Computational techniques have become in_____
problems in transport phenomena.

This book provides a clear, user-oriented introduction to the subject of computational transport phenomena. Each self-contained chapter is devoted to a different topic related to momentum, heat, and mass transport.

Every chapter contains a detailed worked example and a discussion of the problem system equations. Chapters also include an introduction to the numerical methods used to analyze the problem system, a computer code for the solution of the problem system equations, and a discussion of the numerical solution with emphasis on physical interpretation and the analysis of errors. When appropriate, a comparison of the numerical solution with an analytical solution is provided or a discussion of how the numerical solution goes beyond what can be done analytically, especially for nonlinear problems.

Intended for students and a broad range of scientists and engineers, the book includes computer code written in transportable Fortran so that readers can reproduce the numerical solutions and then extend them to other cases. The source code for all of the programs discussed in the book can be downloaded from http://www.lehigh.edu/~wes1/wes1.html. The programs are also available on a DOS-formatted diskette by sending a request and mailing address to wes1@lehigh.edu.

COMPUTATIONAL TRANSPORT PHENOMENA

COMPUTATIONAL TRANSPORT PHENOMENA

Numerical Methods for the Solution of Transport Problems

W. E. SCHIESSER C. A. SILEBI
Lehigh University *Lehigh University*

CAMBRIDGE
UNIVERSITY PRESS

PUBLISHED BY THE PRESS SYNDICATE OF THE UNIVERSITY OF CAMBRIDGE
The Pitt Building, Trumpington Street, Cambridge CB2 1RP, United Kingdom

CAMBRIDGE UNIVERSITY PRESS
The Edinburgh Building, Cambridge CB2 2RU, United Kingdom
40 West 20th Street, New York, NY 10011-4211, USA
10 Stamford Road, Oakleigh, Melbourne 3166, Australia

© Cambridge University Press 1997

This book is in copyright. Subject to statutory exception
and to the provisions of relevant collective licensing agreements,
no reproduction of any part may take place without
the written permission of Cambridge University Press.

First published 1997

Printed in the United States of America

Typeset in Times Roman

Library of Congress Cataloging-in-Publication Data
Schiesser, W. E.
Computational transport phenomena: numerical methods for the
solution of transport problems / W. E. Schiesser. C. A. Silebi.
p. cm.
ISBN 0-521-55378-4 (hardbound). – ISBN 0-521-55653-8 (pbk.)
1. Fluid dynamics. 2. Viscosity. 3. Chemical engineering-
Problems, exercises, etc. I. Silebi, C. A. II. Title
QA929.S35   1997
530.4$'$75$'$015194 – dc20   96-22309
                                          CIP

*A catalog record for this book is available from
the British Library*

ISBN   0 521 55378 4 hardback
ISBN   0 521 55653 8 paperback

To
Gottfried Wilhelm Leibniz

Contents

Preface ix

1 Momentum transport 1
   M1 Laminar boundary layer flow   3
   M2 Unsteady laminar flow in a circular tube   25
   M3 Nonlinear, front-sharpening convective systems   55

2 Heat transport 87
   H1 Heat conduction in a semi-infinite system   89
   H2 One-dimensional heat conduction   107
   H3 Heat transfer in a circular fin   135
   H4 The Graetz problem with constant wall heat flux   167
   H5 The Graetz problem with constant wall temperature   207
   H6 Heat exchanger dynamics   289

3 Mass transport 327
   MA1 A dynamic mass transfer model   329
   MA2 Mass transfer with simultaneous convection and diffusion   370
   MA3 Transient multicomponent diffusion   418

Appendix A Main program RKF45M to call ODE integrator RKF45   437
Appendix B Main program MEM and ODE integrator EULERM   443
Appendix C Subroutine DSS024   449

Index 453

Preface

"Transport phenomena" is the accepted name for the mathematical description of the fundamental phenomena of momentum, energy, and mass transport. This acceptance has developed over some 35 years and was initiated in 1960 by the classic book by Bird, Stewart, and Lightfoot, *Transport Phenomena* (Wiley). Concurrently, the digital computer has evolved as an essential element in the analysis of chemical engineering systems. Yet, rather surprisingly, these two fundamental developments have not come together in a book that presents the computer-based analysis of transport phenomena.

This book is intended to fill the need for an introductory presentation of computational transport phenomena. It is not intended as a replacement for any existing books in transport phenomena, rather, as a supplement to these books. Thus each chapter has the following characteristics: (a) a detailed, worked example, including discussion of the problem system equations; (b) a brief introduction to the numerical methods used to analyze the problem system; (c) a computer code for the solution of the problem system equations (focusing primarily on the routines specific to the problem, with some discussion of the library routines also required to make up a complete code); (d) discussion of the numerical solution, with emphasis on physical interpretation and the analysis of errors; and (e) when appropriate, a comparison of the numerical solution with an analytical solution, or a discussion of how the numerical solution goes beyond what can be done analytically, especially for nonlinear problems.

Each chapter was developed originally during our teaching of a graduate course in transport phenomena. Since each chapter is self-contained, the chapters can be selected in essentially any order and number; we have on occasion used only one chapter during courses in momentum, heat, and mass transfer, applied mathematics, and numerical analysis to illustrate the utility and some of the basic features of the computer analysis of transport problems. Chapter 5 gives the most complete, self-contained discussion of the numerical integration of partial differential equations applied to problems in transport phenomena; this chapter is therefore recommended to the reader who is interested in an introduction to the numerical solution of transport problems.

The intended audience for this book includes chemists, biomedical engineers, chemical engineers, environmental scientists and engineers, mechanical engineers, physicists, and applied mathematicians. The presentation is applied (rather than theoretical, i.e., a "this is how you solve this type of problem" approach illustrated by example). The computer code for each study is written in transportable Fortran so that readers can reproduce the numerical solutions and then extend them to other cases. The source code for all of the programs discussed in the book can be downloaded from http://www.lehigh.edu/~wes1/wes1.html. The programs are also available on a DOS-formatted diskette by sending a request and mailing address to wes1@lehigh.edu.

We have found that the computer-based analysis of transport phenomena extends this important, fundamental field through the detailed study of problems that can also be studied analytically, and by providing a means for the analysis of complex, nonlinear problems that cannot be studied analytically. We think that the numerical approach to the solution of complex transport problems offers the best chance of extending the use of the mathematical description of transport phenomena to new and important application areas and that the use of computer-based analysis is a natural extension of the previously developed analytical methods. We welcome your thoughts, experiences, and insights concerning the use of computational transport phenomena.

W. E. Schiesser
C. A. Silebi

Part one

Momentum transport

M1	Laminar boundary layer flow	3
M2	Unsteady laminar flow in a circular tube	25
M3	Nonlinear, front-sharpening convective systems	55

M1

Laminar boundary layer flow

The equations for incompressible laminar flow over a flat plate with zero pressure gradient can be derived from an order-of-magnitude analysis of the Navier-Stokes equations (Hughes and Brighton 1991). The final equations are
Continuity:

$$\frac{\partial v_x}{\partial x} + \frac{\partial v_y}{\partial y} = 0 \qquad (M1.1)$$

Momentum:

$$v_x \frac{\partial v_x}{\partial x} + v_y \frac{\partial v_x}{\partial y} = \nu \frac{\partial^2 v_x}{\partial y^2} \qquad (M1.2)$$

where

- ν kinematic viscosity $= \mu/\rho$
- μ viscosity
- ρ density

The boundary conditions for eqs. (M1.1) and (M1.2) are

$$v_x(x, 0) = 0 \qquad (M1.3)$$

$$v_y(x, 0) = 0 \qquad (M1.4)$$

$$v_x(x, \infty) = U_s \qquad (M1.5)$$

where U_s is the free steam velocity.

A conventional approach to the solution of eqs. (M1.1) to (M1.5) is to define two new variables, η and ψ, as

$$\eta = y\sqrt{U_s/(\nu x)} \qquad (M1.6)$$

$$\psi = \sqrt{\nu x U_s f(\eta)} \qquad (M1.7)$$

where $f(\eta)$ is a function to be computed. The velocity components are then related to the stream function, ψ, as

$$v_x = \frac{\partial \psi}{\partial y} \tag{M1.8}$$

$$v_y = -\frac{\partial \psi}{\partial x} \tag{M1.9}$$

The change of variables to η and ψ has the advantages that (a) the continuity equation (M1.1) is identically satisfied by ψ, and (b) the similarity transformation of eq. (M1.6) reduces the original problem, eqs. (M1.1) to (M1.5), to the following third-order ordinary differential equation (ODE):

$$f\frac{d^2 f}{d\eta^2} + 2\frac{d^3 f}{d\eta^3} = 0 \tag{M1.10}$$

Equation (M1.10) has the boundary conditions

$$f(0) = 0 \tag{M1.11}$$

$$df(0)/d\eta = 0 \tag{M1.12}$$

$$df(\infty)/d\eta = 1 \tag{M1.13}$$

Compute a solution to eqs. (M1.10) to (M1.13) and compare this numerical solution with the series solution reported by Blasius (Hughes and Brighton 1991, 105–8):

$$f(\eta) = \alpha \sum_{n=0}^{\infty} \left\{\frac{-1}{2}\right\}^2 \frac{\alpha^{n+1} C_n}{(3n+2)!} \eta^{(3n+2)} \tag{M1.14}$$

where $\alpha = 0.3320$, and the first few values of C_n are $C_0 = 1$; $C_1 = 1$; $C_2 = 11$; $C_3 = 375$; $C_4 = 27{,}897$; $C_5 = 3{,}817{,}137$.

Solution

The solution to eq. (M1.10) will be computed by solving a related initial-value ODE problem. In this way, a standard initial-value ODE solver can be used, in this case subroutine RKF45 (Forsythe, Malcolm, and Moler 1977). In order to use RKF45, which can accommodate only first-order ODEs, eq. (M1.13) must be reduced to a system of three, first-order ODEs (this procedure of reducing a nth-order ODE to a system of n first-order ODEs is a straightforward process). First, we define three new variables as

$$g_1 = f \tag{M1.15}$$

$$g_2 = df/d\eta \tag{M1.16}$$

$$g_3 = d^2 f/d\eta 2 \tag{M1.17}$$

then write eqs. (M1.10) to (M1.13) in terms of g_1, g_2, and g_3 as

$$dg_1/d\eta = g_2 \tag{M1.18}$$

$$dg_2/d\eta = g_3 \tag{M1.19}$$

$$g_1 g_3 + 2 dg_3/d\eta = 0 \tag{M1.20}$$

$$g_1(0) = 0 \tag{M1.21}$$

$$g_2(0) = 0 \tag{M1.22}$$

$$g_2(\infty) = 1 \tag{M1.23}$$

Then we can use RKF45 to integrate this system of first-order ODEs, eqs. (M1.18), (M1.19), and (M1.20), subject to the boundary conditions, eqs. (M1.21), (M1.22), and (M1.23). Equations (M1.22), and (M1.23) are boundary conditions since they are specified at two different values of η, $\eta = 0$ and $\eta = \infty$ (which can be considered as physical boundaries of the laminar flow problem). Note also that eqs. (M1.21), (M1.22), and (M1.23) specify one boundary value in g_1 and two boundary values in g_2, and not one boundary value in each of the three state variables, g_1, g_2, and g_3 (this form of the boundary conditions precludes the use of some library ODE boundary-value solvers).

RKF45 is now applied to eqs. (M1.18) to (M1.23). Since eqs. (M1.18) to (M1.23) are an ODE boundary-value problem and RKF45 is an initial-value ODE integrator (i.e., in an initial-value problem the vector of dependent variables, such as g_1, g_2, and g_3, must be specified at only one value of the independent variable, such as $\eta = 0$ or $\eta = \infty$, but not both values), the use of RKF45 must be extended to accommodate the boundary conditions.

The approach we will use is to assume an initial value for g_3,

$$g_3(0) = assumed\ value \tag{M1.24}$$

then integrate until we effectively reach $\eta = \infty$ and observe if boundary condition (M1.23) is satisfied. If it is not (which will be the case when we start this procedure), the assumed value of $g_3(0)$ is adjusted and the calculation is repeated. By a process of trial and error on the assumed value of $g_3(0)$, we eventually arrive at the value that produces a solution that satisfies eq. (M1.23). Note that this procedure, which is called the "shooting method," replaces the boundary-value problem of eqs. (M1.18) to (M1.23) with the initial-value problem consisting of eqs. (M1.18) to (M1.22) and (M1.24).

Subroutine INITAL to define the three initial conditions, eqs. (M1.21), (M1.22), and (M1.24), is listed below:

```
      SUBROUTINE INITAL
      IMPLICIT DOUBLE PRECISION (A-H,O-Z)
      COMMON/T/    ETA,   NSTOP,   NORUN
     +      /Y/    G1,    G2,      G3
     +      /F/    G1E,   G2E,     G3E
     +      /S/    G3S,   DG3S,    FLAGM,   FLAGP
     +      /I/    IP
C...
C...  SET THE INITIAL ESTIMATE OF G3, INCREMENT IN G3,
C...  CONTROL VARIABLES FOR ITERATION ON G3
      IF(NORUN.EQ.1)THEN
         G3S =0.3D0
         DG3S=0.2D0
         FLAGM=-1.0
         FLAGP=-1.0
      END IF
C...
C...  INITIAL CONDITIONS
      G1=0.0D0
      G2=0.0D0
      G3=G3S
C...
C...  INITIAL DERIVATIVES
      CALL DERV
      IP=0
      RETURN
      END
```

We can note the following points about INITAL:

(1) Double precision coding is used. The independent variable, η, is in COMMON/T/, and the dependent variables, g_1, g_2, and g_3, are in COMMON/Y/. The corresponding derivatives, $dg_1/d\eta$, $dg_2/d\eta$, and $dg_3/d\eta$, defined by eqs. (M1.18), (M1.19), and (M1.20) are in COMMON/F/ (i.e., G1E, G2E, and G3E)

```
      IMPLICIT DOUBLE PRECISION (A-H,O-Z)
      COMMON/T/    ETA,   NSTOP,   NORUN
     +      /Y/    G1,    G2,      G3
     +      /F/    G1E,   G2E,     G3E
     +      /S/    G3S,   DG3S,    FLAGM,   FLAGP
     +      /I/    IP
```

Thus, we have the arrangement of the ODE dependent variables in COMMON/Y/, and their initial-value derivatives in COMMON/F/, that is, RKF45, integrates the derivative vector in COMMON/F/ and returns the dependent-variable vector through COMMON/Y/. The additional variables in COMMON/S/ and /I/ are described subsequently.

(2) For the first run (NORUN = 1), the missing initial condition, $g_3(0)$, is assumed to have a value of 0.3 (as discussed previously through eq. (M1.24))

```
C...
C...    SET THE INITIAL ESTIMATE OF G3, INCREMENT IN G3,
C...    CONTROL VARIABLES FOR ITERATION ON G3
        IF(NORUN.EQ.1)THEN
           G3S =0.3D0
           DG3S=0.2D0
           FLAGM=-1.0
           FLAGP=-1.0
        END IF
```

Also, the interval, DG3S, over which an interval-halving search is used to find the missing initial condition, is initialized to 0.2. Two variables to control the interval-halving search, FLAGM and FLAGP, are also initialized.

(3) After the first run, the missing initial condition, $g_3(0)$, is set by the interval-halving search in subroutine PRINT (to be discussed) and returned to INITAL as G3S to start the next run

```
C...
C...    INITIAL CONDITIONS
        G1=0.0D0
        G2=0.0D0
        G3=G3S
```

(4) The initial derivatives are then computed by a call to DERV, and a counter is initialized that is then used in PRINT

```
C...
C...    INITIAL DERIVATIVES
        CALL DERV
        IP=0
```

This call to DERV tests the programming of the derivatives defined by eqs. (M1.18) to (M1.20).

Subroutine DERV is listed below:

```
        SUBROUTINE DERV
        IMPLICIT DOUBLE PRECISION (A-H,O-Z)
        COMMON/T/    ETA,    NSTOP,   NORUN
     +       /Y/     G1,     G2,      G3
     +       /F/     G1E,    G2E,     G3E
```

```
     +          /S/     G3S,    DG3S,   FLAGM,   FLAGP
     +          /I/     IP
C...
C...   ODES
       G1E=G2
       G2E=G3
       G3E=-0.5D0*G1*G3
       RETURN
       END
```

We can note the following points about DERV:

(1) The COMMON area is the same as in INITAL.
(2) The programming of eqs. (M1.18) to (M1.20) is straightforward

```
C...
C...   ODES
       G1E=G2
       G2E=G3
       G3E=-0.5D0*G1*G3
```

Subroutine PRINT is listed below:

```
       SUBROUTINE PRINT(NI,NO)
       IMPLICIT DOUBLE PRECISION (A-H,O-Z)
       COMMON/T/   ETA,    NSTOP,   NORUN
     +        /Y/     G1,     G2,      G3
     +        /F/     G1E,    G2E,     G3E
     +        /S/     G3S,    DG3S,    FLAGM,   FLAGP
     +        /I/     IP
C...
C...   INCREMENT THE COUNTER TO DETERMINE THE END OF EACH RUN
       IP=IP+1
C...
C...   PRINT THE SOLUTION FOR THE FIRST AND LAST RUNS
       IF((NORUN.EQ.1).OR.(NORUN.EQ.20))THEN
C...
C...      PRINT A HEADING
          IF(IP.EQ.1)WRITE(NO,1)NORUN
1         FORMAT(//,' NORUN = ',I3,//,
     +           7X,'eta',8X,'g1',8X,'g2',8X,'g3')
C...
C...      PRINT THE SOLUTION
          WRITE(NO,2)ETA,G1,G2,G3
2         FORMAT(F10.2,3F10.5)
       END IF
C...
```

```
C...   AT THE END OF EACH RUN, INCREMENT G3 DEPENDING ON THE
C...   SIGN DEVIATION FROM THE PRESCRIBED BOUNDARY CONDITION
C...   G2(INF) = 1
       IF(IP.EQ.11)THEN
          CALL SEARCH
C...
C...      MONITOR THE ITERATION
          WRITE(*,3)NORUN,G2,G3S,DG3S,FLAGP,FLAGM
3         FORMAT(I5,3F12.6,2F10.3)
       END IF
       RETURN
       END
```

We can note the following points about PRINT:

(1) Integer counter IP is incremented each time PRINT is called to establish the end of each run, for which it has the value 11, that is, PRINT is called 11 times for each run

```
C...
C...   INCREMENT THE COUNTER TO DETERMINE THE END OF EACH RUN
       IP=IP+1
```

(2) A total of 20 runs is programmed (via the data file to be discussed subsequently). In order to keep the output to a reasonable level, the output for only runs 1 and 20 is printed

```
C...
C...   PRINT THE SOLUTION FOR THE FIRST AND LAST RUNS
       IF((NORUN.EQ.1).OR.(NORUN.EQ.20))THEN
C...
C...      PRINT A HEADING
          IF(IP.EQ.1)WRITE(NO,1)NORUN
1         FORMAT(//,' NORUN = ',I3,//,
      +            7X,'eta',8X,'g1',8X,'g2',8X,'g3')
C...
C...      PRINT THE SOLUTION
          WRITE(NO,2)ETA,G1,G2,G3
2         FORMAT(F10.2,3F10.5)
       END IF
```

Note that g_1, g_2, and g_3 are printed as a function of η.

(3) During the last call to PRINT at the end of each run (IP = 11), subroutine SEARCH is called to set the missing initial condition, $g_3(0)$, for the next run

```
C...
C...   AT THE END OF EACH RUN, INCREMENT G3 DEPENDING ON THE
C...   SIGN DEVIATION FROM THE PRESCRIBED BOUNDARY CONDITION
```

```
C...    G2(INF) = 1
        IF(IP.EQ.11)THEN
            CALL SEARCH
```

(4) The progress of each run is also monitored by printing the solution at the end of each run to unit *, which is usually the screen of the computer

```
C...
C...        MONITOR THE ITERATION
            WRITE(*,3)NORUN,G2,G3S,DG3S,FLAGP,FLAGM
3           FORMAT(I5,3F12.6,2F10.3)
        END IF
```

Subroutine SEARCH, which implements the interval-halving search for the missing initial condition, $g_3(0)$, is listed below:

```
        SUBROUTINE SEARCH
C...
C...    SUBROUTINE SEARCH SEARCHES FOR A ROOT OF A FUNCTION BY
C...    INTERVAL HALVING
C...
        IMPLICIT DOUBLE PRECISION (A-H,O-Z)
        COMMON/T/    ETA,    NSTOP,   NORUN
       +      /Y/    G1,     G2,      G3
       +      /F/    G1E,    G2E,     G3E
       +      /S/    G3S,    DG3S,    FLAGM,   FLAGP
       +      /I/    IP
C...
C...    IF (1 - G2(INF)) IS NONNEGATIVE, INCREASE G3(0)
        IF(FUNC(G2).GE.0.0D0)THEN
C...
C...        IF A SMALLER INCREMENT IN G3(0) IS REQUIRED, HALVE
C...        DG3S
            IF(FLAGM.GE.0.0D0)DG3S=DG3S/2.0D0
C...
C...        INCREMENT G3(0) POSITIVELY FOR THE NEXT RUN
            G3S=G3S+DG3S
            FLAGP=1.0D0
C...
C...    IF (1 - G2(INF)) IS NEGATIVE, DECREASE G3(0)
        ELSE IF(FUNC(G2).LT.0.0D0)THEN
C...
C...        IF A SMALLER INCREMENT IN G3(0) IS REQUIRED, HALVE
C...        DG3S
            IF(FLAGP.GE.0.0D0)DG3S=DG3S/2.0D0
```

```
C...
C...        INCREMENT G3(0) NEGATIVELY FOR THE NEXT RUN
            G3S=G3S-DG3S
            FLAGM=1.0D0
        END IF
        RETURN
        END
```

We can note the following points about SEARCH:

(1) The COMMON area is the same as in INITAL, DERV, and PRINT.

(2) A test function, $f(g_2(\infty)) = 1 - g_2(\infty)$, implemented in function FUNC, is used to determine if boundary condition (M1.23), has been satisfied at the end of the current solution (this function will be zero if boundary condition (M1.24) is satisfied). In general, the test function, computed by a call to function FUNC, will not be zero, and therefore $g_3(0)$ is adjusted for the next run

```
C...
C...    IF (1 - G2(INF)) IS NONNEGATIVE, INCREASE G3(0)
        IF(FUNC(G2).GE.0.0D0)THEN
C...
C...        IF A SMALLER INCREMENT IN G3(0) IS REQUIRED, HALVE
C...        DG3S
            IF(FLAGM.GE.0.0D0)DG3S=DG3S/2.0D0
C...
C...        INCREMENT G3(0) POSITIVELY FOR THE NEXT RUN
            G3S=G3S+DG3S
            FLAGP=1.0D0
```

Specifically, if $f(g_2(\infty)) > 0$, and if FLAGM > 0, the interval over which the search is performed is halved (recall that FLAGM = -1 initially from INITAL, so this interval halving does not take place initially). Then the current value of $g_3(0)$ is increased by the interval, and FLAGP is set to 1. For example, at the end of the first run, $\eta = 10(\approx \infty)$, $g_1(10) = 7.68202$, $g_2(10) = 0.93455$, $g_3(10) = 0.00000$. Therefore $g_2(10)$ is too small (it should be 1 according to boundary condition (M1.23)), and $f(g_2(\infty)) > 0$. Since FLAGM < 0 initially (from INITAL, FLAGM = -1), $g_3(0)$ for the next run is set to $g_3(0)$(new) = $g_3(0)$(old) + 0.2 = 0.3 + 0.2 = 0.5, and FLAGM is set to 1.

Function FUNC, listed below, is self-explanatory:

```
        DOUBLE PRECISION FUNCTION FUNC(G2)
C...
C...    FUNCTION FUNC DEFINES THE FUNCTION FOR WHICH A ZERO IS
C...    TO BE COMPUTED
```

```
C...
      IMPLICIT DOUBLE PRECISION (A-H,O-Z)
      FUNC=1.0D0-G2
      RETURN
      END
```

(3) The search then continues at the end of the second run according to

```
C...
C...  IF (1 - G2(INF)) IS NEGATIVE, DECREASE G3(0)
      ELSE IF(FUNC(G2).LT.0.0D0)THEN
C...
C...     IF A SMALLER INCREMENT IN G3(0) IS REQUIRED, HALVE
C...     DG3S
         IF(FLAGP.GE.0.0D0)DG3S=DG3S/2.0D0
C...
C...     INCREMENT G3(0) NEGATIVELY FOR THE NEXT RUN
         G3S=G3S-DG3S
         FLAGM=1.0D0
      END IF
```

At the end of the second run, $g_2(10) > 1$ so $f(g_2(\infty)) < 0$. The interval is therefore halved from 0.2 to 0.1 (since FLAGP = 1 from the first run) and is then subtracted from $g_3(0)$, that is, $g_3(0)(\text{new}) = g_3(0)(\text{old}) - 0.1 = 0.5 - 0.1 = 0.4$. This process continues through 20 successive runs, which provides enough iterations on $g_3(0)$ to give five-figure convergence of boundary condition (M1.23).

The output from the first and last (20th) runs from PRINT is listed below:

```
RUN NO. -    1  Blasius problem

INITIAL T -  0.000D+00

  FINAL T -  0.100D+02

  PRINT T -  0.100D+01

NUMBER OF DIFFERENTIAL EQUATIONS -    3

MAXIMUM INTEGRATION ERROR -  0.100D-04

NORUN =    1
         eta          g1          g2          g3
        0.00     0.00000     0.00000     0.30000
        1.00     0.14963     0.29814     0.29260
        2.00     0.58845     0.57178     0.24610
```

3.00	1.26948	0.77510	0.15598
4.00	2.10665	0.88436	0.06737
5.00	3.01477	0.92382	0.01875
6.00	3.94445	0.93304	0.00329
7.00	4.87841	0.93442	0.00036
8.00	5.81292	0.93455	0.00002
9.00	6.74747	0.93455	0.00000
10.00	7.68202	0.93455	0.00000

NORUN = 20

eta	g1	g2	g3
0.00	0.00000	0.00000	0.33206
1.00	0.16557	0.32978	0.32301
2.00	0.65002	0.62977	0.26675
3.00	1.39681	0.84604	0.16136
4.00	2.30575	0.95552	0.06423
5.00	3.28327	0.99154	0.01590
6.00	4.27962	0.99897	0.00240
7.00	5.27924	0.99992	0.00022
8.00	6.27921	1.00000	0.00001
9.00	7.27921	1.00000	0.00000
10.00	8.27921	1.00000	0.00000

We can note the following points about this output:

(1) For NORUN = 1, the initial assumed value of $g_3(0)$ is 0.3, as discussed previously, which produces a final value $g_2(10) = 0.93455$.

(2) For NORUN = 20, the initial condition found by SEARCH is $g_3(0) = 0.33206$, giving a final value $g_2(10) = 1.00000$ (i.e., boundary condition (M1.23) is satisfied to five figures).

(3) g_2 is within 1% of its final value given by boundary condition (M1.23) at $\eta = 5$, that is, $g_2(5) = 0.99154$. Therefore, as a rule of thumb, the thickness of the boundary layer is said to correspond to $\eta = 5$.

To complete this discussion of the numerical output, the new initial values of $g_3(0)$ and $f(g_2(\infty)) = 1 - g_2(\infty)$ are listed below:

G3	1 - G2(oo)
0.500000	0.065445
0.400000	-0.313720
0.350000	-0.132131
0.325000	-0.035704
0.337500	0.014221
0.331250	-0.010895
0.334375	0.001624
0.332812	-0.004646

0.332031	−0.001514
0.332422	0.000054
0.332227	−0.000730
0.332129	−0.000338
0.332080	−0.000142
0.332056	−0.000044
0.332068	0.000005
0.332062	−0.000019
0.332059	−0.000007
0.332057	−0.000001
0.332058	0.000002
0.332058	0.000001

Note that $g_3(0)$ has converged to six figures (i.e., 0.332058) and $f(g_2(\infty)) = 1 - g_2(\infty)$ has a residual value of 0.000001. Thus, the interval-halving algorithm was quite effective for this problem. In all of the preceding programming, $f(g_2(\infty)) = 1 - g_2(\infty)$ was computed by function FUNC, listed below:

```
      DOUBLE PRECISION FUNCTION FUNC(G2)
C...
C...  FUNCTION FUNC DEFINES THE FUNCTION FOR WHICH A ZERO IS
C...  TO BE COMPUTED
C...
      IMPLICIT DOUBLE PRECISION (A-H,O-Z)
      FUNC=1.0D0-G2
      RETURN
      END
```

We should also note that subroutine SEARCH is a general interval-halving search routine that, in principle, can be used to find a root of any function defined in FUNC. It has the advantage of not requiring the derivative of the function as does, for example, Newton's method. (It has the disadvantage of most root finders of requiring an initial estimate of the root; in general, if the function has multiple roots, it will converge to the root closest to the initial estimate, i.e., interval halving is a local method as opposed to a global method.)

Finally, to complete the discussion of the computer code, we should note that the programming of the 20 runs was done by a procedure that is a departure from the usual practice of executing multiple runs by using multiple sets of data. Rather, the multiple runs were specified by the single number 20 in the first line of the data file

```
    20
Blasius problem
0.           10.0        1.0
     3                   0.00001
END OF RUNS
```

This data file therefore required modification of main program RKF45M listed in Appendix A, which calls RKF45. Specifically, the reading of the total number of runs (i.e., 20) and reexecution of these runs was accomplished with the programming

```
C...
C...    INITIALIZE THE RUN COUNTER
        NORUN=0
C...
C...    READ TOTAL NUMBER OF RUNS
        READ(NI,1005,END=999)NTOT
1005    FORMAT(I5)
C...
C...    BEGIN A RUN
1       NORUN=NORUN+1
C...
C...    INITIALIZE THE RUN TERMINATION VARIABLE
        NSTOP=0
C...
C...    READ THE FIRST LINE OF DATA
        IF((NORUN-1)*(NORUN-(NTOT+1))).EQ.0)
       +    READ(NI,1000,END=999)(TITLE(I),I=1,20)
C...
C...    TEST FOR  END OF RUNS  IN THE DATA
        DO 2 I=1,3
        IF(TITLE(I).NE.ENDRUN(I))GO TO 3
2       CONTINUE
C...
C...    AN END OF RUNS HAS BEEN READ, SO TERMINATE EXECUTION
999     STOP
C...
C...    READ THE SECOND LINE OF DATA
3       IF(NORUN.EQ.1)READ(NI,1001,END=999)T0,TF,TP
C...
C...    READ THE THIRD LINE OF DATA
        IF(NORUN.EQ.1)READ(NI,1002,END=999)NEQN,ERROR
C...
C...    PRINT A DATA SUMMARY
        IF(NORUN.EQ.1)WRITE(NO,1003)NORUN,(TITLE(I),I=1,20),
       1                            T0,TF,TP,
       2                            NEQN,ERROR
C...
C...    INITIALIZE TIME
        T=T0
```

We can note the following points about this coding:

(1) It is a replacement for the coding in main program RKF45M of Appendix A starting at the lines

```
C...
C...   INITIALIZE THE RUN COUNTER
       NORUN=0
```

and ending at the lines

```
C...
C...   INITIALIZE TIME
       T=T0
```

(2) The total number of runs, for example, 20, is read with an I5 format at the beginning of the first run

```
C...
C...   READ TOTAL NUMBER OF RUNS
       READ(NI,1005,END=999)NTOT
1005   FORMAT(I5)
```

(3) A loop, starting at statement 1, is then used to step through the series of runs

```
C...
C...   BEGIN A RUN
1      NORUN=NORUN+1
C...
C...   INITIALIZE THE RUN TERMINATION VARIABLE
       NSTOP=0
```

(4) For the first and last runs, the data file is read. For the first run, three lines of data define (a) the documentation title: Blasius problem (see the preceding data file); (b) the initial, final, and output intervals of the independent variable, for example, $\eta = 0$, 10, and 1, respectively; and (c) the number of ODEs and the error tolerance to be used by RKF45, for example, 3 and 0.00001, respectively

```
C...
C...   READ THE FIRST LINE OF DATA
       IF((NORUN-1)*(NORUN-(NTOT+1))).EQ.0)
     +    READ(NI,1000,END=999)(TITLE(I),I=1,20)
C...
C...   TEST FOR  END OF RUNS  IN THE DATA
       DO 2 I=1,3
       IF(TITLE(I).NE.ENDRUN(I))GO TO 3
2      CONTINUE
C...
C...   AN END OF RUNS HAS BEEN READ, SO TERMINATE EXECUTION
999    STOP
```

```
C...
C...   READ THE SECOND LINE OF DATA
3      IF(NORUN.EQ.1)READ(NI,1001,END=999)TO,TF,TP
C...
C...   READ THE THIRD LINE OF DATA
       IF(NORUN.EQ.1)READ(NI,1002,END=999)NEQN,ERROR
```

For the last run, the characters END OF RUNS are read to terminate execution of the program.

(5) For the first run, the data are printed in a summary

```
C...
C...   PRINT A DATA SUMMARY
       IF(NORUN.EQ.1)WRITE(NO,1003)NORUN,(TITLE(I),I=1,20),
     1                TO,TF,TP,
     2                NEQN,ERROR
```

The programming then continues on as in main program RKF45M in Appendix A. Again, the usual procedure for specifying multiple runs is to repeat the three basic lines of data, which are read successively (at the beginning of each run) by main program RKF45M.

Finally, we consider the series solution of eq. (M1.14). A program to compute the series for $0 \leq \eta \leq 10$ is listed below:

```
       PROGRAM M1SER
C...
C...   PROGRAM M1SER COMPUTES THE SOLUTION TO THE BLASIUS
C...   EQUATION FROM THE SERIES SOLUTION [1]
C...
C...          INF             (ALPHA**(N+1)CN)
C...   F = SUM(-1/2)**N  ---------------- ETA**(3*N + 2)   (1)
C...          N=0              (3*N + 2)!
C...
C...   WHERE ALPHA = 0.3320, C0 = 1, C1 = 1, C2 = 11,
C...   C3 = 375, C4 = 27,897, C5 = 3,817,137
C...
C...   [1]  HUGHES, W. F., AND J. A. BRIGHTON, THEORY AND
C...   PROBLEMS OF FLUID DYNAMICS, 2ND ED., MCGRAW-HILL,
C...   INC., NEW YORK, 1991
C...
C...   DOUBLE PRECISION CODING IS USED
       IMPLICIT DOUBLE PRECISION (A-H,O-Z)
C...
C...   UPPER LIMIT OF THE SERIES (RATHER THAN INFINITY)
       PARAMETER (NU=5)
```

```fortran
C...
C...   COMMON AREA FOR THE SERIES COEFFICIENTS
       COMMON/COEFF/ ALPHA, C(0:NU)
C...
C...   OPEN OUTPUT FILE
       NO=1
       OPEN(NO,FILE='OUTPUT',STATUS='UNKNOWN')
C...
C...   SET THE COEFFICIENTS IN THE SERIES
       CALL COEFF
C...
C...   WRITE A HEADING FOR THE COMPUTED OUTPUT
       WRITE(NO,5)
5      FORMAT(//,5X,' Series Solution',//,
      +          4X,'eta',9X,'f(eta)',/,
      +          15X,'(series)')
C...
C...   STEP THROUGH A SERIES OF VALUES OF ETA
       DO 1 I=1,11
          ETA=DFLOAT(I-1)
C...
C...      EVALUATE THE SERIES
          SUM=0.0D0
          DO 2 N=0,5
          SUM=SUM+TERM(N,ETA)
2         CONTINUE
C...
C...      PRINT THE SERIES SOLUTION
          WRITE(NO,3)ETA,SUM
3         FORMAT(F7.2,F15.4)
C...
C...      NEXT VALUE OF ETA
1      CONTINUE
C...
C...   PERFORM A HAND CALCULATION FOR ETA = 1 AS A CHECK OF
C...   THE PRECEDING PROGRAMMING
       CALL HAND(F)
       WRITE(NO,4)F
4      FORMAT(//,' Series for eta = 1 by',/,
      +          ' hand calculation = ',F6.4)
C...
C...   CALCULATIONS ARE COMPLETE
       STOP
       END
```

We can note the following points about program M1SER:

(1) Double precision coding is used. The number of terms actually used in the series of eq. (M1.14) is six (note the summation index of the series in eq. (M.14) starts at $n = 0$). The values of α and the coefficients, C_n, in eq. (M1.14) are in COMMON/COEFF/. Output is written to file OUTPUT

```
C...
C...    DOUBLE PRECISION CODING IS USED
        IMPLICIT DOUBLE PRECISION (A-H,O-Z)
C...
C...    UPPER LIMIT OF THE SERIES (RATHER THAN INFINITY)
        PARAMETER (NU=5)
C...
C...    COMMON AREA FOR THE SERIES COEFFICIENTS
        COMMON/COEFF/ ALPHA, C(0:NU)
C...
C...    OPEN OUTPUT FILE
        NO=1
        OPEN(NO,FILE='OUTPUT',STATUS='UNKNOWN')
```

(2) The series is evaluated for 11 values of η over the interval $0 \leq \eta \leq 10$ at intervals of one

```
C...
C...    STEP THROUGH A SERIES OF VALUES OF ETA
        DO 1 I=1,11
            ETA=DFLOAT(I-1)
C...
C...        EVALUATE THE SERIES
            SUM=0.0D0
            DO 2 N=0,5
            SUM=SUM+TERM(N,ETA)
2           CONTINUE
C...
C...        PRINT THE SERIES SOLUTION
            WRITE(NO,3)ETA,SUM
3           FORMAT(F7.2,F15.4)
C...
C...        NEXT VALUE OF ETA
1       CONTINUE
```

For each value of η, six terms in the series are summed in DO loop 2. Individual terms are computed by function TERM. After the summation of the six terms, η and the series are printed. The function DFLOAT is discussed subsequently.

(3) As an independent check of the programming of eq. (M1.14), a detailed "hand" calculation is hard-coded in subroutine HAND for $\eta = 1$

```
C...
C...   PERFORM A HAND CALCULATION FOR ETA = 1 AS A CHECK OF
C...   THE PRECEDING PROGRAMMING
       CALL HAND(F)
       WRITE(NO,4)F
4      FORMAT(//,' Series for eta = 1 by',/,
      +            ' hand calculation = ',F6.4)
```

Subroutine COEFF, listed below, is just straightforward programming of the first six coefficients in eq. (M1.14):

```
       SUBROUTINE COEFF
C...
C...   SUBROUTINE COEFF SETS THE COEFFICIENTS IN THE SERIES
C...   SOLUTION
C...
       IMPLICIT DOUBLE PRECISION (A-H,O-Z)
C...
C...   UPPER LIMIT OF THE SERIES
       PARAMETER (NU=5)
C...
C...   COMMON AREA FOR THE SERIES COEFFICIENTS
       COMMON/COEFF/ ALPHA, C(0:NU)
C...
C...   SET THE COEFFICIENTS
       ALPHA=0.3320D0
       C(0)=1.0D0
       C(1)=1.0D0
       C(2)=11.0D0
       C(3)=375.0D0
       C(4)=27897.0D0
       C(5)=3817137.0D0
       RETURN
       END
```

Function TERM is listed below

```
       DOUBLE PRECISION FUNCTION TERM(N,ETA)
C...
C...   FUNCTION TERM COMPUTES A TERM IN THE SERIES SOLUTION
C...
       IMPLICIT DOUBLE PRECISION (A-H,O-Z)
C...
C...   UPPER LIMIT OF THE SERIES
       PARAMETER (NU=5)
C...
```

```
C...    COMMON AREA FOR THE SERIES COEFFICIENTS
        COMMON/COEFF/ ALPHA, C(0:NU)
C...
C...    TERM IN THE SERIES
        TERM=((-0.5D0)**N)*(ALPHA**(N+1))*C(N)*
     +       (1.0D0/FACT(3*N+2))*(ETA**(3*N+2))
        RETURN
        END
```

The programming of the individual terms in the series of eq. (M1.14) follows directly from the series. FACT is a function to compute the factorial in each of the terms

```
        DOUBLE PRECISION FUNCTION FACT(N)
C...
C...    FUNCTION COMPUTES THE FACTORIAL IN THE SERIES SOLUTION
C...
        IMPLICIT DOUBLE PRECISION (A-H,O-Z)
        FACT=1.0D0
        DO 2 I=1,N
           FACT=FACT*DFLOAT(I)
2       CONTINUE
        RETURN
        END
```

Finally, subroutine HAND contains the detailed arithmetic for six terms in the series at $\eta = 1$

```
        SUBROUTINE HAND
C...
C...    SUBROUTINE HAND DOES A 'HAND' CALCULATION OF THE
C...    SERIES FOR ETA = 1 TO CHECK THE PROGRAMMING OF THE
C...    SERIES
C...
C...         INF           (ALPHA**(N+1)CN)
C...    F = SUM(-1/2)**N  ---------------- ETA**(3*N + 2)    (1)
C...         N=0              (3*N + 2)!
C...
C...    WHERE ALPHA =  0.3320, C0 = 1, C1 = 1, C2 = 11,
C...    C3 = 375, C4 = 27,897, C5 = 3,817,137
C...
        IMPLICIT DOUBLE PRECISION (A-H,O-Z)
C...
C...    COMPUTE SIX TERMS IN THE SERIES
C...
C...    N=0
        TERM0=0.3320D0/2.0D0
```

```
C...
C...   N=1
       FACT5=1.0D0*2.0D0*3.0D0*4.0D0*5.0D0
       TERM1=-0.5D0*(0.3320**2)/FACT5
C...
C...   N=2
       FACT8=FACT5*6.0D0*7.0D0*8.0D0
       TERM2=(-0.5D0**2)*(0.3320**3)*11.0D0/FACT8
C...
C...   N=3
       FACT11=FACT8*9.0D0*10.0D0*11.0D0
       TERM3=(-0.5D0**3)*(0.3320**4)*375.0D0/FACT11
C...
C...   N=4
       FACT14=FACT11*12.0D0*13.0D0*14.0D0
       TERM4=(-0.5D0**4)*(0.3320**5)*27897.0D0/FACT14
C...
C...   N=5
       FACT17=FACT14*15.0D0*16.0D0*17.0D0
       TERM5=(-0.5D0**5)*(0.3320**6)*3817137.0D0/FACT17

C...   SUM SIX TERMS
       F=TERM0+TERM1+TERM2+TERM3+TERM4+TERM5
C...
C...   HAND CALCULATION IS COMPLETE
       RETURN
       END
```

In main program M1SER and function FACT, a function, DFLOAT, is used to convert a single precision integer to a double precision, floating-point number. Some Fortran compilers do not include DFLOAT, and therefore we include a function for this purpose

```
       DOUBLE PRECISION FUNCTION DFLOAT(I)
C...
C...   FUNCTION DFLOAT CONVERTS A SINGLE PRECISION INTEGER
C...   INTO A DOUBLE PRECISION FLOATING POINT. THIS FUNCTION
C...   IS PROVIDED IN CASE THE USER'S FORTRAN COMPILER DOES
C...   NOT INCLUDE DFLOAT.
C...
       DFLOAT=DBLE(FLOAT(I))
       RETURN
       END
```

The output from program M1SER is listed below:

Series Solution

eta	f(eta) (series)
0.00	0.0000
1.00	0.1655
2.00	0.6499
3.00	1.3966
4.00	2.3034
5.00	3.1429
6.00	0.1407
7.00	-63.1697
8.00	-741.6028
9.00	-5998.2526
10.00	-38049.8006

Series for eta = 1 by hand calculation = 0.1655

We note that the series gives values of $f(\eta)$, which are close to the numerical solution, $g_1(\eta)$ for $\eta \leq 5$ (i.e., over the "boundary layer" discussed previously for which $g_2(5) \approx 0.99$):

$$f(1) = 0.1655 \qquad g_1(1) = 0.16557$$
$$f(2) = 0.6499 \qquad g_1(2) = 0.65002$$
$$f(3) = 1.3966 \qquad g_1(3) = 1.39681$$
$$f(4) = 2.3034 \qquad g_1(4) = 2.30575$$
$$f(5) = 3.1429 \qquad g_1(5) = 3.28327$$

For $\eta > 5$, the series diverges, for example,

$$f(6) = 0.1407 \qquad g_1(6) = 4.27962$$
$$f(7) = -63.1697 \qquad g_1(7) = 5.27924$$

Thus, in this sense, the numerical solution is better than the analytical solution, that is, the numerical solution approaches more closely the infinite value of η in boundary condition (M1.23).

To summarize, we have considered the solution of a boundary-value problem in ODEs, defined by eqs. (M1.18) to (M1.23), by solving iteratively a related initial-value ODE problem. The shooting method worked satisfactorily for this problem, but in general it has significant limitations: (a) Frequently, the solution at the end of the interval(s), for example, $\eta = 10$, is so sensitive to the assumed initial condition(s) that convergence to the required solution cannot be achieved, and (b) If n initial conditions are missing, then an n-dimensional search must be conducted for these missing initial conditions, and the feasibility of such a search drops off rather sharply with increasing n. Therefore, the shooting method is not recommended for these more demanding boundary-value ODE problems, and more general methods should be considered.

References

Forsythe, G. E., M. A. Malcolm, and C. B. Moler. 1977. *Computer Methods for Mathematical Computations*. Englewood Cliffs, NJ: Prentice-Hall.

Hughes, W. F., and J. A. Brighton. 1991. *Theory and Problems of Fluid Dynamics*. 2d ed. New York: McGraw-Hill, 105–8.

M2

Unsteady laminar flow in a circular tube

Consider the partial differential equation (PDE) for unsteady, laminar flow in a circular tube presented by Bird, Stewart and Lightfoot (BSL) (1960, 126, eq. (4.1-21)):

$$\frac{\partial \phi}{\partial \tau} = 4 + \frac{1}{\xi}\frac{\partial}{\partial \xi}\left\{\xi \frac{\partial \phi}{\partial \xi}\right\} = 4 + \frac{\partial^2 \phi}{\partial \xi^2} + \frac{1}{\xi}\frac{\partial \phi}{\partial \xi} \quad \text{(M2.1)}$$

where

$$\phi = \frac{v_z}{(p_0 - p_L)R^2/(4\mu L)} = \frac{v_z}{v_{\max}}, \quad \xi = \frac{r}{R}, \quad \tau = \frac{\nu t}{R^2}$$

Compute a numerical solution to eq. (M2.1), $\phi(\tau, \xi)$, subject to the initial and boundary conditions

$$\phi(0, \xi) = 0 \quad \text{(M2.2)}$$

$$\phi(\tau, 1) = 0 \quad \text{(M2.3)}$$

$$\frac{\partial \phi(\tau, 0)}{\partial \xi} = 0 \quad \text{(M2.4)}$$

Use 31 grid points in the ξ direction. Compare your numerical solution with the values computed from BSL, eq. (4.1-4), for $\xi = 0$.

Solution

The method of lines (MOL) solution (Schiesser 1991) to eqs. (M2.1) to (M2.4) is programmed in the following subroutines INITAL, DERV, PRINT, and subordinate routines. Initial condition (M2.2) is implemented in subroutine INITAL:

```
SUBROUTINE INITAL
IMPLICIT DOUBLE PRECISION (A-H,O-Z)
PARAMETER(NR=31)
COMMON/T/     TAU,     NSTOP,     NORUN
```

```
     +          /Y/    PHI(NR)
     +          /F/    PHIT(NR)
     +          /C/         DR,          DRS,       R(NR)
     +          /I/         NG,          IP
C...
C...    SET NUMBER OF RADIAL GRID POINTS
        IF(NORUN.EQ.1)NG=11
        IF(NORUN.EQ.2)NG=21
        IF(NORUN.EQ.3)NG=31
C...
C...    SPATIAL GRID
        DR=1.0D0/DFLOAT(NG-1)
        DRS=DR**2
        DO 1 I=1,NG
           R(I)=DFLOAT(I-1)*DR
1       CONTINUE
C...
C...    INITIAL CONDITION
        DO 2 I=1,NG
           PHI(I)=0.0D0
2       CONTINUE
C...
C...    INITIAL DERIVATIVES
        CALL DERV
        IP=0
        RETURN
        END
```

We can note the following points about subroutine INITAL:

(1) ϕ in eq. (M2.1) is defined on a spatial grid in array PHI(NR) in COMMON/Y/. The derivative $\partial \phi/\partial \tau$ in eq. (M2.1) is stored in array PHIT(NR) in COMMON/F/. The derivatives in PHIT(NR) are integrated forward in τ ($=$ TAU in COMMON/T/) by an ordinary differential equation (ODE) integrator, in this case RKF45 (Forsythe, Malcolm, and Moler 1977) to produce PHI(NR)

```
        PARAMETER(NR=31)
        COMMON/T/        TAU,       NSTOP,      NORUN
     +          /Y/    PHI(NR)
     +          /F/    PHIT(NR)
     +          /C/         DR,          DRS,       R(NR)
     +          /I/         NG,          IP
```

The spatial grid in ξ is stored in array R(NR), which is in COMMON/C/ (note that in the programming, R is used to denote the dimensionless radius, ξ, of eqs. (M2.1) to (M2.4)).

(2) Three runs are programmed for spatial grids of NG = 11, 21, and 31 points so that the effect of varying the grid spacing can be observed

```
C...
C...   SET NUMBER OF RADIAL GRID POINTS
       IF(NORUN.EQ.1)NG=11
       IF(NORUN.EQ.2)NG=21
       IF(NORUN.EQ.3)NG=31
```

(3) The spatial grid is then defined with spacing DR for $0 \leq \xi \leq 1$

```
C...
C...   SPATIAL GRID
       DR=1.0D0/DFLOAT(NG-1)
       DRS=DR**2
       DO 1 I=1,NG
          R(I)=DFLOAT(I-1)*DR
1      CONTINUE
```

Function DFLOAT, listed below, is included because some Fortran compilers do not have this function as an intrinsic

```
       DOUBLE PRECISION FUNCTION DFLOAT(I)
C...
C...   FUNCTION DFLOAT CONVERTS A SINGLE PRECISION INTEGER
C...   INTO A DOUBLE PRECISION FLOATING POINT NUMBER
C...
       IMPLICIT DOUBLE PRECISION (A-H,O-Z)
       DFLOAT=DBLE(FLOAT(I))
       RETURN
       END
```

The statement

```
       DFLOAT=DBLE(FLOAT(I))
```

first converts "I" to a single precision variable by a call to FLOAT, which is part of standard Fortran (and therefore is included in most Fortran compilers). This single precision result is then converted to double precision by a call to DBLE, which again is part of standard Fortran.

(4) Initial condition (M2.2) is programmed in DO loop 2

```
C...
C...   INITIAL CONDITION
       DO 2 I=1,NG
          PHI(I)=0.0D0
2      CONTINUE
```

(5) Finally, initial derivatives are computed and a counter for plotting (used in subroutine PRINT) is initialized

```
C...
C...   INITIAL DERIVATIVES
       CALL DERV
       IP=0
```

Subroutine DERV defines the eq. (M2.1), that is, the NG elements of PHIT(NR) are computed

```
       SUBROUTINE DERV
       IMPLICIT DOUBLE PRECISION (A-H,O-Z)
       PARAMETER(NR=31)
       COMMON/T/       TAU,     NSTOP,     NORUN
      +       /Y/  PHI(NR)
      +       /F/  PHIT(NR)
      +       /C/       DR,       DRS,     R(NR)
      +       /I/       NG,        IP
C...
C...   STEP THROUGH NG GRID POINTS
       DO 1 I=1,NG
C...
C...      R = 0
          IF(I.EQ.1)THEN
              PHIT(1)=4.0D0+4.0D0*(PHI(2)-PHI(1))/DRS
C...
C...      R EQ 1
          ELSE IF(I.EQ.NG)THEN
              PHI(NG)=0.0D0
              PHIT(NG)=0.0D0
C...
C...      R NE 0 OR 1
          ELSE
              PHIT(I)=4.0D0+(PHI(I+1)-2.0D0*PHI(I)+PHI(I-1))/DRS
      +              +(1.0D0/R(I))*(PHI(I+1)-PHI(I-1))/(2.0D0*DR)
          END IF
1      CONTINUE
       RETURN
       END
```

We can note the following points about subroutine DERV:

(1) The derivatives in PHIT are programmed by stepping through NG grid points via DO loop 1

```
C...
C...    STEP THROUGH NG GRID POINTS
        DO 1 I=1,NG
```

(2) At $\xi = 0$, the right hand side (RHS) of eq. (M2.1) is indeterminate (as a consequence of boundary condition (M2.4)), and we have from l'Hospital's rule

$$\lim_{\xi \to \infty} \frac{\partial^2 \phi}{\partial \xi^2} + \frac{1}{\xi}\frac{\partial \phi}{\partial \xi} = 2\frac{\partial^2 \phi}{\partial \xi^2} \quad \text{(M2.5)}$$

Thus, eq. (M2.1) at $\xi = 0$ becomes

$$\frac{\partial \phi}{\partial \tau} = 4 + 2\frac{\partial^2 \phi}{\partial \xi^2} \quad \text{(M2.6)}$$

Equation (M2.6) is programmed as

```
C...
C...        R = 0
            IF(I.EQ.1)THEN
            PHIT(1)=4.0D0+4.0D0*(PHI(2)-PHI(1))/DRS
```

Here we have used in eq. (M2.6) the finite difference approximation

$$2\frac{\partial^2 \phi}{\partial \xi^2} \approx 2\frac{\phi(2) - 2\phi(1) + \phi(0)}{\Delta \xi^2} = 4\frac{\phi(2) - \phi(1)}{\Delta \xi^2} \quad \text{(M2.7)}$$

The RHS term of eq. (M2.7) follows from the approximation of boundary condition (M2.4) as

$$\frac{\partial \phi(\tau, 0)}{\partial \xi} \approx \frac{\phi(2) - \phi(0)}{2\Delta \xi} = 0 \quad \text{(M2.8)}$$

or $\phi(0) = \phi(2)$. In other words, at $\xi = 0$ we have programmed the MOL approximation of eq. (M2.6) as the ODE

$$\frac{d\phi(1)}{d\tau} = 4 + 4\frac{\phi(2) - \phi(1)}{\Delta \xi^2} \quad \text{(M2.9)}$$

(3) Boundary condition (M2.3) is programmed as

```
C...
C...        R EQ 1
            ELSE IF(I.EQ.NG)THEN
            PHI(NG)=0.0D0
            PHIT(NG)=0.0D0
```

Thus, $\phi(1, \tau)$ ($=$ PHI(NG)) is set to one and its temporal derivative ($=$ PHIT(NG)) is set to zero since the derivative of the constant boundary value is zero.

(4) For $0 < \xi < 1$, the MOL approximation of eq. (M2.1) is

$$\frac{d\phi(i)}{d\tau} = 4 + \frac{\phi(i+1) - 2\phi(i) + \phi(i-1)}{\Delta\xi^2} + \frac{1}{\xi(i)} \frac{\phi(i+1) - \phi(i-1)}{2\Delta\xi} \quad (M2.10)$$

which is programmed in DO loop 1 as

```
C...
C...       R NE 0 OR 1
           ELSE
             PHIT(I)=4.0D0+(PHI(I+1)-2.0D0*PHI(I)+PHI(I-1))/DRS
     +              +(1.0D0/R(I))*(PHI(I+1)-PHI(I-1))/(2.0D0*DR)
           END IF
1       CONTINUE
```

This completes the programming of all NG ODEs as the MOL approximation of eqs. (M2.1) to (M2.4).

The solution is printed in subroutine PRINT

```
        SUBROUTINE PRINT(NI,NO)
        IMPLICIT DOUBLE PRECISION (A-H,O-Z)
        PARAMETER(NR=31)
        COMMON/T/      TAU,     NSTOP,      NORUN
     +        /Y/  PHI(NR)
     +        /F/ PHIT(NR)
     +        /C/       DR,       DRS,       R(NR)
     +        /I/       NG,        IP
C...
C...  MONITOR SOLUTION
      WRITE(*,*)TAU,PHI(1)
C...
C...  OPEN A FILE FOR MATLAB PLOTTING OF THE SOLUTION FOR
C...  THE 31 POINT GRID (NORUN = 31)
      IF((TAU.LT.0.001D0).AND.(NORUN.EQ.3))
     +   OPEN(1,FILE='m2.out')
C...
C...  WRITE SOLUTION
      IF(NORUN.EQ.1)NGI=1
      IF(NORUN.EQ.2)NGI=2
      IF(NORUN.EQ.3)NGI=3
      WRITE(NO,1)TAU,(PHI(I),I=1,NG,NGI)
1     FORMAT(/,' TAU = ',F6.2,/,'    R',
     + 3X,'  0',3X,'0.1',3X,'0.2',3X,'0.3',3X,'0.4',
     + 3X,'0.5',3X,'0.6',3X,'0.7',3X,'0.8',3X,'0.9',
     + 3X,'1.0',/,' PHI',11F6.3,/)
C...
```

```
C...     WRITE CENTERLINE VELOCITY TO MORE FIGURES FOR
C...     COMPARISON OF THE SOLUTIONS
         WRITE(NO,2)PHI(1)
2        FORMAT(' PHI(0) = ',F8.5,//)
C...
C...     WRITE THE SOLUTION FOR MATLAB PLOTTING
         IP=IP+1
         IF((NORUN.EQ.3).AND.
       +    ((IP.EQ. 1).OR.(IP.EQ. 2).OR.(IP.EQ. 3).OR.
       +     (IP.EQ. 4).OR.(IP.EQ. 5).OR.(IP.EQ. 7).OR.
       +     (IP.EQ. 9).OR.(IP.EQ.11).OR.(IP.EQ.21)))THEN
         DO 3 I=1,NG
            WRITE(1,4)R(I),PHI(I)
4           FORMAT(F10.3,F10.5)
3        CONTINUE
         END IF
         RETURN
         END
```

We can note the following points about PRINT:

(1) The COMMON area is the same as in INITAL and DERV.

(2) The progress of the solution is monitored on the screen. At the beginning of the solution for the third run (NORUN = 3), corresponding to 31 radial grid points, file m2.out is opened for Matlab plotting

```
C...
C...     MONITOR SOLUTION
         WRITE(*,*)TAU,PHI(1)
C...
C...     OPEN A FILE FOR MATLAB PLOTTING OF THE SOLUTION FOR
C...     THE 31 POINT GRID (NORUN = 31)
         IF((TAU.LT.0.001D0).AND.(NORUN.EQ.3))
       +     OPEN(1,FILE='m2.out')
```

Any plotting system that uses either Fortran-callable subroutines or writes a file that is then subsequently plotted can be used at this point.

(3) The solution is printed during each call to PRINT, for varying numbers of grid points

```
C...
C...     WRITE SOLUTION
         IF(NORUN.EQ.1)NGI=1
         IF(NORUN.EQ.2)NGI=2
         IF(NORUN.EQ.3)NGI=3
         WRITE(NO,1)TAU,(PHI(I),I=1,NG,NGI)
```

```
1       FORMAT(/,' TAU = ',F6.2,/,'   R',
     +  3X,'  0',3X,'0.1',3X,'0.2',3X,'0.3',3X,'0.4',
     +  3X,'0.5',3X,'0.6',3X,'0.7',3X,'0.8',3X,'0.9',
     +  3X,'1.0',/,' PHI',11F6.3,/)
```

Thus, by using the increment NGI, the values of $\phi(\xi, \tau)$ are printed for the same values of ξ, that is, $\xi = 0, 0.1, 0.2, \ldots, 1.0$, for 11, 21, and 31 grid points used in the three runs corresponding to NORUN $= 1, 2$, and 3, respectively.

(4) At the end of each call to PRINT, the centerline velocity is printed to five significant figures for comparison with the analytical solution that is discussed subsequently

```
C...
C...    WRITE CENTERLINE VELOCITY TO MORE FIGURES FOR
C...    COMPARISON OF THE SOLUTIONS
        WRITE(NO,2)PHI(1)
2       FORMAT(' PHI(0) = ',F8.5,//)
```

(5) The numerical solution is stored during calls to PRINT corresponding to $\tau = 0, 0.05, 0.1, 0.15, 0.2, 0.3, 0.4, 0.5$, and 1

```
C...
C...    WRITE THE SOLUTION FOR MATLAB PLOTTING
        IP=IP+1
        IF((NORUN.EQ.3).AND.
     +     ((IP.EQ. 1).OR.(IP.EQ. 2).OR.(IP.EQ. 3).OR.
     +      (IP.EQ. 4).OR.(IP.EQ. 5).OR.(IP.EQ. 7).OR.
     +      (IP.EQ. 9).OR.(IP.EQ.11).OR.(IP.EQ.21)))THEN
        DO 3 I=1,NG
            WRITE(1,4)R(I),PHI(I)
4           FORMAT(F10.3,F10.5)
3       CONTINUE
        END IF
```

PRINT is called with an interval in τ of 0.05 as defined in the following data file. The Matlab plot of the solution is given in Figure M2.1 (the "m" file used to draw Figure M2.1 is available as explained in the preface). Note in particular that boundary conditions (M2.3) and (M2.4) are satisfied by the numerical solution.

The data file read by the main program that calls integrator RKF45 (program RKF45M listed in Appendix A) is listed below:

```
BSL, EX4P1-2, NG = 11
0.          1.0         0.05
   11                       0.0001
```

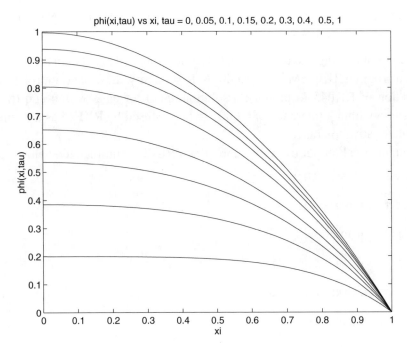

Figure M2.1. Velocity Profiles Computed by the Method of Lines Solution of eqs. (M2.1) to (M2.4)

```
BSL, EX4P1-2, NG = 21
0.          1.0          0.05
   21                      0.0001
BSL, EX4P1-2, NG = 31
0.          1.0          0.05
   31                      0.0001
END OF RUNS
```

We can note the following points about this data file:

(1) Three sets of data are provided corresponding to the three runs for NG = 11, 21, and 31 (NORUN = 1, 2, and 3 in the preceding COMMON/T/). The number of ODEs in each case is designated as the first entry in the third line of data, read by an I5 format; the second number in each of these lines is the error tolerance for the ODE integration, in this case 0.0001, which is read by a 20X,E10.0 format and is used by RKF45 to determine the integration step.

(2) The second line of data in each of the three sets contains the initial value of the independent variable, in the present case $\tau (= 0.0)$, the final value $(= 1.0)$, and the output interval $(= 0.05)$ for the solution (the interval in τ for which subroutine PRINT is called to print the numerical solution), read by a 3E10.0 format.

34   *Momentum transport*

(3) The last line of data, containing the characters END OF RUNS, terminates execution of the main program.

The details of reading and using the data file are explained in detail in the comments of main program RKF45M in Appendix A. In summary, these data control the ODE integration of RKF45 to produce a solution, $\phi(\tau)$, of eqs. (M2.9) and (M2.10) accurate to within a tolerance of 0.0001 (which is used by RKF45 as a composite absolute/relative tolerance).

The numerical output from the subroutine PRINT is summarized below:

```
RUN NO. -    1   BSL, EX4P1-2, NG = 11

INITIAL T -   0.000D+00

  FINAL T -   0.100D+01

  PRINT T -   0.500D-01

NUMBER OF DIFFERENTIAL EQUATIONS -   11

MAXIMUM INTEGRATION ERROR -   0.100D-03

TAU =    0.00
  R      0    0.1   0.2   0.3   0.4   0.5   0.6   0.7   0.8   0.9   1.0
PHI  0.000 0.000 0.000 0.000 0.000 0.000 0.000 0.000 0.000 0.000 0.000


PHI(0) =   0.00000

TAU =    0.05
  R      0    0.1   0.2   0.3   0.4   0.5   0.6   0.7   0.8   0.9   1.0
PHI  0.199 0.199 0.199 0.197 0.194 0.188 0.178 0.158 0.126 0.075 0.000


PHI(0) =   0.19939

TAU =    0.10
  R      0    0.1   0.2   0.3   0.4   0.5   0.6   0.7   0.8   0.9   1.0
PHI  0.384 0.382 0.377 0.368 0.354 0.331 0.299 0.253 0.191 0.107 0.000


PHI(0) =   0.38391

TAU =    0.15
  R      0    0.1   0.2   0.3   0.4   0.5   0.6   0.7   0.8   0.9   1.0
PHI  0.534 0.531 0.520 0.502 0.475 0.437 0.385 0.319 0.234 0.129 0.000


PHI(0) =   0.53416

TAU =    0.20
  R      0    0.1   0.2   0.3   0.4   0.5   0.6   0.7   0.8   0.9   1.0
PHI  0.650 0.645 0.630 0.604 0.566 0.515 0.449 0.367 0.266 0.144 0.000
```

PHI(0) = 0.64972

              .   .
                .   .
              .   .

TAU = 1.00
  R   0   0.1   0.2   0.3   0.4   0.5   0.6   0.7   0.8   0.9   1.0
 PHI 0.996 0.987 0.957 0.907 0.837 0.748 0.638 0.509 0.359 0.190 0.000

PHI(0) = 0.99648

RUN NO. -   2 BSL, EX4P1-2, NG = 21

INITIAL T - 0.000D+00

  FINAL T - 0.100D+01

  PRINT T - 0.500D-01

NUMBER OF DIFFERENTIAL EQUATIONS -   21

MAXIMUM INTEGRATION ERROR - 0.100D-03

TAU = 0.00
  R   0   0.1   0.2   0.3   0.4   0.5   0.6   0.7   0.8   0.9   1.0
 PHI 0.000 0.000 0.000 0.000 0.000 0.000 0.000 0.000 0.000 0.000 0.000

PHI(0) = 0.00000

TAU = 0.05
  R   0   0.1   0.2   0.3   0.4   0.5   0.6   0.7   0.8   0.9   1.0
 PHI 0.200 0.199 0.199 0.198 0.195 0.189 0.178 0.159 0.127 0.076 0.000

PHI(0) = 0.19954

TAU = 0.10
  R   0   0.1   0.2   0.3   0.4   0.5   0.6   0.7   0.8   0.9   1.0
 PHI 0.385 0.383 0.378 0.369 0.355 0.332 0.300 0.254 0.191 0.108 0.000

PHI(0) = 0.38486

TAU = 0.15
  R   0   0.1   0.2   0.3   0.4   0.5   0.6   0.7   0.8   0.9   1.0
 PHI 0.536 0.532 0.522 0.503 0.476 0.438 0.386 0.319 0.235 0.129 0.000

PHI(0) = 0.53556

TAU = 0.20
  R   0   0.1   0.2   0.3   0.4   0.5   0.6   0.7   0.8   0.9   1.0
 PHI 0.651 0.646 0.631 0.605 0.567 0.516 0.450 0.368 0.266 0.144 0.000

PHI(0) = 0.65127

.   .
.   .
.   .

TAU = 1.00
 R   0   0.1   0.2   0.3   0.4   0.5   0.6   0.7   0.8   0.9   1.0
PHI 0.997 0.987 0.957 0.907 0.837 0.748 0.638 0.509 0.359 0.190 0.000

PHI(0) = 0.99654

RUN NO. -   3 BSL, EX4P1-2, NG = 31

INITIAL T - 0.000D+00

  FINAL T - 0.100D+01

  PRINT T - 0.500D-01

NUMBER OF DIFFERENTIAL EQUATIONS - 31

MAXIMUM INTEGRATION ERROR - 0.100D-03

TAU = 0.00
 R   0   0.1   0.2   0.3   0.4   0.5   0.6   0.7   0.8   0.9   1.0
PHI 0.000 0.000 0.000 0.000 0.000 0.000 0.000 0.000 0.000 0.000 0.000

PHI(0) = 0.00000

TAU = 0.05
 R   0   0.1   0.2   0.3   0.4   0.5   0.6   0.7   0.8   0.9   1.0
PHI 0.200 0.199 0.199 0.198 0.195 0.189 0.178 0.159 0.127 0.076 0.000

PHI(0) = 0.19960

TAU = 0.10
 R   0   0.1   0.2   0.3   0.4   0.5   0.6   0.7   0.8   0.9   1.0
PHI 0.385 0.384 0.379 0.370 0.355 0.332 0.300 0.254 0.191 0.108 0.000

PHI(0) = 0.38503

TAU = 0.15
 R   0   0.1   0.2   0.3   0.4   0.5   0.6   0.7   0.8   0.9   1.0
PHI 0.536 0.532 0.522 0.504 0.476 0.438 0.386 0.320 0.235 0.129 0.000

PHI(0) = 0.53585

TAU = 0.20
 R   0   0.1   0.2   0.3   0.4   0.5   0.6   0.7   0.8   0.9   1.0
PHI 0.652 0.647 0.631 0.605 0.567 0.516 0.450 0.368 0.266 0.145 0.000

```
PHI(0) =   0.65156
```

                    .   .
                           .
                    .   .

```
TAU =   1.00
R      0    0.1   0.2   0.3   0.4   0.5   0.6   0.7   0.8   0.9   1.0
PHI 0.997 0.987 0.957 0.907 0.837 0.748 0.638 0.509 0.359 0.190 0.000

PHI(0) =   0.99660
```

We can note the following points about the preceding output:

(1) The velocity starts out for $\tau = 0$ at the initial condition given by eq. (M2.2)

```
TAU =   0.00
R      0    0.1   0.2   0.3   0.4   0.5   0.6   0.7   0.8   0.9   1.0
PHI 0.000 0.000 0.000 0.000 0.000 0.000 0.000 0.000 0.000 0.000 0.000

PHI(0) =   0.00000
```

(2) The velocity profile then develops with increasing τ, as expected. For example, at $\tau = 0.3$ with NG = 11 (NORUN = 1), the velocity profile is

```
TAU =   0.30
R      0    0.1   0.2   0.3   0.4   0.5   0.6   0.7   0.8   0.9   1.0
PHI 0.803 0.796 0.774 0.738 0.686 0.618 0.533 0.430 0.307 0.164 0.000

PHI(0) =   0.80272
```

(3) The centerline velocity, $\phi(\tau, 0)$, approaches one for large τ, which is in accordance with the definition of the velocity $\phi = v_z/v_{max}$. For example, at $\tau = 1$ for NG = 11, the centerline velocity is

```
PHI(0) =   0.99648
```

(4) Convergence of the solution to three figures using 31 radial grid points can be inferred from a comparison of the numerical solutions for 11, 21, and 31 grid points. For example, for $\tau = 0.2$, the three centerline velocities are

```
NG = 11
PHI(0) =   0.64972

NG = 21
PHI(0) =   0.65127

NG = 31
PHI(0) =   0.65156
```

Thus, the use of 31 radial grid points appears to ensure reasonable engineering accuracy. Note that in general this type of check for spatial convergence is essential in order to assess the accuracy of the numerical solution since the method of lines as implemented with a fixed number of grid points does not provide spatial step size adjustment in ξ as RKF45 does in τ.

We can now consider a comparison of the preceding numerical solution with the analytical solution (BSL 1960, eq. (4.1–40))

$$\phi(\xi, \tau) = (1 - \xi^2) - 8 \sum_{n=1}^{\infty} \frac{J_0(\alpha_n \xi)}{\alpha_n^3 J_1(\alpha_n)} e^{-\alpha_n^2 \tau} \quad \text{(M2.11)}$$

where

J_0 Bessel function of the first kind of order zero
J_1 Bessel function of the first kind of order one
α_n roots of $J_0(\alpha) = 0, n = 1, 2, 3, \ldots$

The series of eq. (M2.11) converges rapidly because of the $1/\alpha_n^3$ and $e^{-\alpha_n^2 \tau}$ terms. In the following Fortran program, 20 terms are summed, which provides five-figure accuracy, except for $\tau = 0$.

A main program to evaluate ϕ from eq. (M2.11) for $\xi = 0$ is listed below (note that $J_0(0) = 1$):

```
      PROGRAM M2CL
C...
C...  PROGRAM M2CL EVALUATES THE INFINITE SERIES
C...
C...                    INF           1
C...     PHI(0) = 1 - 8 SUM-----------------*EXP(-E(N)**2*T)
C...                    N=1(E(N)**3)*J1(E(N))
C...
C...  WHICH IS EQ. (4.1-40), BIRD ET AL (1960), WITH R = 0
C...  (FOR THE CENTERLINE DIMENSIONLESS VELOCITY).  HERE
C...
C...     E(N)    NTH ROOT OF J0(X)
C...
C...     J1(X)   BESSEL FUNCTION OF THE FIRST KIND
C...
C...  BIRD, R. B., W. E. STEWART AND E. N. LIGHTFOOT (1960),
C...  TRANSPORT PHENOMENA, JOHN WILEY AND SONS, NEW YORK,
C...  P 129
C...
C...  DOUBLE PRECISION CODING IS USED
      IMPLICIT DOUBLE PRECISION (A-H,O-Z)
C...
C...  INPUT/OUTPUT UNIT NUMBERS
```

```
      COMMON/IO/ NI, NO
C...
C...  NUMBERS OF TERMS IN THE SERIES, NUMBER OF VALUES
C...  OF TAU
      PARAMETER(NTERMS=20, NTAU=9)
C...
C...  DIMENSION ARRAYS
      DIMENSION TAU(NTAU), E(NTERMS), BESSJ1(NTERMS)
C...
C...  DIMENSIONLESS TIMES
      DATA TAU/    0.0D0,     0.05D0,    0.1D0,     0.15D0,
     +             0.2D0,     0.3D0,     0.4D0,     0.5D0,
     +             2.0D0/
C...
C...  OUTPUT FILE
      NO=1
      OPEN(NO,FILE='output')
C...
C...  COMPUTE E(N), J1(E(N))
      CALL ROOTJ0(NTERMS,E,BESSJ1)
C...
C...  PRINT HEADING FOR SERIES SOLUTION
      WRITE(NO,1)
1     FORMAT(//,' EQ. (4.1-40), BSL',/)
C...
C...  EVALUATE THE SERIES FOR NTAU VALUES OF TAU
      DO 3 NT=1,NTAU
C...
C...     INITIALIZE SERIES
         SERIES=1.0D0
C...
C...     SUM SERIES
         DO 4 N=1,NTERMS
         SERIES=SERIES-8.0D0/((E(N)**3)*BESSJ1(N))
     +          *DEXP(-(E(N)**2)*TAU(NT))
4        CONTINUE
C...
C...  WRITE SERIES
      WRITE(NO,2)NT,TAU(NT),SERIES
2     FORMAT('NT = ',I5,'   TAU = ',F6.2,'   PHI(0) = ',F10.5)
C...
C...  NEXT VALUE OF TAU
3     CONTINUE
C...
C...  CALCULATIONS ARE COMPLETE
      END
```

We can note the following points about main program M2CL:

(1) The dimensioning of the arrays is for 20 terms in the series of eq. (M2.11) and nine values of τ: 0, 0.05, 0.1, 0.15, 0.2, 0.3, 0.4, 0.5, and 2.0

```
C...
C...   NUMBERS OF TERMS IN THE SERIES, NUMBER OF VALUES
C...   OF TAU
       PARAMETER(NTERMS=20, NTAU=9)
C...
C...   DIMENSION ARRAYS
       DIMENSION TAU(NTAU), E(NTERMS), BESSJ1(NTERMS)
C...
C...   DIMENSIONLESS TIMES
       DATA TAU/    0.0D0,       0.05D0,      0.1D0,       0.15D0,
      +            0.2D0,        0.3D0,       0.4D0,        0.5D0,
      +            2.0D0/
```

(2) The eigenvalues, α_n, and the corresponding values of $J_1(\alpha_n)$, $n = 1, 2, 3, \ldots,$ 20 are computed by a call to subroutine ROOTJ0 and are returned in arrays E and BESSJ1, respectively.

```
C...
C...   COMPUTE E(N), J1(E(N))
       CALL ROOTJ0(NTERMS,E,BESSJ1)
```

Subroutine ROOTJ0 is discussed subsequently.

(3) After printing a heading for the computed results, the series in eq. (M2.11) is summed (in DO loop 4) and printed (in DO loop 3) for the nine values of τ using the previously calculated values of α_n and $J_1(\alpha_n)$

```
C...
C...   EVALUATE THE SERIES FOR NTAU VALUES OF TAU
       DO 3 NT=1,NTAU
C...
C...      INITIALIZE SERIES
          SERIES=1.0D0
C...
C...      SUM SERIES
          DO 4 N=1,NTERMS
             SERIES=SERIES-8.0D0/((E(N)**3)*BESSJ1(N))
      +            *DEXP(-(E(N)**2)*TAU(NT))
4         CONTINUE
C...
C...      WRITE SERIES
          WRITE(NO,2)NT,TAU(NT),SERIES
```

```
2       FORMAT('NT = ',I5,'   TAU = ',F6.2,'   PHI(0) = ',F10.5)
C...
C...    NEXT VALUE OF TAU
3       CONTINUE
```

Note in particular that the programming in DO loop 4 follows directly from eq. (M2.11).

Subroutine ROOTJ0 is listed below:

```
        SUBROUTINE ROOTJ0(NROOTS,ROOTS,BFJ1)
C...
C...    SUBROUTINE ROOTJ0 COMPUTES THE ROOTS OF J0(X), AND
C...    EVALUATES J1(X) AT THESE ROOTS
C...
C...    ARGUMENTS
C...
C...       NROOTS    NUMBER OF SUCESSIVE ROOTS OF J0(X) TO BE
C...                 COMPUTED STARTING WITH THE FIRST ROOT
C...                 (INPUT)
C...
C...       ROOTS     ARRAY WITH THE NROOTS COMPUTED ROOTS
C...                 (OUTPUT)
C...
C...       BFJ1      ARRAY WITH NROOTS VALUES OF J1(X)
C...                 EVALUATED AT THE NROOTS OF J0(X) (OUTPUT)
C...
C...    DOUBLE PRECISION CODING IS USED
        IMPLICIT DOUBLE PRECISION (A-H,O-Z)
C...
C...    INPUT/OUTPUT UNIT NUMBERS
        COMMON/IO/ NI, NO
C...
C...    ARRAYS FOR ROOTS AND J1
        DIMENSION ROOTS(NROOTS), BFJ1(NROOTS)
C...
C...    STEPS THROUGH NROOTS ROOTS
        DO 1 N=1,NROOTS
C...
C...       COMPUTE ROOT OF J0(X)
           ROOTS(N)=ZERO(N)
C...
C...       EVALUATE J1(X) AT THE ROOT
           BFJ1(N)=BESSJ1(ROOTS(N))
C...
C...       WRITE RESULTS TO OUTPUT FILE WITH LABELS
           WRITE(NO,2)N,ROOTS(N),BFJ1(N)
```

```
2         FORMAT(' N = ',I4,
     +             '    Z FOR J0(Z) = 0',F15.10,
     +             '    J1(Z)',F14.10,/)
C...
C...  NEXT ROOT
1         CONTINUE
          RETURN
          END
```

We can note the following points about subroutine ROOTJ0:

(1) NROOTS (= 20) roots of $J_0(\alpha_n)$ are computed by DO loop 1, that is, $\alpha_1, \alpha_2, \alpha_3, \ldots, \alpha_{20}$. During each pass through DO loop 1, the root is calculated by a call to function ZERO

```
C...
C...  STEPS THROUGH NROOTS ROOTS
      DO 1 N=1,NROOTS
C...
C...      COMPUTE ROOT OF J0(X)
          ROOTS(N)=ZERO(N)
```

Function ZERO is discussed subsequently.

(2) $J_1(\alpha_n)$ (= BFJ1(N)) is then evaluated at the previously computed root by a call to function BESSJ1

```
C...
C...      EVALUATE J1(X) AT THE ROOT
          BFJ1(N)=BESSJ1(ROOTS(N))
```

Function BESSJ1 is discussed subsequently.

(3) The root, α_n, and the corresponding value of $J_1(\alpha_n)$ are printed at the end of DO loop 1

```
C...
C...      WRITE RESULTS TO OUTPUT FILE WITH LABELS
          WRITE(NO,2)N,ROOTS(N),BFJ1(N)
2         FORMAT(' N = ',I4,
     +             '    Z FOR J0(Z) = 0',F15.10,
     +             '    J1(Z)',F14.10,/)
C...
C...  NEXT ROOT
1         CONTINUE
```

Function ZERO is listed below:

```
      DOUBLE PRECISION FUNCTION ZERO(N)
C...
C...  FUNCTION ZERO COMPUTES THE ZEROS OF J0(Z) FROM A
C...  FORMULA GIVEN BY WATSON, G. N., A TREATISE ON THE
C...  THEORY OF BESSEL FUNCTIONS, CAMBRIDGE UNIVERSITY
C...  PRESS, SECOND EDITION, 1966, PAGE 506.
C...
C...  THE FORMULA FROM WATSON IS FOR "LARGE ZEROS".  FOR THE
C...  FIRST FIVE "SMALL ZEROS", THE VALUES ARE SET DIRECTLY
C...  (RATHER THAN BY THE APPROXIMATION)
C...
C...  DOUBLE PRECISION CODING IS USED
      IMPLICIT DOUBLE PRECISION (A-H,O-Z)
C...
C...  THE FIRST FIVE ROOTS OF J0(X) ARE SET EXPLICITLY
      IF(N.EQ.1)ZERO=2.4048255577D0
      IF(N.EQ.2)ZERO=5.5200781103D0
      IF(N.EQ.3)ZERO=8.6537279129D0
      IF(N.EQ.4)ZERO=11.7915344391D0
      IF(N.EQ.5)ZERO=14.9309177086D0
      IF(N.LE.5)RETURN
C...
C...  FOR N GT 5, THE WATSON FORMULA IS USED
      PI=4.0D0*DATAN(1.0D0)
      B=(DFLOAT(N)-0.25D0)*PI
      ZERO=B+1.0D0/((2.0D0**3)*B)-31.0D0/(3.0D0*(2.0D0**7)*
                 (B**3))
     1     +3779.0D0/(15.0D0*(2.0D0**15)*(B**5))
     2     -6277237.0D0/(105.0D0*(2.0D0**15)*(B**7))
      RETURN
      END
```

We can note the following points about ZERO:

(1) The first five roots of $J_0(\alpha)$ are defined explicitly, that is, $\alpha_1 = 2.4048\ldots$, $\alpha_2 = 5.5200\ldots$, $\alpha_3 = 8.6537\ldots$, $\alpha_4 = 11.7915\ldots$, $\alpha_5 = 14.9309\ldots$

```
C...
C...  THE FIRST FIVE ROOTS OF J0(X) ARE SET EXPLICITLY
      IF(N.EQ.1)ZERO=2.4048255577D0
      IF(N.EQ.2)ZERO=5.5200781103D0
      IF(N.EQ.3)ZERO=8.6537279129D0
      IF(N.EQ.4)ZERO=11.7915344391D0
      IF(N.EQ.5)ZERO=14.9309177086D0
      IF(N.LE.5)RETURN
```

This explicit definition of the first several (five) roots was used because of the apparent lack of an accurate formula for the beginning roots of $J_0(\alpha)$.

(2) Beyond the fifth root, a formula was used from the classic book on Bessel functions by Watson (1966, 506)

```
C...
C...   FOR N GT 5, THE WATSON FORMULA IS USED
       PI=4.0D0*DATAN(1.0D0)
       B=(DFLOAT(N)-0.25D0)*PI
       ZERO=B+1.0D0/((2.0D0**3)*B)-31.0D0/(3.0D0*(2.0D0**7)*
                     (B**3))
     1       +3779.0D0/(15.0D0*(2.0D0**15)*(B**5))
     2       -6277237.0D0/(105.0D0*(2.0D0**15)*(B**7))
```

The function DFLOAT, discussed previously, is used to convert the single precision number of the root, n in α_n ($=$ N), to a double precision variable ($=$ B) for use in the Watson formula.

Function BESSJ1 to evaluate $J_1(\alpha)$ at the roots of $J_0(\alpha)$ is listed below:

```
       DOUBLE PRECISION FUNCTION BESSJ1(X)
C...
C...   FUNCTION BESSJ1 COMPUTES THE BESSEL FUNCTION J1(X).
C...   BESSJ1 IS TAKEN FROM W. H. PRESS, B. P. FLANNERY,
C...   S. A. TEUKOLSKY AND V. T. VETTERLING, (1985),
C...   NUMERICAL RECIPES, CAMBRIDGE UNIVERSITY PRESS,
C...   CAMBRIDGE, CHAPTER 6, WITH CONVERSION TO DOUBLE
C...   PRECISION AND EDITING BY W. E. SCHIESSER.
C...
C...   DOUBLE PRECISION CODING IS USED
       IMPLICIT DOUBLE PRECISION (A-H,O-Z)
       DATA           R1,              R2,             R3,
     +                R4,              R5,             R6
     +   /72362614232.D0,    -7895059235.D0,    242396853.1D0,
     +     -2972611.439D0,     15704.48260D0,    -30.16036606D0/
       DATA           S1,              S2,             S3,
     +                S4,              S5,             S6
     +   /144725228442.D0,    2300535178.D0,    18583304.74D0,
     +        99447.43394D0,   376.9991397D0,           1.0D0/
       DATA           P1,              P2,             P3,
     +                P4,              P5/
     +               1.D0,      0.183105D-2,   -0.3516396496D-4,
     +     0.2457520174D-5,  -0.240337019D-6/
       DATA           Q1,              Q2,             Q3,
     +                Q4,              Q5/
     +     0.04687499995D0, -0.2002690873D-3,   0.8449199096D-5,
     +       -0.88228987D-6,   0.105787412D-6/
```

```
      IF(DABS(X).LT.8.0D0)THEN
         Y=X**2
         BESSJ1=X*(R1+Y*(R2+Y*(R3+Y*(R4+Y*(R5+Y*R6)))))
     +          /(S1+Y*(S2+Y*(S3+Y*(S4+Y*(S5+Y*S6)))))
      ELSE
         AX=DABS(X)
         Z=8.0D0/AX
         Y=Z**2
         XX=AX-2.356194491D0
         BESSJ1=DSQRT(0.636619772D0/AX)*(DCOS(XX)*(P1+Y*
     +          (P2+Y*(P3+Y*+(P4+Y*P5))))-Z*DSIN(XX)*(Q1+Y*
     +          (Q2+Y*(Q3+Y*(Q4+Y*Q5)))))*DSIGN(1.0D0,X)
      ENDIF
      RETURN
      END
```

We can note the following points about function BESSJ1:

(1) Constants for use in the formulas for $J_1(\alpha)$ are first evaluated in a series of DATA statements. These constants were converted to double precision format from the single precision constants of the original subroutine BESSJ1 by Press et al. (1985).

(2) Two sets of formulas are used to compute BESSJ1, depending on whether the argument of $J_1(\alpha)$ (= X) corresponds to $\alpha < 8$ or $\alpha \geq 8$.

A summary of the output from the preceding main program and subordinate routines is listed below:

```
N =  1    Z FOR J0(Z) = 0    2.4048255577    J1(Z)   0.5191474974

N =  2    Z FOR J0(Z) = 0    5.5200781103    J1(Z)  -0.3402648073

N =  3    Z FOR J0(Z) = 0    8.6537279129    J1(Z)   0.2714522999

N =  4    Z FOR J0(Z) = 0   11.7915344391    J1(Z)  -0.2324598313

N =  5    Z FOR J0(Z) = 0   14.9309177086    J1(Z)   0.2065464330

N =  6    Z FOR J0(Z) = 0   18.0710638439    J1(Z)  -0.1877288043

N =  7    Z FOR J0(Z) = 0   21.2116365743    J1(Z)   0.1732658946

N =  8    Z FOR J0(Z) = 0   24.3524715029    J1(Z)  -0.1617015508

N =  9    Z FOR J0(Z) = 0   27.4934791169    J1(Z)   0.1521812138
```

```
N = 10    Z FOR J0(Z) = 0   30.6346064596    J1(Z) -0.1441659777
                    .                                 .
                    .                                 .
                    .                                 .
N = 20    Z FOR J0(Z) = 0   62.0484691900    J1(Z) -0.1012934989
```

EQ. (4.1-40), BSL

```
NT =  1    TAU =   0.00    PHI(0) =    0.00015
NT =  2    TAU =   0.05    PHI(0) =    0.19962
NT =  3    TAU =   0.10    PHI(0) =    0.38519
NT =  4    TAU =   0.15    PHI(0) =    0.53607
NT =  5    TAU =   0.20    PHI(0) =    0.65180
NT =  6    TAU =   0.30    PHI(0) =    0.80455
NT =  7    TAU =   0.40    PHI(0) =    0.89038
NT =  8    TAU =   0.50    PHI(0) =    0.93852
NT =  9    TAU =   2.00    PHI(0) =    0.99999
```

We can note the following points about this output:

(1) The series of eq. (M2.11) converges to about four-figure accuracy for $\tau = 0$

```
NT =     1    TAU =    0.00    PHI(0) =      0.00015
```

(the correct value is $\phi(0, 0) = 0$ according to initial condition (M2.2)) because only the term $1/\alpha_n^3$ decreases with increasing n. For $\tau > 0$, the series converges to five-figure accuracy with 20 terms because of the additional term $e^{(-\alpha_n^2 \tau)}$.

(2) The series of eq. (M2.11) agrees with the preceding numerical solution to about three figures for NG = 31 (31 radial grid points). This is demonstrated by the following results at $\tau = 0.2$

(Numerical solution)

```
NG = 11
PHI(0) =  0.64972

NG = 21
PHI(0) =  0.65127

NG = 31
PHI(0) =  0.65156
```

(Series solution, eq. (M2.11))

```
NT =     5    TAU =   0.20    PHI(0) =     0.65180
```

Unsteady laminar flow in a circular tube

These results again indicate that the numerical solution is accurate to three-figures for NG = 31 (since the analytical solution is presumably correct if enough terms are used in the series of eq. (M2.11)).

Finally, we consider another approach to the numerical solution of eqs. (M2.1) to (M2.4) based on the use of finite differences for both the initial-value derivative, $\partial \phi / \partial \tau$, and the boundary-value derivatives, $\partial^2 \phi / \partial \xi^2 + (1/\xi) \partial \phi / \partial \xi$, in eq. (M2.1). Actually, we used finite differences for both groups of derivatives in the preceding method of lines solution, but the finite difference approximation of $\partial \phi / \partial \tau$, and the integration of this derivative with respect to τ, were performed within subroutine RKF45, so we did not have to be concerned with the details of this integration (since they were essentially handled automatically by RKF45). Now, we consider the explicit programming of the finite difference approximation of $\partial \phi / \partial \tau$ and the corresponding integration with respect to τ. However, the programming is set up in a format that closely resembles the previous MOL code, that is, we retain the use of the same subroutines INITAL, DERV, and PRINT.

The main program for the explict programming of the finite differences for the derivative $\partial \phi / \partial \tau$ is listed below:

```
      PROGRAM M2EX
C...
C...  PROGRAM M2EX COMPUTES AN EXPLICIT FINITE DIFFERENCE
C...  SOLUTION TO THE UNSTEADY POISEUILLE FLOW PDE IN A
C...  CIRCULAR TUBE.  THIS IS EQUIVALENT TO AN EXPLICIT
C...  EULER INTEGRATION
C...
      IMPLICIT DOUBLE PRECISION (A-H,O-Z)
      PARAMETER(NR=31)
      COMMON/T/     TAU,     NSTOP,     NORUN
     +       /Y/  PHI(NR)
     +       /F/  PHIT(NR)
     +       /C/       DR,       DRS,    R(NR)
     +       /I/       NG
C...
C...  COMMON AREA FOR INPUT/OUTPUT UNIT NUMBERS
      COMMON/IO/ IO, NO
C...
C...  OPEN OUTPUT FILE
      NO=1
      OPEN(NO,FILE='output')
C...
C...  SET INITIAL CONDITIONS
      TAU=0.0D0
      CALL INITAL
C...
```

```
C...   OUTPUT INTERVAL FOR SOLUTION
       TAUP=0.05D0
C...
C...   UPDATE THE DERIVATIVE VECTOR BY A CALL TO DERV AND
C...   PRINT SOLUTION (INITIAL CONDITION THE FIRST CALL TO
C...   PRINT)
1      CALL DERV
       CALL PRINT(NI,NO)
C...
C...   CHECK FOR END OF SOLUTION
       IF(TAU.GT.0.99D0)STOP
C...
C...   TAKE NSTEPS EULER STEPS, EACH OF LENGTH DTAU
       NSTEPS=200
       DTAU=TAUP/DFLOAT(NSTEPS)
       DO 2 NS=1,NSTEPS
C...
C...      DERIVATIVE VECTOR
          CALL DERV
C...
C...      EULER STEP
          DO 3 I=1,NG
          PHI(I)=PHI(I)+PHIT(I)*DTAU
3         CONTINUE
          TAU=TAU+DTAU
C...
C...      NEXT EULER STEP
2      CONTINUE
C...
C...   NSTEPS COMPLETED.  PRINT SOLUTION
       GO TO 1
       END
```

We can note the following points about main program M2EX (which is essentially a replacement for main program RKF45M listed in Appendix A):

(1) The COMMON area is essentially the same as in the preceding MOL code

```
       PARAMETER(NR=31)
       COMMON/T/      TAU,      NSTOP,     NORUN
      +       /Y/  PHI(NR)
      +       /F/ PHIT(NR)
      +       /C/        DR,       DRS,    R(NR)
      +       /I/       NG
```

The only change in the COMMON area is the removal of the counter IP, since Matlab plotting of the numerical solution is not included in subroutine PRINT.

(2) COMMON/IO/ is used to share the input/output unit numbers between subroutines (which is the same arrangement used in main program RKF45M in Appendix A)

```
C...
C...    COMMON AREA FOR INPUT/OUTPUT UNIT NUMBERS
        COMMON/IO/ IO, NO
C...
C...    OPEN OUTPUT FILE
        NO=1
        OPEN(NO,FILE='output')
```

Also, an output file, output, is again produced by execution of M2EX.

(3) The initial conditions for the 31 ODEs are set by a call to INITAL. However, rather than read the initial value of τ from a data file, as was done in main program RKF45M, the initial value of τ is set just before the call to INITAL

```
C...
C...    SET INITIAL CONDITIONS
        TAU=0.0D0
        CALL INITAL
```

Note that as a result of this call to INITAL, the initial conditions are returned to main program M2EX from INITAL via COMMON/Y/.

(4) The output interval in ($\tau = 0.05$) is then set (rather than reading it from a data file)

```
C...
C...    OUTPUT INTERVAL FOR SOLUTION
        TAUP=0.05D0
```

(5) The initial value integration of the 31 ODEs with respect to τ is then initiated by a call to subroutine DERV, which sets the initial value of the derivative $\partial\phi/\partial\tau$.

```
C...
C...    UPDATE THE DERIVATIVE VECTOR WITH BY A CALL TO DERV
C...    AND PRINTSOLUTION (INITIAL CONDITION THE FIRST CALL
C...    TO PRINT)
1       CALL DERV
        CALL PRINT(NI,NO)
```

Note that the 31 initial derivatives are returned from DERV to M2EX through COMMON/F/. These initial derivatives are then printed by a call to PRINT (since they are also available to PRINT through COMMON/F/).

(6) A check is then made for whether the solution is complete, that is, whether τ has reached the final value $\tau = 1$. Of course, since the solution has just started,

this final value has not yet been reached. However, when it is eventually reached, a STOP statement is executed

```
C...
C...    CHECK FOR END OF SOLUTION
        IF(TAU.GT.0.99D0)STOP
```

(7) To initiate the integration of $\partial\phi/\partial\tau$, 200 integration steps are specified (to cover the interval $0 \leq \tau \leq 0.05$, i.e., to move to the first output point) and the integration step is computed as $\Delta\tau = 0.05/200 = 0.00025$

```
C...
C...    TAKE NSTEPS EULER STEPS, EACH OF LENGTH DTAU
        NSTEPS=200
        DTAU=TAUP/DFLOAT(NSTEPS)
```

The choice of this intergration interval, $\Delta\tau = 0.00025$, is discussed subsequently.

(8) The numerical integration from $\tau = 0$ to $\tau = 0.05$ is then performed by DO loop 2 (note the NSTEPS = 200 integration steps as DO loop 2 is executed).

```
        DO 2 NS=1,NSTEPS
C...
C...        DERIVATIVE VECTOR
            CALL DERV
C...
C...        EULER STEP
            DO 3 I=1,NG
            PHI(I)=PHI(I)+PHIT(I)*DTAU
3           CONTINUE
            TAU=TAU+DTAU
C...
C...        NEXT EULER STEP
2       CONTINUE
```

The essential steps of the numerical integration are

(8.1) Subroutine DERV is called to evaluate the derivative vector at the starting point of each integration step, that is, at $\tau = 0, 0.00025, 0.00050, \ldots, 0.05$

```
C...
C...        DERIVATIVE VECTOR
            CALL DERV
```

(8.2) A step is then taken along the solution using Euler's method

```
C...
C...        EULER STEP
            DO 3 I=1,NG
```

```
           PHI(I)=PHI(I)+PHIT(I)*DTAU
3          CONTINUE
           TAU=TAU+DTAU
```

The programming in DO loop 3 is an implementation of the basic Euler integration formula for ODEs,

$$\phi_{j+1} = \phi_j + \frac{d\phi_j}{d\tau}\Delta\tau \qquad (M2.12)$$

where j is an index denoting the position along the numerical solution, that is, $j = 0$ corresponds to $\tau = 0$, $j = 1$ corresponds to $\tau = 0.00025$, etc. Note that DO loop 3 applies eq. (M2.12) to all NG ($= 31$) ODEs. Note also that the independent variable, τ, is updated when the Euler step is completed, that is, $\tau_{i+1} = \tau_i + \Delta\tau$.

(8.3) After a Euler step of 0.00025 is taken for the NG ODEs by DO loop 3, the next Euler step is taken by the outer DO loop 2

```
C...
C...       NEXT EULER STEP
2          CONTINUE
```

Thus, through the 200 steps of DO loop 2, the solution for the interval $0 \leq \tau \leq 0.05$ is computed.

(9) The solution over one output interval, 0.05, has now been computed, so a branch to statement 1 (the call to DERV) computes the derivative vector at the new output point to produce the derivative vector, $d\phi_j/d\tau$, required by eq. (M2.12) for the next Euler step (and also, the derivative vector will be updated in case it is printed by the subsequent call to PRINT).

```
C...
C...       NSTEPS COMPLETED.  PRINT SOLUTION
           GO TO 1
```

Then subroutine PRINT is called to print the numerical solution (through COMMON/Y/) and possibly the derivative vector (through COMMON/F/).

(10) The test is again made for the end of the solution, and if the final value of τ ($= 1$) has not been reached, the entire procedure is repeated to move the solution to the next output point. This numerical integration is continued until the final value of τ is reached, at which point the Fortran STOP statement is executed. The numerical integration starting at statement 1 will be repeated 20 times in going from $\tau = 0$ to $\tau = 1$ in steps of 0.05.

The output from the preceding main program M2EX, along with subroutines INITAL, DERV, and PRINT considered previously for the MOL solution, is summarized below:

```
TAU =   0.00
  R     0    0.1   0.2   0.3   0.4   0.5   0.6   0.7   0.8   0.9   1.0
PHI 0.000 0.000 0.000 0.000 0.000 0.000 0.000 0.000 0.000 0.000 0.000

PHI(0) =  0.00000

TAU =   0.05
  R     0    0.1   0.2   0.3   0.4   0.5   0.6   0.7   0.8   0.9   1.0
PHI 0.200 0.199 0.199 0.198 0.195 0.189 0.179 0.159 0.127 0.076 0.000

PHI(0) =  0.19962

TAU =   0.10
  R     0    0.1   0.2   0.3   0.4   0.5   0.6   0.7   0.8   0.9   1.0
PHI 0.385 0.384 0.379 0.370 0.355 0.333 0.300 0.254 0.191 0.108 0.000

PHI(0) =  0.38523

TAU =   0.15
  R     0    0.1   0.2   0.3   0.4   0.5   0.6   0.7   0.8   0.9   1.0
PHI 0.536 0.533 0.522 0.504 0.476 0.438 0.387 0.320 0.235 0.129 0.000

PHI(0) =  0.53612

TAU =   0.20
  R     0    0.1   0.2   0.3   0.4   0.5   0.6   0.7   0.8   0.9   1.0
PHI 0.652 0.647 0.632 0.606 0.568 0.516 0.451 0.368 0.267 0.145 0.000

PHI(0) =  0.65185

                       .              .
                       .              .
                       .              .

TAU =   1.00
  R     0    0.1   0.2   0.3   0.4   0.5   0.6   0.7   0.8   0.9   1.0
PHI 0.997 0.987 0.957 0.907 0.837 0.748 0.638 0.509 0.359 0.190 0.000

PHI(0) =  0.99659
```

We note that the finite difference solution essentially agrees with the MOL solution and the series of eq. (M2.11)

```
(MOL numerical solution)

NG = 11

PHI(0) =   0.64972

NG = 21
```

```
PHI(0)  =    0.65127

NG  =   31

PHI(0)  =    0.65156
```

(Series solution, eq. (M2.11))

```
NT  =     5    TAU  =    0.20    PHI(0)  =       0.65180
```

(Finite difference numerical solution)

```
PHI(0)  =    0.65185
```

In fact, the finite difference solution is closer to the series solution than the MOL solution for NG = 31. This closer agreement is probably due to the large number of steps used in the finite difference solution (of course, we do not know how many steps RKF45 used as it adjusted the integration step in τ to meet the error criterion of 0.0001). In any case, we see that a choice of the integration step, in this case $\Delta \tau = 0.00025$, is required in the finite difference method. This step must be selected carefully in accordance with several requirements:

(1) $\Delta \tau$ must be selected small enough to achieve the required accuracy in the numerical solution. In this case, the numerical solution was apparently accurate to at least four figures. However, we had two other independent solutions to gauge the accuracy of the finite difference solution, that is, the MOL solution and the series solution. In general, we will not have other solutions to evaluate a numerical solution, and therefore some experimentation with $\Delta \tau$ would be required, that is, we could reduce $\Delta \tau$ and observe the effect on the solution.

(2) $\Delta \tau$ should not be taken to be excessively small since this would (a) waste computer time and (b) possibly produce unexpected inaccuracy through roundoff error accumulated over a large number of integration steps.

(3) $\Delta \tau$ must be selected small enough to ensure stability of the numerical solution. In general the stability requirement for an explicit finite difference solution like the one programmed in M2EX is of the form

$$\Delta \tau / \Delta \xi^2 < C \quad (M2.13)$$

where C is a constant for a particular finite difference formulation, for example, $C = 1/2$ is typical for explicit finite difference formulations. Thus, as we decrease $\Delta \xi$ by taking more radial grid points to improve the accuracy in the finite difference approximation of the radial group $\partial^2 \phi / \partial \xi^2 + (1/\xi) \partial \phi / \partial \xi$, for example, NG = 11, 21, and 31, the maximum value of $\Delta \tau$ to maintain stability in the numerical solution decreases according to inequality (M2.13). In general, we must be sure that $\Delta\tau$ is small enough to ensure stability for whatever value of $\Delta\xi$ we use. If the stability criterion such as inequality (M2.13) requires an excessively small interval in the initial-value variable (e.g., $\Delta\tau$), possibly because a very small interval in the boundary value (e.g., $\Delta\xi$) is required to achieve the required spatial resolution in the solution, then an implicit initial-value integration should be used rather than an explicit integration such as eq. (M2.12). For example, if the implicit Euler method

$$\phi_{j+1} = \phi_j + \frac{d\phi_{j+1}}{d\tau}\Delta\tau \qquad (M2.14)$$

is used, then the resulting finite difference equations are generally unconditionally stable for linear PDEs, that is, inequality (M2.13) is replaced with

$$\Delta\tau/\Delta\xi^2 < \infty \qquad (M2.15)$$

However, in order to use implicit integration such as eq. (M2.14), solution of systems of simultaneous algebraic equations will be required, that is, we cannot step directly from τ_j to τ_{j+1} as we did in using the explicit Euler method of eq. (M2.12) in DO loop 3 of program M2EX; rather, we must solve systems of algebraic equations to step from τ_j to τ_{j+1} (note that ϕ_{j+1} appears in both sides of eq. (M2.14)).

In summary, we have considered two numerical methods for the solution of PDEs, the method of lines and finite differences, and an analytical method, an infinite series produced through the use of separation of variables and orthogonal functions. We have observed that the three methods give consistent numerical results. Generally, numerical and analytical solutions, when available, should be used in combination as checks on computed results.

References

Bird, R. B., W. E. Stewart, and E. N. Lightfoot. 1960. *Transport Phenomena*. New York: John Wiley and Sons.
Forsythe, G. E., M. A. Malcolm, and C. B. Moler. 1977. *Computer Methods for Mathematical Computations*. Englewood Cliffs, NJ: Prentice-Hall.
Press, W. H., B. P. Flannery, S. A. Teukolsky, and W. T. Vetterling. 1985. *Numerical Recipes*. Cambridge: Cambridge University Press.
Schiesser, W. E. 1991. *The Numerical Method of Lines Integration of Partial Differential Equations*. San Diego: Academic Press.
Watson, G. N. 1966. *A Treatise on the Theory of Bessel Functions*. 2d ed. Cambridge: Cambridge University Press.

M3

Nonlinear, front-sharpening convective systems

We consider here some fundamental characteristics of convective systems that complicate the calculation of numerical solutions of the descriptive partial differential equations (PDEs) for these systems. We start with the momentum balance, the precursor of the Navier-Stokes equations (Bird, Stewart and Lightfoot (BSL) 1960, eq. (3.2-20)):

$$\frac{\partial}{\partial t}(\rho v_x) + \frac{\partial}{\partial x}(\rho v_x v_x) + \frac{\partial}{\partial y}(\rho v_y v_x) + \frac{\partial}{\partial z}(\rho v_z v_x) + \frac{\partial}{\partial x}(\tau_{xx}) + \frac{\partial P}{\partial x} - \rho g_x = 0 \quad \text{(M3.1)}$$

where

v_x, v_y, v_z	components of the velocity vector density
ρ	density
P	pressure
τ_{xx}	normal component of the stress tensor
g_x	x-component of the gravitational acceleration vector

Equation (M3.1) is the x-component of the vector momentum balance (expressing Newton's second law in the x direction).

To focus on the terms of particular interest in eq. (M3.1), we consider one-dimensional flow ($v_y = v_z = 0$) and drop the pressure and gravitational terms

$$\frac{\partial}{\partial t}(\rho v_x) + \frac{\partial}{\partial x}(\rho v_x v_x) + \frac{\partial}{\partial x}(\tau_{xx}) = 0 \quad \text{(M3.2)}$$

Also, if the shear is expressed in terms of a viscosity (through Newton's law for fluids), eq. (M3.2) becomes

$$\frac{\partial}{\partial t}(\rho v_x) + \frac{\partial}{\partial x}(\rho v_x v_x) = \mu \frac{\partial^2 v_x}{\partial x^2} \quad \text{(M3.3)}$$

The next step that is usually taken in analyzing eq. (M3.3) is to expand the left hand side (LHS)

$$\rho \frac{\partial v_x}{\partial t} + v_x \frac{\partial \rho}{\partial t} + \rho v_x \frac{\partial v_x}{\partial x} + v_x \frac{\partial}{\partial x}(\rho v_x) = \mu \frac{\partial^2 v_x}{\partial x^2}$$

or

$$\rho \frac{\partial v_x}{\partial t} + \rho v_x \frac{\partial v_x}{\partial x} + v_x \left\{ \frac{\partial \rho}{\partial t} + \frac{\partial}{\partial x}(\rho v_x) \right\} = \mu \frac{\partial^2 v_x}{\partial x^2} \quad (M3.4)$$

The term in brackets is zero from the continuity equation (mass balance)

$$\frac{\partial \rho}{\partial t} + \frac{\partial}{\partial x}(\rho v_x) = 0 \quad (M3.5)$$

and therefore eq. (M3.4) reduces to

$$\frac{\partial v_x}{\partial t} + v_x \frac{\partial v_x}{\partial x} = \nu \frac{\partial^2 v_x}{\partial x^2} \quad (M3.6)$$

where $\nu = \mu/\rho$ (the kinematic viscosity).

Equation (M3.6) is the well-known Burgers' equation, which has been studied extensively, both analytically and numerically (thus, we might consider Burgers' equation as a special case of the Navier-Stokes equations).

For the purpose of the subsequent analysis, we consider the case $\nu = 0$, and in conformity with the usual convention of the numerical analysis literature, we denote the dependent variable as u, that is, $u \Leftrightarrow v_x$. The equation that is the starting point for the subsequent analysis, the inviscid Burgers' equation, is therefore from eq. (M3.6)

$$\frac{\partial u}{\partial t} + u \frac{\partial u}{\partial x} = 0 \quad (M3.7)$$

Clearly eq. (M3.7) is a highly simplified form of the general momentum balance, eq. (M3.1). However, for the purpose of the subsequent analysis, it has the nonlinear advection group, $\partial u/\partial t + u \partial u/\partial x$, which is of primary interest (and, of course, appears in computational fluid dynamics problems in general, when both the dynamic momentum and convective momentum terms, e.g., $\partial u/\partial t$ and $u \partial u/\partial x$, are taken into consideration).

The intention of the following analysis is to

(1) Compare analytical and numerical solutions to eq. (M3.7) in order to illustrate and evaluate the numerical methods

(2) Illustrate some important properties of the nonlinear advection group, for example, front flattening and front sharpening, which can have a major effect on the performance of numerical methods used to integrate eq. (M3.7).

Thus, we proceed first to an analytical solution of eq. (M3.7).

If we assume a product solution of the form

$$u(x, t) = f(x)g(t) \quad (M3.8)$$

substitution in eq. (M3.7) gives

$$f(x)g'(t) + f(x)g(t)^2 f'(x) = 0 \qquad (M3.9)$$

which can be separated into

$$g'(t)/g^2(t) = -f'(x) = \lambda \qquad (M3.10)$$

where λ is a separation constant (to be determined). Thus, we have two ordinary differential equations (ODEs)

$$g'(t) - \lambda g^2(t) = 0 \qquad (M3.11)$$

$$f'(x) + \lambda = 0 \qquad (M3.12)$$

Equation (M3.11) has a solution

$$g(t) = \frac{1}{a + bt} \qquad (M3.13)$$

where a and b are constants (to be determined) and $\lambda = -b$. Then, from eq. (M3.12)

$$f(x) = bx + c \qquad (M3.14)$$

where c is a constant (to be determined). Thus, a solution to eq. (M3.7) is

$$u(x, t) = \frac{c + bx}{a + bt} \qquad (M3.15)$$

In the subsequent analysis, we take the one initial condition and the one boundary condition required by eq. (M3.7) to be consistent with eq. (M3.15).

We now consider a method of lines (MOL) solution to eq. (M3.7), subject to the initial and boundary conditions from eq. (M3.15). The first case is for (1) $a > 0$ and $b > 0$ and (2) $a > 0$ and $b < 0$, that is, two executions or "runs" of the program. Subroutine INITAL to define the initial condition for eq. (M3.7) (from eq. (M3.15), $u(x, 0) = (c + bx)/a$) is listed below:

```
      SUBROUTINE INITAL
      IMPLICIT DOUBLE PRECISION (A-H,O-Z)
      COMMON/T/    T,   NSTOP,   NORUN
     +      /I/    N,   NCASE,   IP,     NT
C...
C...  SELECT CASE
      NCASE=1
C...
C...  SELECT INITIALIZATION ROUTINE
      IF(NCASE.EQ.1)CALL INIT1
      IF(NCASE.EQ.2)CALL INIT2
      IF(NCASE.EQ.3)CALL INIT3
      RETURN
      END
```

We can note in subroutine INITAL that three cases are programmed, NCASE = 1, 2, or 3, for which three initialization routines are called, INIT1 for NCASE = 1, etc. In the present INITAL, NCASE = 1.

Subroutine INIT1 is listed below:

```fortran
      SUBROUTINE INIT1
      IMPLICIT DOUBLE PRECISION (A-H,O-Z)
      PARAMETER(NX=101)
      COMMON/T/     T,   NSTOP,   NORUN
     +      /Y/  U(NX)
     +      /F/  UT(NX)
     +      /S/  UX(NX)
     +      /X/  X(NX)
     +      /C/       A,       B,       C,     XL,      DX
     +      /I/       N,   NCASE,      IP,     NT
C...
C...  SELECT THE SOLUTION PARAMETERS
C...
C...     FRONT FLATTENING
         IF(NORUN.EQ.1)THEN
         A=1.0D0
         B=1.0D0
         C=0.0D0
         NT=11
C...
C...     FRONT SHARPENING
         ELSE
     +   IF(NORUN.EQ.2)THEN
         A= 1.01D0
         B=-1.0D0
         C= 1.0D0
         NT=11
         END IF
C...
C...  NUMBER OF GRID POINTS
      N=11
C...
C...  TOTAL LENGTH IN X, GRID SPACING
      XL=1.0D0
      DX=XL/DFLOAT(N-1)
C...
C...  INITIAL CONDITION
      DO 1 I=1,N
         X(I)=XL*DFLOAT(I-1)/DFLOAT(N-1)
         U(I)=ANAL(X(I),0.0D0)
1     CONTINUE
```

```
C...
C...   INITIAL DERIVATIVES
       CALL DERV
       IP=0
       RETURN
       END
```

We can note the following points about subroutine INIT1:

(1) The coding is in double precision. A maximum of 101 grid points in x is specified

```
       IMPLICIT DOUBLE PRECISION (A-H,O-Z)
       PARAMETER(NX=101)
       COMMON/T/     T,    NSTOP,   NORUN
      +       /Y/  U(NX)
      +       /F/  UT(NX)
      +       /S/  UX(NX)
      +       /X/  X(NX)
      +       /C/     A,       B,       C,      XL,      DX
      +       /I/     N,   NCASE,      IP,      NT
```

The following variables are used throughout the code:

U(NX)	$u(x,t)$ in eq. (M3.7)
UT(NX)	$\partial u/\partial t$ in eq. (M3.7), subsequently expressed in subscript notation as u_t
UX(NX)	$\partial u/\partial x$ in eq. (M3.7), in subscript notation, u_x
X(NX)	spatial variable in eq. (M3.7), defined on a grid of NX points
A,B,C	a, b, and c in eq. (M3.15)
XL	length of the x axis
DX	interval used in the finite difference approximation of $\partial u/\partial x$ in eq. (M3.7)
N	number of spatial grid points actually used ($\leq$ NX)
NCASE	case implemented in the computer code, subsequently explained
IP	integer index used in the plotting routine PLOTM
NT	total number of output points used in PLOTM

(2) Two runs are programmed

```
C...
C...   SELECT THE SOLUTION PARAMETERS
C...
C...       FRONT FLATTENING
           IF(NORUN.EQ.1)THEN
           A=1.0D0
           B=1.0D0
```

```
              C=0.0D0
              NT=11
C...
C...          FRONT SHARPENING
              ELSE
     +        IF(NORUN.EQ.2)THEN
              A= 1.01D0
              B=-1.0D0
              C= 1.0D0
              NT=11
              END IF
```

For the first run (NORUN = 1), $a = 1$, $b = 1$, and $c = 0$ in eq. (M3.15). For the second run (NORUN = 2), $a = 1.01$, $b = -1$, and $c = 1$ in eq. (M3.15).

(3) In both runs, the number of spatial grid points, N, is 11

```
C...
C...      NUMBER OF GRID POINTS
          N=11
```

A small number of grid points (i.e., N≪ NX) can be used since the solution is quite smooth for both runs (as discussed subsequently).

(4) The total length of the system, XL, is unity, and the grid spacing for the finite difference approximation of $\partial u/\partial x$ in eq. (M3.7), DX, is then computed for this value of the length

```
C...
C...      TOTAL LENGTH IN X, GRID SPACING
          XL=1.0D0
          DX=XL/DFLOAT(N-1)
```

(5) The spatial grid, X(I), and the initial condition from eq. (M3.15), implemented via function ANAL, are defined in DO loop 1

```
C...
C...      INITIAL CONDITION
          DO 1 I=1,N
             X(I)=XL*DFLOAT(I-1)/DFLOAT(N-1)
             U(I)=ANAL(X(I),0.0D0)
1         CONTINUE
C...
C...      INITIAL DERIVATIVES
```

Function ANAL is a straightforward implementation of eq. (M3.15)

```
          DOUBLE PRECISION FUNCTION ANAL(X,T)
          IMPLICIT DOUBLE PRECISION (A-H,O-Z)
```

```
      COMMON/C/      A,      B,      C,      XL,     DX
      ANAL=(C+B*X)/(A+B*T)
      RETURN
      END
```

Note the use of *a*, *b*, and *c* in eq. (M3.15) available through COMMON/C/.

(6) A call to DERV then computes the initial derivatives in eq. (M3.7). Finally, the index for plotting used in subroutine PLOTM is intialized

```
C...
C...  INITIAL DERIVATIVES
      CALL DERV
      IP=0
```

Subroutine DERV for the calculation of the derivatives in eq. (M3.7) is listed below:

```
      SUBROUTINE DERV
      IMPLICIT DOUBLE PRECISION (A-H,O-Z)
      COMMON/T/     T,   NSTOP,   NORUN
     +      /I/     N,   NCASE,      IP,      NT
C...
C...  SELECT DERIVATIVE SUBROUTINE
      IF(NCASE.EQ.1)CALL DERV1
      IF(NCASE.EQ.2)CALL DERV2
      IF(NCASE.EQ.3)CALL DERV3
      RETURN
      END
```

Note the COMMON area is the same as in INITAL, and three derivative subroutines are called, depending on NCASE (currently NCASE = 1, so DERV1 is called).

Subroutine DERV1 is listed below:

```
      SUBROUTINE DERV1
      IMPLICIT DOUBLE PRECISION (A-H,O-Z)
      PARAMETER(NX=101)
      COMMON/T/      T,   NSTOP,   NORUN
     +      /Y/  U(NX)
     +      /F/ UT(NX)
     +      /S/ UX(NX)
     +      /X/  X(NX)
     +      /C/      A,       B,       C,      XL,      DX
     +      /I/      N,   NCASE,      IP,      NT
C...
C...  BOUNDARY CONDITION AT X = 0
      U(1)=ANAL(0.0D0,T)
      UT(1)=0.0D0
```

```
C...
C...   PDE
       DO 1 I=2,N
          UT(I)=-U(I)*(U(I)-U(I-1))/DX
1      CONTINUE
       RETURN
       END
```

We can note the following points about DERV1:

(1) The COMMON area is the same as in INIT1.

(2) The boundary condition for eq. (M3.7), $u(0, t)$, is defined by analytical solution eq. (M3.15) via function ANAL

```
C...
C...   BOUNDARY CONDITION AT X = 0
       U(1)=ANAL(0.0D0,T)
       UT(1)=0.0D0
```

Also, since $u(0, t)$ is defined, its temporal derivative, $\partial u(0, t)/\partial t$, is set to zero so that the ODE integrator does not move $u(0, t)$ away from its prescribed value. Of course, mathematically this is not correct since this temporal derivative is not zero (from eq. (M3.15), $\partial u(0, t)/t = -bc/(a + bt)^2$). However, the next time DERV1 is called, the boundary value, $u(0, t)$, will again be reset according to eq. (M3.15), via the call to ANAL. By zeroing the derivative, $\partial u(0, t)/\partial t$, the ODE integration at $x = 0$ is facilitated. We should note, however, that this procedure may not work correctly with some ODE integrators, for example, the backward differentiation formulas (BDF) integrators of ODEPACK such as LSODE and LSODES and the differential algebraic equations (DAE) integrator DASSL. For such integrators the correct temporal derivative at $x = 0$ may be required, and the setting of the dependent variable at $x = 0$ would then be dropped. In other words, the following alternative programming could be used:

```
C...
C...   BOUNDARY CONDITION AT X = 0
       UT(1)=ANALT(0.0D0,T)
```

where the $\partial u(x, t)/\partial t$ is programmed in function ANALT as

```
       DOUBLE PRECISION FUNCTION ANALT(X,T)
       IMPLICIT DOUBLE PRECISION (A-H,O-Z)
       COMMON/C/      A,      B,      C,      XL,      DX
       ANALT=-B*(C+B*X)/(A+B*T)**2
       RETURN
       END
```

Now we are actually integrating the correct ODE at $x = 0$, and the initial condition for this ODE, $u(0, 0)$, would be set in INIT1 as $u(0, 0) = c/a$ (from eq. (M3.15)).

(3) The MOL approximation of eq. (M3.7) is finally programmed over points 2 to N

```
C...
C...    PDE
        DO 1 I=2,N
          UT(I)=-U(I)*(U(I)-U(I-1))/DX
1       CONTINUE
```

Here we have used the first-order, backward finite difference approximation

$$\frac{\partial u(x,t)}{\partial x} \approx \frac{u(x,t) - u(x - \Delta x, t)}{\Delta x} \Leftrightarrow (U(I) - U(I-1))/DX \quad \text{(M3.16)}$$

This finite difference representation of the spatial derivative in eq. (M3.7) is usually termed a two-point, upwind approximation since the "upwind" point, $u(x - \Delta x, t)$, is used. Note also that this approximation is first-order correct, that is,

$$\frac{\partial u(x,t)}{\partial x} \approx \frac{u(x,t) - u(x - \Delta x, t)}{\Delta x} + O(\Delta x) \quad \text{(M3.17)}$$

where the capital letter "O" is interpreted as "of order." In other words, the approximation of eq. (M3.17) is exact, or "superaccurate", for constant and linear functions (in x), as we will observe in the subsequent numerical solutions.

Subroutine DERV1 defines the temporal derivatives for all N (= 11) MOL ODEs. This derivative vector, UT(N), is sent through COMMON/F/ to an ODE integrator, in this case RKF45 (Forsythe, Malcolm, and Moler 1977). The ODE integrator then returns the dependent variable vector, U(N), through COMMON/Y/ for the programming of the temporal derivatives in DERV1. The dependent variable vector, U(N), which is the numerical solution to eq. (M3.7), is printed via subroutine PRINT, listed below:

```
        SUBROUTINE PRINT(NI,NO)
        IMPLICIT DOUBLE PRECISION (A-H,O-Z)
        COMMON/T/    T,  NSTOP,   NORUN
       +      /I/    N,  NCASE,   IP,    NT
C...
C...    SELECT PRINT ROUTINE
        IF(NCASE.EQ.1)CALL PRINT1(NI,NO)
        IF(NCASE.EQ.2)CALL PRINT2(NI,NO)
        IF(NCASE.EQ.3)CALL PRINT3(NI,NO)
        RETURN
        END
```

PRINT is similar to INITAL and DERV. For NCASE = 1, PRINT1 is called to print and plot the numerical solution resulting from the integration of the ODEs programmed in DERV1. PRINT1 is listed below:

```
      SUBROUTINE PRINT1(NI,NO)
      IMPLICIT DOUBLE PRECISION (A-H,O-Z)
      PARAMETER(NX=101)
      COMMON/T/     T,   NSTOP,   NORUN
     +      /Y/    U(NX)
     +      /F/    UT(NX)
     +      /S/    UX(NX)
     +      /X/    X(NX)
     +      /C/         A,     B,      C,     XL,      DX
     +      /I/         N,     NCASE,  IP,    NT
C...
C...  MONITOR OUTPUT ON SCREEN
      WRITE(*,*)NORUN,T
C...
C...  PRINT NUMERICAL AND ANALYTICAL SOLUTIONS
      WRITE(NO,3)T
3     FORMAT(/,' t = ',F5.2,/,9X,'x',4X,'u(x,t)',
     +                  3X,'ua(x,t)')
      DO 1 I=1,N
      UA=ANAL(X(I),T)
      WRITE(NO,2)X(I),U(I),UA
2     FORMAT(F10.2,2F10.5)
1     CONTINUE
C...
C...  PLOT THE SOLUTION
      IP=IP+1
      CALL PLOTM
      RETURN
      END
```

We can note the following points about subroutine PRINT1:

(1) The COMMON area is the same as in INIT1 and DERV1.

(2) The numerical solution, UA, is computed via function ANAL, then printed along with the numerical solution, U(I), as a function of x (at the values of t corresponding to when PRINT1 is called, as defined by the data file discussed subsequently)

```
C...
C...  PRINT NUMERICAL AND ANALYTICAL SOLUTIONS
      WRITE(NO,3)T
3     FORMAT(/,' t = ',F5.2,/,9X,'x',4X,'u(x,t)',
     +                  3X,'ua(x,t)')
      DO 1 I=1,N
      UA=ANAL(X(I),T)
```

```
              WRITE(NO,2)X(I),U(I),UA
2             FORMAT(F10.2,2F10.5)
1         CONTINUE
```

(3) Finally, the counter IP is incremented each time PRINT1 is called to control the storing and plotting of the solution by subroutine PLOTM

```
C...
C...   PLOT THE SOLUTION
       IP=IP+1
       CALL PLOTM
```

PLOTM writes a computer file m3.out with the solution to be plotted. This file is then read by Matlab to draw the plots discussed subsequently.

```
       SUBROUTINE PLOTM
       IMPLICIT DOUBLE PRECISION(A-H,O-Z)
       PARAMETER(NX=101)
       COMMON/T/     T,   NSTOP,   NORUN
      +       /Y/   U(NX)
      +       /F/   UT(NX)
      +       /S/   UX(NX)
      +       /X/   X(NX)
      +       /C/       A,      B,      C,     XL,     DX
      +       /I/       N,  NCASE,     IP,     NT
C...
C...   OPEN A FILE FOR MATLAB PLOTTING
       IF((IP.EQ.1).AND.(NORUN.EQ.1))THEN
          OPEN(1,FILE='m3.out')
       END IF
C...
C...   WRITE THE NUMERICAL SOLUTION FOR SUBSEQUENT MATLAB
C...   PLOTTING
       DO 2 I=1,N
          WRITE(1,3)X(I),U(I)
3         FORMAT(F10.3,F10.5)
2      CONTINUE
       RETURN
       END
```

The data file read by the main program that calls ODE integrator RKF45 is listed below:

```
Inviscid Burgers equation, front flattening
0.         1.0         0.1
   11                  0.0001
Inviscid Burgers equation, front sharpening
```

```
0.          1.0         0.1
  11                    0.0001
END OF RUNS
```

This file defines

(1) Line 1: A title for the run.

(2) Line 2: The initial, final, and print interval values of t. Thus, PRINT1 is called at $t = 0, 0.1, 0.2, \ldots, 1$ (i.e., 11 times).

(3) Line 3: 11 ODEs to be integrated by RKF45 with an accuracy of 0.0001.

Three lines of data are read for each run. At the end of the second run, the characters END OF RUNS are read to terminate execution of the main program.

Abbreviated output from the main program and subroutine PRINT1 is listed below:

```
RUN NO. -    1  Inviscid Burgers equation, front flattening

INITIAL T -   0.000D+00

  FINAL T -   0.100D+01

  PRINT T -   0.100D+00

NUMBER OF DIFFERENTIAL EQUATIONS -   11

MAXIMUM INTEGRATION ERROR -   0.100D-03

t =  0.00
          x       u(x,t)     ua(x,t)
        0.00     0.00000     0.00000
        0.10     0.10000     0.10000
        0.20     0.20000     0.20000
        0.30     0.30000     0.30000
        0.40     0.40000     0.40000
        0.50     0.50000     0.50000
        0.60     0.60000     0.60000
        0.70     0.70000     0.70000
        0.80     0.80000     0.80000
        0.90     0.90000     0.90000
        1.00     1.00000     1.00000

t =  0.10
          x       u(x,t)     ua(x,t)
        0.00     0.00000     0.00000
        0.10     0.09091     0.09091
        0.20     0.18182     0.18182
```

	0.30	0.27273	0.27273
	0.40	0.36364	0.36364
	0.50	0.45455	0.45455
	0.60	0.54545	0.54545
	0.70	0.63636	0.63636
	0.80	0.72727	0.72727
	0.90	0.81818	0.81818
	1.00	0.90909	0.90909

.
.
.

t = 0.90

x	u(x,t)	ua(x,t)
0.00	0.00000	0.00000
0.10	0.05263	0.05263
0.20	0.10526	0.10526
0.30	0.15789	0.15789
0.40	0.21053	0.21053
0.50	0.26316	0.26316
0.60	0.31579	0.31579
0.70	0.36842	0.36842
0.80	0.42105	0.42105
0.90	0.47368	0.47368
1.00	0.52632	0.52632

t = 1.00

x	u(x,t)	ua(x,t)
0.00	0.00000	0.00000
0.10	0.05000	0.05000
0.20	0.10000	0.10000
0.30	0.15000	0.15000
0.40	0.20000	0.20000
0.50	0.25000	0.25000
0.60	0.30000	0.30000
0.70	0.35000	0.35000
0.80	0.40000	0.40000
0.90	0.45000	0.45000
1.00	0.50000	0.50000

RUN NO. -   2   Inviscid Burgers equation, front sharpening

INITIAL T -   0.000D+00

  FINAL T -   0.100D+01

  PRINT T -   0.100D+00

NUMBER OF DIFFERENTIAL EQUATIONS -   11

MAXIMUM INTEGRATION ERROR -   0.100D-03

```
t =   0.00
         x      u(x,t)    ua(x,t)
       0.00    0.99010    0.99010
       0.10    0.89109    0.89109
       0.20    0.79208    0.79208
       0.30    0.69307    0.69307
       0.40    0.59406    0.59406
       0.50    0.49505    0.49505
       0.60    0.39604    0.39604
       0.70    0.29703    0.29703
       0.80    0.19802    0.19802
       0.90    0.09901    0.09901
       1.00    0.00000    0.00000

t =   0.10
         x      u(x,t)    ua(x,t)
       0.00    1.09890    1.09890
       0.10    0.98900    0.98901
       0.20    0.87915    0.87912
       0.30    0.76921    0.76923
       0.40    0.65934    0.65934
       0.50    0.54945    0.54945
       0.60    0.43956    0.43956
       0.70    0.32967    0.32967
       0.80    0.21978    0.21978
       0.90    0.10989    0.10989
       1.00    0.00000    0.00000
               .
               .
               .

t =   0.90
         x      u(x,t)    ua(x,t)
       0.00    9.09091    9.09091
       0.10    8.18169    8.18182
       0.20    7.27312    7.27273
       0.30    6.36346    6.36364
       0.40    5.45462    5.45455
       0.50    4.54549    4.54545
       0.60    3.63640    3.63636
       0.70    2.72731    2.72727
       0.80    1.81822    1.81818
```

```
      0.90    0.90911    0.90909
      1.00    0.00000    0.00000

t =   1.00
        x     u(x,t)     ua(x,t)
      0.00  100.00000  100.00000
      0.10   89.99937   90.00000
      0.20   80.00154   80.00000
      0.30   69.99988   70.00000
      0.40   60.00046   60.00000
      0.50   50.00040   50.00000
      0.60   40.00043   40.00000
      0.70   30.00044   30.00000
      0.80   20.00045   20.00000
      0.90   10.00042   10.00000
      1.00    0.00000    0.00000
```

We can note the following points about this output:

(1) Considering first the output from the first run, the initial condition of eq. (M3.15) with $a = b = 1$, $c = 0$ (set in INIT1) is

$$u(x, 0) = x \tag{M3.18}$$

This linear function is apparent in the output at $t = 0$

```
t =   0.00
        x     u(x,t)    ua(x,t)
      0.00   0.00000    0.00000
      0.10   0.10000    0.10000
      0.20   0.20000    0.20000
      0.30   0.30000    0.30000
      0.40   0.40000    0.40000
      0.50   0.50000    0.50000
      0.60   0.60000    0.60000
      0.70   0.70000    0.70000
      0.80   0.80000    0.80000
      0.90   0.90000    0.90000
      1.00   1.00000    1.00000
```

(Checking the initial condition is, of course, always important in developing a numerical PDE solution; also, the output of the solution is in accordance with the data file, i.e., the output is at $t = 0, 0.1, 0.2, \ldots, 1.$)

(2) The evolution of the solution is in accordance with eq. (M3.15)

$$u(x, t) = \frac{x}{1+t} \tag{M3.19}$$

Note that at $t = 1$, the analytical and numerical solutions are still in agreement to five figures, that is, numerical integration errors did not accumulate significantly as the solution evolved. This agreement is not unexpected since the solution is linear in x (from eq. (M3.19)) and therefore the finite difference approximation of eq. (M3.17) is exact. Also, the integration in t by RKF45 surpassed the error tolerance of 0.0001 (RKF45 uses this number in a combined absolute/error tolerance). Of course we cannot generally expect an accurate numerical solution to a PDE to be computed so easily; for example, the first-order approximation of eq. (M3.17) will typically give solutions of limited accuracy since the variation of the solution with x will usually depart substantially from a linear relationship.

(3) Continuing on to the solution for the second run, we see again that the numerical solution is in five-figure agreement with the analytical solution (with $a = 1.01, b = -1$, and $c = 1$ in eq. (M3.15))

$$u(x, t) = \frac{1 - x}{1.01 - t} \quad \quad (M3.20)$$

Note in particular that the initial condition, $u(x, 0) = (1 - x)/1.01$, is again linear in x

```
t =   0.00
       x       u(x,t)    ua(x,t)
      0.00    0.99010    0.99010
      0.10    0.89109    0.89109
      0.20    0.79208    0.79208
      0.30    0.69307    0.69307
      0.40    0.59406    0.59406
      0.50    0.49505    0.49505
      0.60    0.39604    0.39604
      0.70    0.29703    0.29703
      0.80    0.19802    0.19802
      0.90    0.09901    0.09901
      1.00    0.00000    0.00000
```

and at $t = 1$, the solution has increased to a maximum value of 100

```
t =   1.00
       x       u(x,t)      ua(x,t)
      0.00  100.00000   100.00000
      0.10   89.99937    90.00000
      0.20   80.00154    80.00000
      0.30   69.99988    70.00000
      0.40   60.00046    60.00000
      0.50   50.00040    50.00000
      0.60   40.00043    40.00000
      0.70   30.00044    30.00000
```

```
      0.80    20.00045    20.00000
      0.90    10.00042    10.00000
      1.00     0.00000     0.00000
```

The five-figure agreement at $t = 1$ is particularly interesting since the solution of eq. (M3.20) has a singularity at $t = 1.01$ (from eq. (M3.20)).

The solutions for the two runs are plotted in Figures M3.1a and M3.1b. The linear character of these solutions is apparent (which also suggests that good accuracy in the solutions could have been achieved using only two points in the spatial grid in x). Also, the front flattening and front sharpening of the two solutions is clear. In the first solution, the front flattening occurs because the velocity, the multiplying u in the nonlinear advective group $u\partial u/\partial x$ in eq. (M3.7), is relatively small at small values of x, that is, the left end is flowing more slowly than the right end. In the second solution, the front sharpens because the velocity is relatively large at small values of x, the left end is flowing more rapidly than the right end. The "m" file used to draw Figures M3.1a and M3.1b is available as explained in the preface.

The preceding solutions are not typical of PDE solutions in the sense that they were easily calculated to good accuracy with a small number of spatial grid points. To investigate a more typical PDE problem, we repeat the preceding example but keep the boundary value, $u(0, t)$, fixed at its initial value, $u(0, 0)$. Physically, this might represent an entering velocity that does not change with time, that is, a constant entering flow rate. The preceding programming is modified in a minor way to include this revised boundary condition. However, to clarify just how this case was programmed, we now consider subroutines INIT2, DERV2, and PRINT2, which are called from INITAL, DERV, and PRINT, respectively, by setting NCASE $= 2$ in INITAL.

Subroutine INIT2 is listed below:

```
      SUBROUTINE INIT2
      IMPLICIT DOUBLE PRECISION (A-H,O-Z)
      PARAMETER(NX=101)
      COMMON/T/      T,   NSTOP,   NORUN
     +       /Y/  U(NX)
     +       /F/  UT(NX)
     +       /S/  UX(NX)
     +       /X/  X(NX)
     +       /C/      A,       B,       C,      XL,      DX
     +       /I/      N,   NCASE,      IP,      NT
C...
C...  SELECT THE SOLUTION PARAMETERS
C...
C...      FRONT FLATTENING
          IF(NORUN.EQ.1)THEN
```

72 *Momentum transport*

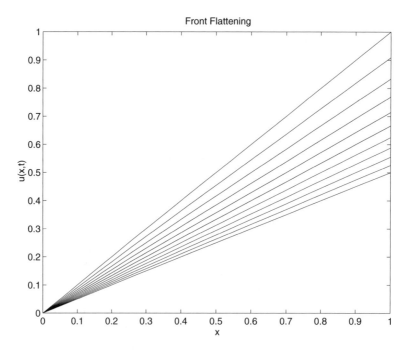

Figure M3.1a. Front-Flattening Solution of Eq. (M3.7) with Initial and Boundary Conditions from Eq. (M3.15)

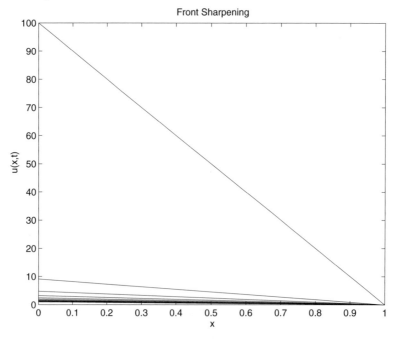

Figure M3.1b. Front-Sharpening Solution of Eq. (M3.7) with Initial and Boundary Conditions from Eq. (M3.15)

```
              A=1.0D0
              B=1.0D0
              C=0.0D0
              NT=11
C...
C...          FRONT SHARPENING
              ELSE
     +        IF(NORUN.EQ.2)THEN
              A= 1.0D0
              B=-1.0D0
              C= 1.0D0
              NT=13
              END IF
C...
C...     NUMBER OF GRID POINTS
         N=101
C...
C...     TOTAL LENGTH IN X, GRID SPACING
         XL=1.0D0
         DX=XL/DFLOAT(N-1)
C...
C...     INITIAL CONDITION
         DO 1 I=1,N
            X(I)=XL*DFLOAT(I-1)/DFLOAT(N-1)
            U(I)=ANAL(X(I),0.0D0)
1        CONTINUE
C...
C...     INITIAL DERIVATIVES
         CALL DERV
         IP=0
         RETURN
         END
```

INIT2 differs from INIT1 in only two ways:

(1) The number of spatial grid points has been increased from 11 to 101 to improve the spatial resolution, which is required because the solution is no longer necessarily linear in x.

(2) The value of a in the second run was changed from 1.01 to 1 and the number of output points, NT, was increased from 11 to 13 (so that PRINT2 is called at $t = 0, 0.1, 0.2, \ldots, 1.2$). Note that this does not cause a problem with the singularity of eq. (M3.20) (at $t = 1.01$) since eq. (M3.20) is not the analytical solution to this problem (in fact, the analytical solution is unknown). However, we still use eq. (M3.15) to define the initial condition in DO loop 1.

Subroutine DERV2 is listed below:

```
      SUBROUTINE DERV2
      IMPLICIT DOUBLE PRECISION (A-H,O-Z)
      PARAMETER(NX=101)
      COMMON/T/      T,  NSTOP,   NORUN
     +      /Y/ U(NX)
     +      /F/ UT(NX)
     +      /S/ UX(NX)
     +      /X/ X(NX)
     +      /C/      A,     B,      C,    XL,      DX
     +      /I/      N, NCASE,     IP,    NT
C...
C... BOUNDARY CONDITION AT X = 0
      U(1)=ANAL(0.0D0,0.0D0)
      UT(1)=0.0D0
C...
C... PDE
      DO 1 I=2,N
         UT(I)=-U(I)*(U(I)-U(I-1))/DX
1     CONTINUE
      RETURN
      END
```

Subroutine DERV2 differs from subroutine DERV1 in only one way: The call to ANAL to define the boundary condition at $x = 0$ is

```
U(1)=ANAL(0.0D0,0.0D0)
```

rather than

```
U(1)=ANAL(0.0D0,T)
```

Thus, the boundary value, $u(0, t)$, remains at the initial value, $u(0, 0)$.
Subroutine PRINT2 is identical to PRINT1 and therefore is not listed.

The data file for NCASE = 2 is

```
Inviscid Burgers equation, front flattening, n = 101
0.          1.0         0.1
  101                        0.0001
Inviscid Burgers equation, front sharpening, n = 101
0.          1.2         0.1
  101                        0.0001
END OF RUNS
```

There are two differences in this data file with respect to the NCASE = 1 data file:

(1) The number of ODEs is now 101.

(2) For the second solution, t runs to 1.2.

Abbreviated output from these the two runs is listed below:

```
RUN NO. -    1  Inviscid Burgers equation, front flattening,
                n = 101

INITIAL T -  0.000D+00

  FINAL T -  0.100D+01

  PRINT T -  0.100D+00

NUMBER OF DIFFERENTIAL EQUATIONS - 101

MAXIMUM INTEGRATION ERROR -   0.100D-03

t =  0.00
          x      u(x,t)
        0.00    0.00000
        0.01    0.01000
        0.02    0.02000
        0.03    0.03000
        0.04    0.04000
        0.05    0.05000
          .        .
          .        .
          .        .
        0.95    0.95000
        0.96    0.96000
        0.97    0.97000
        0.98    0.98000
        0.99    0.99000
        1.00    1.00000

t =  0.10
          x      u(x,t)
        0.00    0.00000
        0.01    0.00909
        0.02    0.01818
        0.03    0.02727
        0.04    0.03636
        0.05    0.04545
          .        .
          .        .
          .        .
        0.95    0.86364
```

```
          0.96    0.87273
          0.97    0.88182
          0.98    0.89091
          0.99    0.90000
          1.00    0.90909
            .
            .
            .

t =    1.00
           x    u(x,t)
          0.00    0.00000
          0.01    0.00500
          0.02    0.01000
          0.03    0.01500
          0.04    0.02000
          0.05    0.02500
            .      .
            .      .
            .      .
          0.95    0.47500
          0.96    0.48000
          0.97    0.48500
          0.98    0.49000
          0.99    0.49500
          1.00    0.50000
```

RUN NO. -   2 Inviscid Burgers equation, front sharpening, n = 101

INITIAL T -   0.000D+00

  FINAL T -   0.120D+01

  PRINT T -   0.100D+00

NUMBER OF DIFFERENTIAL EQUATIONS - 101

MAXIMUM INTEGRATION ERROR -   0.100D-03

```
t =    0.00
           x    u(x,t)
          0.00    1.00000
          0.01    0.99000
          0.02    0.98000
          0.03    0.97000
```

0.04	0.96000
0.05	0.95000
.	
.	
.	
0.95	0.05000
0.96	0.04000
0.97	0.03000
0.98	0.02000
0.99	0.01000
1.00	0.00000
.	
.	
.	

t = 1.20

x	u(x,t)
0.00	1.00000
0.01	1.00000
0.02	1.00000
0.03	1.00000
0.04	1.00000
0.05	1.00000
.	.
.	.
.	.
0.95	0.99951
0.96	0.99935
0.97	0.99909
0.98	0.99876
0.99	0.99833
1.00	0.00000

We can note the following points about the preceding output:

(1) The first solution demonstrates front flattening and remains linear throughout the interval $0 \leq t \leq 1$; these characteristics are apparent in the plot of the solution in Figure M3.2a.

(2) The second solution demonstrates front sharpening; by the time it has progressed to $t = 1.2$, the solution decreases from 0.99833 to 0.00000 within an interval $\Delta x = 0.01$. This front sharpening is apparent in the plot of the solution in Figure M3.2b and indicates the requirement for a finer spatial grid, for example, 101 grid points rather than 11. The performance of the first-order approximation of eq. (M3.17) is surprisingly good considering the sharp spatial variation in the solution; in fact, we cannot usually expect such good results from a low-order approximation.

78   *Momentum transport*

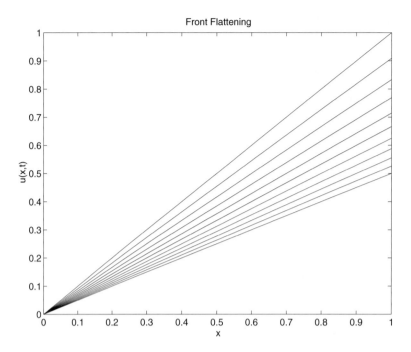

Figure M3.2a. Front-Flattening Solution of Eq. (M3.7) for Constant Boundary Condition Using Upwind Approximations

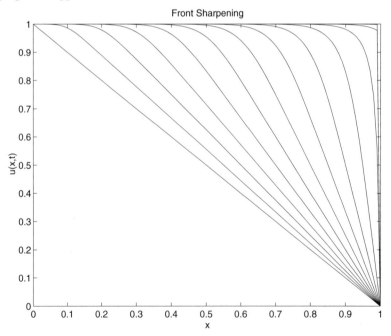

Figure M3.2b. Front-Sharpening Solution of Eq. (M3.7) for Constant Boundary Condition Using Upwind Approximations

(3) An interesting feature of the second solution is $u(1, t) = 0$. In other words, the front does not "flow out of the system" at $x = 1$, as indicated in Figure M3.2b, but rather, merely sharpens at $x = 1$. The reason for this can be seen in considering DO loop 2 of DERV2

```
C...
C...    PDE
        DO 1 I=2,N
            UT(I)=-U(I)*(U(I)-U(I-1))/DX
1       CONTINUE
```

Note that initially $u(1, 0) = 0$. Therefore, at $t = 0$, U(N) = 0, and from DO loop 2, UT(N) = 0. Since the velocity at $x = 1$, U(N), starts at zero, and its temporal derivative therefore is zero, it cannot change with increasing t, that is, it remains at zero, which is what we observe in the numerical solution. Thus, the case $u(1, 0) \neq 0$ should be interesting and can easily be run by a small change in the initial condition set in INIT2.

To investigate the effect of the order of the approximation for the spatial derivative, $\partial u / \partial x$, in eq. (M3.7), we now consider using a second-order, centered approximation

$$\frac{\partial u(x, t)}{\partial x} \approx \frac{u(x + \Delta x, t) - u(x - \Delta x, t)}{2\Delta x} + O(\Delta x^2) \tag{M3.21}$$

Intuitively it would seem that the centered approximation of eq. (M3.21) should produce a better solution than the upwind approximation of eq. (M3.17). To investigate this point, we again consider the preceding INITAL, DERV, and PRINT, with NCASE = 3, for which INIT3, DERV3, and PRINT3 are called. INIT3 is essentially the same as INIT2, so it is not listed here. The only difference is that the second solution is terminated at $t = 1$ (NT = 11 in INIT3) rather than at $t = 1.2$ (NT = 13 in INIT2); the reason for the shorter run with INIT3 is explained subsequently. DERV3 is listed below:

```
        SUBROUTINE DERV3
        IMPLICIT DOUBLE PRECISION (A-H,O-Z)
        PARAMETER(NX=101)
        COMMON/T/      T,  NSTOP,   NORUN
       +       /Y/  U(NX)
       +       /F/  UT(NX)
       +       /S/  UX(NX)
       +       /X/  X(NX)
       +       /C/     A,       B,       C,     XL,     DX
       +       /I/     N,  NCASE,      IP,     NT
C...
C...    BOUNDARY CONDITION AT X = 0
```

```
      U(1)=ANAL(0.0D0,0.0D0)
      UT(1)=0.0D0
C...
C...  PDE
C...  X NE 0 OR XL
      DO 1 I=2,N-1
         UT(I)=-U(I)*(U(I+1)-U(I-1))/(2.0D0*DX)
1     CONTINUE
C...
C...  X = XL
      UT(N)=-U(N)*(U(N)-U(N-1))/DX
      RETURN
      END
```

We can note the following points about DERV3:

(1) The boundary condition at $x = 0$ is the same as in DERV2

```
C...
C...  BOUNDARY CONDITION AT X = 0
      U(1)=ANAL(0.0D0,0.0D0)
      UT(1)=0.0D0
```

(2) The temporal derivative $\partial u(1,t)/\partial t$ is computed with the approximation of eq. (M3.17) in order to avoid the use of the fictitious value $u(1 + \Delta x, t)$

```
C...
C...  X = XL
      UT(N)=-U(N)*(U(N)-U(N-1))/DX
```

(3) The interior temporal derivatives are computed with the centered approximation of eq. (M3.21)

```
C...
C...  PDE
C...  X NE 0 OR XL
      DO 1 I=2,N-1
         UT(I)=-U(I)*(U(I+1)-U(I-1))/(2.0D0*DX)
1     CONTINUE
```

Subroutine PRINT3 is the same as PRINT2, so it is not listed here.

The data file for the two solutions computed with DERV3 is listed below:

```
Inviscid Burgers equation, front flattening, n = 101
0.          1.0         0.1
  101                      0.0001
Inviscid Burgers equation, front sharpening, n = 101
0.          1.0         0.1
```

```
     101                   0.0001
END OF RUNS
```

Note that the only difference between this data file and that for NCASE = 2 is the termination of the second run at $t = 1$ rather than 1.2.

Abbreviated output from INIT3, DERV3, and PRINT3 is listed below:

```
RUN NO. -    1   Inviscid Burgers equation, front flattening,
                 n = 101

INITIAL T -  0.000D+00

  FINAL T -  0.100D+01

  PRINT T -  0.100D+00

NUMBER OF DIFFERENTIAL EQUATIONS - 101

MAXIMUM INTEGRATION ERROR -   0.100D-03

t =  0.00
            x      u(x,t)
         0.00    0.00000
         0.01    0.01000
         0.02    0.02000
         0.03    0.03000
         0.04    0.04000
         0.05    0.05000
           .        .
           .        .
           .        .
         0.95    0.95000
         0.96    0.96000
         0.97    0.97000
         0.98    0.98000
         0.99    0.99000
         1.00    1.00000

t =  0.10
            x      u(x,t)
         0.00    0.00000
         0.01    0.00909
         0.02    0.01818
         0.03    0.02727
         0.04    0.03636
         0.05    0.04545
```

```
            .    .
            .    .
            .    .
         0.95   0.86364
         0.96   0.87273
         0.97   0.88182
         0.98   0.89091
         0.99   0.90000
         1.00   0.90909
            .
            .
            .

t =   1.00
          x    u(x,t)
       0.00   0.00000
       0.01   0.00500
       0.02   0.01000
       0.03   0.01500
       0.04   0.02000
       0.05   0.02500
          .
          .
          .
       0.95   0.47495
       0.96   0.47996
       0.97   0.48504
       0.98   0.49004
       0.99   0.49499
       1.00   0.49998
```

RUN NO. -   2   Inviscid Burgers equation, front sharpening, n = 101

INITIAL T -   0.000D+00

  FINAL T -   0.100D+01

  PRINT T -   0.100D+00

NUMBER OF DIFFERENTIAL EQUATIONS - 101

MAXIMUM INTEGRATION ERROR -   0.100D-03

```
t =   0.00
        x      u(x,t)
       0.00   1.00000
       0.01   0.99000
       0.02   0.98000
       0.03   0.97000
       0.04   0.96000
       0.05   0.95000
         .       .
         .       .
         .       .
       0.95   0.05000
       0.96   0.04000
       0.97   0.03000
       0.98   0.02000
       0.99   0.01000
       1.00   0.00000

t =   0.10
        x      u(x,t)
       0.00   1.00000
       0.01   1.00023
       0.02   1.00060
       0.03   0.99974
       0.04   0.99861
       0.05   0.99900
         .       .
         .       .
         .       .
       0.95   0.05556
       0.96   0.04444
       0.97   0.03333
       0.98   0.02222
       0.99   0.01111
       1.00   0.00000
         .
         .
         .

t =   1.00
        x      u(x,t)
       0.00   1.00000
       0.01   1.00000
       0.02   1.00000
       0.03   1.00000
```

```
0.04    1.00000
0.05    1.00001
  .       .
  .       .
  .       .
0.95    1.01462
0.96    1.02332
0.97    1.03823
0.98    1.07794
0.99    1.34886
1.00    0.00000
```

We can note the following points about this output:

(1) The first solution is essentially the same as with NCASE = 2, which is not unexpected since the solution is linear, and we would therefore expect the second-order approximation of eq. (M3.21) to perform as well as the first-order approximation of eq. (M3.17). In fact, the only difference between the NCASE = 2 and NCASE = 3 solutions is due to the mixed use of eq. (M3.17) (at $x = 1$) and eq. (M3.21) (at $x \neq 1$) in DERV3. The linear character of the first solution is apparent in Figure M3.3a.

(2) The second solution has unrealistic oscillations (e.g., note $u(0.99, 1) = 1.34886$), and these oscillations become larger with increasing t (which is the reason for terminating the second solution at $t = 1$; beyond $t = 1$, the oscillations become so large they cannot be plotted). The character of the second solution is evident in Figure M3.3b.

From these results, we arrive at the following conclusions:

(1) Centered approximations do not work well for the dynamic analysis of strongly convective systems.

(2) Some form of upwinding is essential for the dynamic analysis of strongly convective systems.

Of course, we have not proven either of these conclusions rigorously, but rather, we infer them from the results of the one test problem based on eq. (M3.7). More generally, these conclusions have proven to be correct based on accumulated experience.

Conclusion (2) has an important implication. In using upwinding, the direction of flow must be known (in order to know the upwind direction for the approximation). But, in general, during the analysis of a convective system, we may not know the direction of flow intuitively. Therefore, some test or analytical method must be included in the analysis to indicate the correct direction for upwinding. If this is done incorrectly and downwinding is actually used, the solution will in general be unstable.

Nonlinear, front-sharpening convective systems

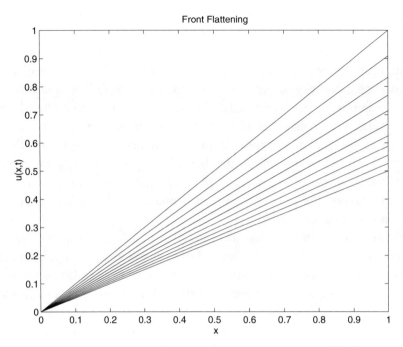

Figure M3.3a. Front-Flattening Solution of Eq. (M3.7) for Constant Boundary Condition Using Centered Approximations

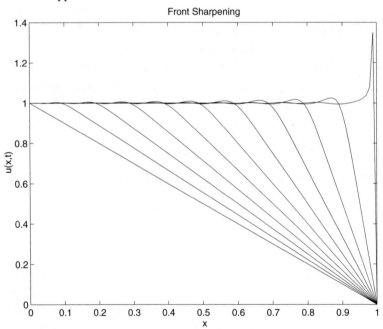

Figure M3.3b. Front-Sharpening Solution of Eq. (M3.7) for Constant Boundary Condition Using Centered Approximations

Finally, we mention one other generalization based essentially on experience. If a system of PDEs for a strongly convective system is to be integrated numerically, upwinding in all of the PDEs may not be required. For example, if the continuity, momentum, and energy equations are to be integrated, upwinding of the continuity and energy equations may be sufficient to compute a solution. The momentum balance can be approximated by centered differences, which is an important advantage since the momentum balance in effect provides the fluid velocity (i.e., the direction of flow is not required in approximating the momentum balance, and the fluid velocity that comes from the momentum balance can be used with appropriate upwinding in the continuity and energy balances). Again, we emphasize that there is apparently no proof of this statement, and some analysis to at least support this conclusion would be reassuring.

We should also appreciate that the features we have noted in the preceding solutions, for example, front-flattening and front-sharpening, are nonlinear effects (due to the nonlinear convective term $u\partial u/\partial x$ that occurs frequently in a momentum balance). In contrast, a linear advection term such as $v\partial u/\partial x$ (with v constant) would not produce such unsymmetrical effects, for example, $u(x, t)$ increasing or decreasing with x would produce solutions with essentially the same features.

In other words, nonlinear terms in a PDE can often produce unexpected, and even surprising, effects in the solution. For example, front-sharpening can progress to the point that a shock developes. Under these circumstances a numerical method will fail unless special precautions are taken to "capture the shock" (the jargon used for this type of problem and associated numerical methods). However, in the preceding examples, we took special precautions to ensure that the numerical methods did not fail outright, that is, they performed well enough that the numerical output could at least be plotted and discussed.

References

Bird, R. B., W. E. Stewart, and E. N. Lightfoot. 1960. *Transport Phenomena*. New York: John Wiley and Sons.

Forsythe, G. E., M. A. Malcolm, and C. B. Moler. 1977. *Computer Methods for Mathematical Computations*. Englewood Cliffs, NJ: Prentice-Hall.

Part two

Heat transport

H1	Heat conduction in a semi-infinite system	89
H2	One-dimensional heat conduction	107
H3	Heat transfer in a circular fin	135
H4	The Graetz problem with constant wall heat flux	167
H5	The Graetz problem with constant wall temperature	207
H6	Heat exchanger dynamics	289

H1

Heat conduction in a semi-infinite system

Fourier's second law for one-dimensional heat conduction in Cartesian coordinates is

$$\frac{\partial u}{\partial t} = \alpha \frac{\partial^2 u}{\partial x^2}$$

or in the usual subscript notation for partial derivatives

$$u_t = \alpha u_{xx} \tag{H1.1}$$

where we have followed the convention of the numerical analysis literature by taking the dependent variable as u (rather than T, for example, to designate temperature). t and x are space and time, respectively, or in mathematical terms, initial- and boundary-value independent variables. α is the thermal diffusivity.

Derive a solution to eq. (H1.1) subject to the initial condition

$$u(x, 0) = u_0 \tag{H1.2}$$

and boundary conditions

$$u(0, t) = u_s \tag{H1.3}$$

$$u(\infty, t) = u_0 \tag{H1.4}$$

where u_0 and u_s are prescribed initial and surface temperatures, respectively. Note that this problem is defined on a semi-infinite interval in x as a consequence of boundary conditions (H1.3) and (H1.4) (defined at $x = 0$ and $x = \infty$, respectively). The solution to eqs. (H1.1) to (H1.4) is sketched in Figure H1.1.

Solution

We first derive an analytical solution to this problem by two standard methods. Then the analytical solution is evaluated numerically. To facilitate the subsequent

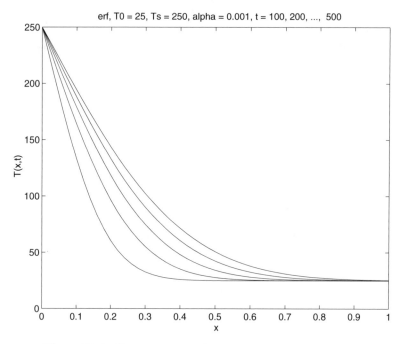

Figure H1.1. Temperature Profiles in a Semi-infinite System

analysis, a dimensionless temperature, u', is defined as

$$u' = \frac{u - u_s}{u_0 - u_s} \tag{H1.5}$$

Finally, we drop the prime of eq. (H1.5), agreeing that we are now considering a dimensionless temperature. Then eqs. (H1.1) to (H1.4) can be written as

$$u_t = \alpha u_{xx} \tag{H1.6}$$

$$u(x, 0) = 1 \tag{H1.7}$$

$$u(0, t) = 0 \tag{H1.8}$$

$$u(\infty, t) = 1 \tag{H1.9}$$

Now we proceed to the analytical solution of eqs. (H1.6) to (H1.9).

(1) Solution by a similarity transformation

If we define a new variable, η, by combining the independent variables x and t through a similarity transformation (Bird, Stewart, and Lightfoot (BSL) 1960, 124–6), defined as

$$\eta = \frac{x}{\sqrt{4\alpha t}} \tag{H1.10}$$

then $u = u(\eta)$, that is, eq. (H1.6) becomes an ordinary differential equation (ODE) in η as we shall see. The partial derivatives in eq. (H1.6) become

$$u_t = (du/d\eta)\eta_t = (du/d\eta)\left[(-1/2)\frac{x}{\sqrt{4\alpha t}}t^{-(3/2)}\right] = (du/d\eta)[-(1/2)\eta t^{-1}]$$

$$u_x = (du/d\eta)\eta_x = (du/d\eta)\left[\frac{1}{\sqrt{4\alpha t}}\right]$$

$$u_{xx} = \{d\left[(du/d\eta)\eta_x\right]/d\eta\}\eta_x = d^2u/d\eta^2\left[\frac{1}{\sqrt{4\alpha t}}\right]^2 = (d^2u/d\eta^2)(\eta^2/x^2)$$

Substitution of u_t and u_{xx} in eq. (H1.6) gives

$$(du/d\eta)[-(1/2)\eta t^{-1}] = \alpha(d^2u/d\eta^2)(\eta^2/x^2)$$

or

$$d^2u/d\eta^2 + (2/\eta)\left[\frac{x}{\sqrt{4\alpha t}}\right]^2 (du/d\eta) = 0$$

and finally

$$d^2u/d\eta^2 + 2\eta(du/d\eta) = 0 \tag{H1.11}$$

Equation (H1.11) is to be integrated subject to the boundary conditions

$$u(0) = 0 \tag{H1.12}$$

$$u(\infty) = 1 \tag{H1.13}$$

that follow directly from eqs. (H1.7), (H1.8), (H1.9), and (H1.10).

If we define $\phi = du/d\eta$, eq. (H1.11) can be written as a first order ODE

$$d\phi/d\eta + 2\phi\eta = 0 \tag{H1.14}$$

which is then separable into

$$d\phi/\phi = -2\eta d\eta \tag{H1.15}$$

Integration of eq. (H1.15) gives

$$ln(\phi) = -\eta^2 + C_1$$

or

$$\phi = C_1 e^{-\eta^2}$$

A second integration then gives

$$u(\eta) = \int_0^\eta \phi(\lambda)d\lambda + C_2 = C_1\int_0^\eta e^{-\lambda^2}d\lambda + C_2 \tag{H1.16}$$

From boundary condition (H1.12), $C_2 = 0$. From boundary condition (H1.13), eq. (H1.16) gives

$$u(\infty) = C_1 \int_0^\infty e^{-\lambda^2} d\lambda = 1 \tag{H1.17}$$

Since

$$\int_0^\infty e^{-\lambda^2} d\lambda = \sqrt{\pi}/2$$

from eq. (H1.17), we have $C_1 = 2/\sqrt{\pi}$, and the solution to eqs. (H1.6) to (H1.9) is

$$u(\eta) = 2/\sqrt{\pi} \int_0^\eta e^{-\lambda^2} d\lambda = erf(\eta) \tag{H1.18}$$

with η given by eq. (H1.10). The convention is to write this solution as *erf*, which denotes the error function. We will consider the numerical evaluation of this solution subsequently (since the integral of eq. (H1.18) cannot be evaluated analytically for finite η; note, however, that *erf* does have the properties $erf(0) = 0$ and $erf(\infty) = 1$, in accordance with boundary conditions (H1.12) and (H1.13)). Before proceeding to the numerical solution, we derive the analytical solution by a second method.

(2) Solution by the Laplace transform

If the Laplace transform of $u(x, t)$ defined by eqs. (H1.6) to (H1.9) with respect to t is defined as

$$L\{u(x,t)\} = \int_0^\infty u(x,t)e^{-st} dt = \bar{u}(x,s) \tag{H1.19}$$

eqs. (H1.6) to (H1.9) transform to

$$s\bar{u} - 1 = \alpha \frac{d^2 \bar{u}}{dx^2} \tag{H1.20}$$

$$\bar{u}(0,s) = 0 \tag{H1.21}$$

$$\bar{u}(\infty,s) = 1/s \tag{H1.22}$$

Equations (H1.20) to (H1.22) constitute a linear, constant coefficient ODE problem, which has a solution consisting of the sum of a homogeneous solution and a particular solution. The homogeneous solution is of the form

$$\bar{u}_h = Ce^{\lambda x}$$

which, when substituted in the homogeneous form of eq. (H1.20), gives

$$\alpha \lambda^2 C e^{\lambda x} - sCe^{\lambda x} = 0$$

so that the *characteristic equation* is

$$\lambda^2 - s/\alpha = 0 \tag{H1.23}$$

The eigenvalues are therefore $\lambda_1 = \sqrt{s/\alpha}$ and $\lambda_2 = -\sqrt{s/\alpha}$, and the homogeneous solution is a superposition of solutions (because eq. (H1.20) is linear)

$$\bar{u}_h = C_1 e^{\lambda_1 x} + C_2 e^{\lambda_2 x} \tag{H1.24}$$

where constants C_1 and C_2 will be determined from boundary conditions (H1.21) and (H1.22).

First, however, a particular solution must be added to $\bar{u}_h$ of eq. (H1.24). We take for the particular solution

$$\bar{u}_p = K$$

which, when substituted in eq. (H1.20), gives $K = 1/s$ or

$$\bar{u}_p = 1/s \tag{H1.25}$$

Therefore, the general solution to eq. (H1.20) is (from summing eqs. (H1.24) and (H1.25))

$$\bar{u}(x,s) = C_1 e^{\lambda_1 x} + C_2 e^{\lambda_2 x} + 1/s \tag{H1.26}$$

Boundary condition (H1.22) requires $C_1 = 0$, where we have taken

$$\lim_{x \to \infty} e^{\lambda_2 x} = 0$$

since $C_2 = -\sqrt{s/\alpha}$. (This would clearly be true if s is positive and real, but actually s is a complex variable, so the situation is somewhat more complicated; of course, α is positive and real.) Then, from boundary condition (H1.21), $C_1 = -1/s$. The solution to this point is therefore (from eq. (H1.26))

$$\bar{u}(x,s) = 1/s\{1 - e^{\lambda_2 x}\} = 1/s\{1 - e^{-\sqrt{s/\alpha}x}\} \tag{H1.27}$$

The inversion of eq. (H1.27) to obtain $u(x,t)$ is a rather complicated procedure if we use the definition of the inverse Laplace transform. Fortunately, the inverse transform is tabulated (Spiegel 1991, 169), so we can write down immediately from eq. (H1.27)

$$u(x,t) = L_t^{-1}\{\bar{u}(x,s)\} = erf\left(\frac{x}{2\sqrt{\alpha t}}\right) \tag{H1.28}$$

which is the same result as eq. (H1.18), keeping in mind eq. (H1.10).

(3) Numerical solution

We now consider the calculation of $u(x,t)$ from eqs. (H1.10) and (H1.28), which requires the evaluation of the integral in these equations. We will use Simpson's rule:

$$I(a,b) = \int_a^b f(x)dx \approx (h/3)\{f(a) + 4f(a+h) + 2f(a+2h)$$
$$+ 4(a+3h) + \cdots + 4f(b-h) + f(b)\} \tag{H1.29}$$

where $h = (b - a)/n$ and n is the number of intervals or "panels" in the approximation (which must be even). The value of n is varied in the following program to observe the variation in the value of the computed integral with the number of points used in the approximation of the integrand, $f(x)$.

A main program to perform the integration according to eq. (H1.29) is listed below:

```
      PROGRAM H1
C...
C...  DOUBLE PRECISION CODING IS USED
      IMPLICIT DOUBLE PRECISION (A-H,O-Z)
C...
C...  EXTERNAL FUNCTIONS TO CALCULATION INTEGRANDS FOR
C...  VARIOUS CASES
C...
C...     F3    THIRD ORDER POLYNOMIAL    (CUBIC FUNCTION)
C...
C...     F4   FOURTH ORDER POLYNOMIAL (QUARTIC FUNCTION)
C...
C...     FE   INTEGRAND OF ERF(X)
C...
      EXTERNAL F3, F4, FE
C...
C...  OPEN OUTPUT FILE
      NO=7
      OPEN(NO,FILE='output')
C...
C...  INTEGRATION LIMITS
      A=0.0D0
      B=1.0D0
C...
C...  STEP THROUGH A SERIES OF RUNS FOR POLYNOMIALS WITH
C...  DIFFERING NUMBERS OF PANELS
      DO 2 NORUN=1,4
      IF(NORUN.EQ.1)NP=2
      IF(NORUN.EQ.2)NP=4
      IF(NORUN.EQ.3)NP=10
      IF(NORUN.EQ.4)NP=100
C...
C...  INTEGRATION OF THIRD ORDER POLYNOMIAL
      CI3=SIMP(F3,A,B,NP)
C...
C...  INTEGRATION OF FOURTH ORDER POLYNOMIAL
      CI4=SIMP(F4,A,B,NP)
```

Heat conduction in a semi-infinite system

```fortran
C...
C...   PRINT NUMERICAL INTEGRALS
       WRITE (NO,1)NP,
      +              A,B,CI3,
      +              A,B,CI4
1      FORMAT(' NP = ',I3,//,
      +       ' A = ',F6.2,'   B = ',F6.2,'   I3 = ',F10.7,/,
      +       ' A = ',F6.2,'   B = ',F6.2,'   I4 = ',F10.7,/)
2      CONTINUE
C...
C...   HEAT CONDUCTION PROBLEM
C...
C...   THREE RUNS FOR T = 50, 100, 200 S (SECONDS)

       DO 3 NORUN=1,3
       IF(NORUN.EQ.1)T= 50.0D0
       IF(NORUN.EQ.2)T=100.0D0
       IF(NORUN.EQ.3)T=200.0D0
C...
C...   THERMAL DIFFUSIVITY, INITIAL, SURFACE TEMPERATURES
       ALPHA=0.0001D0
       T0= 25.0D0
       TS=250.0D0
C...
C...   PI
       PI=4.0D0*DATAN(1.0D0)
C...
C...   X COORDINATES
       X1=0.25D0
       X2=1.00D0
C...
C...   LIMITS
       A=0.0D0
       B1=X1/DSQRT(4.0D0*ALPHA*T)
       B2=X2/DSQRT(4.0D0*ALPHA*T)
C...
C...   EVALUATION OF ERF(X)
       NP=100
       ERFB1=SIMP(FE,A,B1,NP)
       ERFB2=SIMP(FE,A,B2,NP)
C...
C...   INCLUDE FACTOR 2/SQRT(PI)
       ERFB1=2.0D0/DSQRT(PI)*ERFB1
       ERFB2=2.0D0/DSQRT(PI)*ERFB2
C...
C...   TEMPERATURES AT X1, X2
```

```
            T1=TS+(T0-TS)*ERFB1
            T2=TS+(T0-TS)*ERFB2
C...
C...    PRINT NUMERICAL RESULTS
            WRITE(NO,4)NP,T,A,B1,ERFB1,T1,A,B2,ERFB2,T2
4           FORMAT(//,' NP = ',I3,' T = ',F6.1,//,
           +'     A = ',F6.2,'     B1 = ',F6.2,
           +'     ERF(B1) = ',F10.7,'     T1 = ',F10.7,/,
           +'     A = ',F6.2,'     B2 = ',F6.2,
           +'     ERF(B2) = ',F10.7,'     T2 = ',F10.7,/)
C...
C...    ERF(X) FROM HASTINGS APPROXIMATION
            ERFB1=ERF(B1)
            ERFB2=ERF(B2)
            WRITE(NO,5)B1,ERFB1,B2,ERFB2
5           FORMAT(//,' ERF FROM HASTINGS APPROXIMATION',/,
           +        ' B1 = ',F6.2,'     ERF(B1) = ',F10.7,/,
           +        ' B2 = ',F6.2,'     ERF(B2) = ',F10.7,/)
C...
C...    NEXT TIME
3           CONTINUE
C...
C...    CALCULATIONS COMPLETE
            STOP
            END
```

We can note the following points about program H1:

(1) Double precision coding is used. Three external functions are then declared, which compute the integrand in eq. (H1.29) for three cases

```
C...
C...    EXTERNAL FUNCTIONS TO CALCULATION INTEGRANDS FOR
C...    VARIOUS CASES
C...
C...        F3    THIRD ORDER POLYNOMIAL    (CUBIC FUNCTION)
C...
C...        F4    FOURTH ORDER POLYNOMIAL (QUARTIC FUNCTION)
C...
C...        FE    INTEGRAND OF ERF(X)
C...
            EXTERNAL F3, F4, FE
```

Simpson's rule is first applied to a third-order polynomial using function F3. Then a fourth-order polynomial is integrated using function F4. Finally, the integrand of the error function in eq. (H1.18) is integrated using function FE. The third- and fourth-order polynomials are included in the analysis to observe the performance of Simpson's rule in computing integrals for which exact values are known (through analytical integration). The calls to F3, F4, and FE are considered subsequently.

(2) After defining an output file, OUTPUT, for saving the computed results, the limits of the integrals in the case of the third- and fourth-order polynomials are set, that is, the variable of integration covers the interval $a \leq x \leq b$ or in this case $0 \leq x \leq 1$

```
C...
C...   INTEGRATION LIMITS
       A=0.0D0
       B=1.0D0
```

Then four integrations are performed using eq. (H1.29) with $n = 2, 4, 10$, and 100. Of course, we would expect in general that the computed integral would become more accurate with increasing numbers of points

```
C...
C...   STEP THROUGH A SERIES OF RUNS FOR POLYNOMIALS WITH
C...   DIFFERING NUMBERS OF PANELS
       DO 2 NORUN=1,4
       IF(NORUN.EQ.1)NP=2
       IF(NORUN.EQ.2)NP=4
       IF(NORUN.EQ.3)NP=10
       IF(NORUN.EQ.4)NP=100
```

(3) The Simpson's rule integration is then performed on the third- and fourth-order polynomials by calls to a function SIMP (to be discussed subsequently)

```
C...
C...   INTEGRATION OF THIRD ORDER POLYNOMIAL
       CI3=SIMP(F3,A,B,NP)
C...
C...   INTEGRATION OF FOURTH ORDER POLYNOMIAL
       CI4=SIMP(F4,A,B,NP)
```

Note that SIMP has four arguments defining (a) the function to compute the integrand (externals F3 and F4), (b) the limits of integration (A and B), and (c) the number of panels (intervals) (NP).

(4) The computed integrals are then printed

```
C...
C...   PRINT NUMERICAL INTEGRALS
       WRITE (NO,1)NP,
      +            A,B,CI3,
      +            A,B,CI4
```

```
1       FORMAT(' NP = ',I3,//,
        +         '  A = ',F6.2,'  B = ',F6.2,'  I3 = ',F10.7,/,
        +         '  A = ',F6.2,'  B = ',F6.2,'  I4 = ',F10.7,/)
2       CONTINUE
```

We now consider the output from the program to this point before continuing on with a discussion of the remainder of the main program, H1. The output for the integrals of the third- and fourth-order polynomials (from the preceding WRITE statement) is listed below:

```
NP =    2

A =    0.00   B =   1.00   I3 =   0.2500000
A =    0.00   B =   1.00   I4 =   0.2083333

NP =    4

A =    0.00   B =   1.00   I3 =   0.2500000
A =    0.00   B =   1.00   I4 =   0.2005208

NP =   10

A =    0.00   B =   1.00   I3 =   0.2500000
A =    0.00   B =   1.00   I4 =   0.2000133

NP =  100

A =    0.00   B =   1.00   I3 =   0.2500000
A =    0.00   B =   1.00   I4 =   0.2000000
```

We can note the following points about this output:

(1) The integral of the third-order polynomial remains constant at 0.2500000, irrespective of the number of intervals used in eq. (H1.29), and in fact, this is the correct answer, that is,

$$I_3(0, 1) = \int_0^1 x^3 dx = 1/4 \tag{H1.30}$$

Thus, we come to the conclusion that Simpson's rule is exact for third-order polynomials, which is a surprising result since Simpson's rule is derived by approximating the integrand with a second-order polynomial. In other words, Simpson's rule will integrate a third-order polynomial correctly using only three values of the polynomial (or two intervals in eq. (H1.29)). Again, this is an unexpected result since in general four values of a third-order polynomial are required just to evaluate the four constants in a third-order polynomial.

(2) The integral of the fourth-order polynomial changes with the number of intervals and appears to converge to a value of 0.2000000, which is the correct value

$$I_4(0, 1) = \int_0^1 x^4 dx = 1/5 \tag{H1.31}$$

Thus, we can conclude that Simpson's rule is not exact for fourth-order polynomials. But, from this example we can infer that the accuracy of the computed integral will improve with increasing numbers of points (if the integrand can be more closely approximated by a second-order polynomial over a smaller interval; in this case, an interval $h = (1 - 0)/100 = 0.01$ produced a numerical integral correct to seven figures).

To conclude this discussion of the numerical integration of polynomials, we consider the routines called by the main program. The externals, F3 and F4, are listed below:

```
DOUBLE PRECISION FUNCTION F3(X)
IMPLICIT DOUBLE PRECISION (A-H,O-Z)
F3=X**3
RETURN
END

DOUBLE PRECISION FUNCTION F4(X)
IMPLICIT DOUBLE PRECISION (A-H,O-Z)
F4=X**4
RETURN
END
```

Note that these are straightforward implementations of third- and fourth-order polynomials. The value of X is an input from function SIMP that implements eq. (H1.29).

Function SIMP is listed below:

```
      DOUBLE PRECISION FUNCTION SIMP(F,A,B,NP)
C...
C...  FUNCTION SIMP COMPUTES A SIMPSON'S RULE QUADRATURE
C...
C...  ARGUMENT LIST
C...
C...     F       EXTERNAL FUNCTION TO DEFINE THE INTEGRAND
C...             (INPUT)
C...
C...     A       LOWER LIMIT OF THE INTEGRAL (INPUT)
C...
C...     B       UPPER LIMIT OF THE INTEGRAL (INPUT)
```

```
C...
C...      NP       NUMBER OF PANELS (MUST BE EVEN) (INPUT)
C...
C...   DOUBLE PRECISION CODING IS USED
       IMPLICIT DOUBLE PRECISION (A-H,O-Z)
C...
C...   EXTERNAL FUNCTION FOR INTEGRAND
       EXTERNAL F
C...
C...   CHECK FOR AN EVEN NUMBER OF PANELS
       IF((NP/2*2).NE.NP)THEN
          WRITE(*,2)NP
2         FORMAT(//,' NP = ',I3,' (NOT EVEN)')
          SIMP=0.0D0
          RETURN
       END IF
C...
C...   INTEGRATION INTERVAL
       DX=(B-A)/DFLOAT(NP)
C...
C...   INITIALIZE SUM
       SIMP=(F(A)+4.0D0*F(A+DX)+F(B))
C...
C...   CONTINUE SUM
       IF(NP.GT.2)THEN
          DO 1 I=2,NP-1,2
             SIMP=SIMP
     +             +2.0D0*F(A+DFLOAT(I  )*DX)
     +             +4.0D0*F(A+DFLOAT(I+1)*DX)

1         CONTINUE
       END IF
C...
C...   INTEGRAL
       SIMP=(DX/3.0D0)*SIMP
       RETURN
       END
```

We can note the following points about function SIMP:

(1) The first argument is an external to define the integrand in eq. (H1.29), as discussed previously

```
C...
C...   EXTERNAL FUNCTION FOR INTEGRAND
       EXTERNAL F
```

(2) A check is then made for an even number of intervals (eq. (H1.29) is valid only for even n; note that in making this check, we are making use of some properties of integer arithmetic)

```
C...
C...   CHECK FOR AN EVEN NUMBER OF PANELS
       IF((NP/2*2).NE.NP)THEN
          WRITE(*,2)NP
2         FORMAT(//,' NP = ',I3,' (NOT EVEN)')
          SIMP=0.0D0
          RETURN
       END IF
```

(3) The panel width, h, in eq. (H1.29) is then computed (DX), and the sum is initiated as

$$\{f(a) + 4f(a+h) + f(b)\}$$

```
C...
C...   INTEGRATION INTERVAL
       DX=(B-A)/DFLOAT(NP)
C...
C...   INITIALIZE SUM
       SIMP=(F(A)+4.0D0*F(A+DX)+F(B))
```

Note in particular the calls to function F to compute the integrand.

(4) The sum in eq. (H1.29) is then continued with alternating weighting factors of 2 and 4

```
C...
C...   CONTINUE SUM
       IF(NP.GT.2)THEN
          DO 1 I=2,NP-1,2
             SIMP=SIMP
     +            +2.0D0*F(A+DFLOAT(I  )*DX)
     +            +4.0D0*F(A+DFLOAT(I+1)*DX)

1         CONTINUE
       END IF
```

(5) After the sum is completed, the integral is computed according to eq. (H1.29)

```
C...
C...   INTEGRAL
       SIMP=(DX/3.0D0)*SIMP
```

Finally, we should note that function SIMP is a general-purpose quadrature (integration) routine for any integrand programmed in external function F. Its principal limitation is a lack of error control, that is, the accuracy of the computed integral will depend, in general, on the value of NP used as the fourth argument, so that usually a repeated calculation with different values of NP is required to infer the accuracy of the numerical integral. More sophisticated quadrature routines are available that vary the number of points (the width h) to achieve a user-specified maximum error in the computed integral; see, for example, subroutine QUANC8 (Forsythe, Malcolm, and Moler 1977).

We now return to the remainder of main program H1 for the calculation of the solution of eqs. (H1.18) and (H1.28).

(6) Three runs are programmed for $t = 50, 100,$ and 200

```
C...
C...   HEAT CONDUCTION PROBLEM
C...
C...   THREE RUNS FOR T = 50, 100, 200 S (SECONDS)

       DO 3 NORUN=1,3
       IF(NORUN.EQ.1)T= 50.0D0
       IF(NORUN.EQ.2)T=100.0D0
       IF(NORUN.EQ.3)T=200.0D0
```

The thermal diffusivity is set as $\alpha = 0.001$, and the initial and surface temperatures are $u_o = 25$ and $u_s = 250$

```
C...
C...   THERMAL DIFFUSIVITY, INITIAL, SURFACE TEMPERATURES
       ALPHA=0.0001D0
       T0= 25.0D0
       TS=250.0D0
C...
C...   PI
       PI=4.0D0*DATAN(1.0D0)
C...
C...   X COORDINATES
       X1=0.25D0
       X2=1.00D0
```

π used in eqs. (H1.18) and (H1.28) is then computed, and two values of x for which the temperature, $u(x, t)$, is to be computed are set, that is, we will compute $u(0.25, 50)$, $u(0.25, 100)$, $u(0.25, 200)$, $u(1, 50)$, $u(1, 100)$, and $u(1, 200)$.

(7) Now that $x, t,$ and α are set, η can be computed (from eq. (H1.10))

```
C...
C...   LIMITS
       A=0.0D0
```

Heat conduction in a semi-infinite system

```
      B1=X1/DSQRT(4.0D0*ALPHA*T)
      B2=X2/DSQRT(4.0D0*ALPHA*T)
```

B1 and B2 are then the upper limits of the integral in eq. (H1.18).

(8) The integral in eqs. (H1.18) and (H1.28) is then computed with 100 intervals by calls to SIMP; the factor $2/\sqrt{\alpha}$ is included in the integral to arrive at *erf*

```
C...
C...  EVALUATION OF ERF(X)
      NP=100
      ERFB1=SIMP(FE,A,B1,NP)
      ERFB2=SIMP(FE,A,B2,NP)
C...
C...  INCLUDE FACTOR 2/SQRT(PI)
      ERFB1=2.0D0/DSQRT(PI)*ERFB1
      ERFB2=2.0D0/DSQRT(PI)*ERFB2
```

External FE follows directly from the integrand of eq. (H1.18)

```
      DOUBLE PRECISION FUNCTION FE(X)
      IMPLICIT DOUBLE PRECISION (A-H,O-Z)
      FE=DEXP(-X**2)
      RETURN
      END
```

(9) The actual temperatures are computed from eq. (H1.5) and printed

```
C...
C...  TEMPERATURES AT X1, X2
      T1=TS+(T0-TS)*ERFB1
      T2=TS+(T0-TS)*ERFB2
C...
C...  PRINT NUMERICAL RESULTS
      WRITE(NO,4)NP,T,A,B1,ERFB1,T1,A,B2,ERFB2,T2
4     FORMAT(//,' NP = ',I3,' T = ',F6.1,//,
     +  '    A = ',F6.2,'    B1 = ',F6.2,
     +  '    ERF(B1) = ',F10.7,'    T1 = ',F10.7,/,
     +  '    A = ',F6.2,'    B2 = ',F6.2,
     +  '    ERF(B2) = ',F10.7,'    T2 = ',F10.7,/)
```

(10) Finally, as an independent check of the *erf* computed from SIMP, an approximation for *erf* reported by Hastings (1955) is computed by calls to function ERF

```
C...
C...  ERF(X) FROM HASTINGS APPROXIMATION
      ERFB1=ERF(B1)
      ERFB2=ERF(B2)
      WRITE(NO,5)B1,ERFB1,B2,ERFB2
```

```
5       FORMAT(//,' ERF FROM HASTINGS APPROXIMATION',/,
      +        '    B1 = ',F6.2,'    ERF(B1) = ',F10.7,/,
      +        '    B2 = ',F6.2,'    ERF(B2) = ',F10.7,/)
C...
C...    NEXT TIME
3       CONTINUE
```

The values of *erf* (ERFB1 and ERFB2) from ERF for the two values of B1 and B2 are printed for comparison with the values computed by SIMP (for tabulated functions including *erf*, see also Abramowitz and Stegun 1964; Luke 1975)

Function ERF is listed below:

```
        DOUBLE PRECISION FUNCTION ERF(X)
C...
C...    FUNCTION ERF COMPUTES THE ERROR FUNCTION ACCORDING TO
C...    THE HASTINGS APPROXIMATION [GREENBERG (1978, P11)]
C...
        IMPLICIT DOUBLE PRECISION (A-H,O-Z)
C...
C...    THE FOLLOWING CODING IS A STRAIGHTFORWARD
C...    IMPLEMENTATION OF THE HASTINGS APPROXIMATION
C...    CITED ABOVE
        T=  1.0D0/(1.0D0+0.3275911D0*X)
        A1= 0.254829592D0
        A2=-0.284496736D0
        A3= 1.421413741D0
        A4=-1.453152027D0
        A5= 1.061405429D0
        ERF=1.0D0-(A1*T+A2*T**2+A3*T**3+A4*T**4+A5*T**5)
     +       *DEXP(-X**2)
        RETURN
        END
```

The programming in function ERF follows directly from the Hastings approximation.

To conclude the discussion of the programming, function DFLOAT is listed below:

```
        DOUBLE PRECISION FUNCTION DFLOAT(I)
C...
C...    FUNCTION DFLOAT CONVERTS A SINGLE PRECISION INTEGER
C...    INTO A DOUBLE PRECISION FLOATING POINT. THIS FUNCTION
C...    IS PROVIDED IN CASE THE USER'S FORTRAN COMPILER DOES
C...    NOT INCLUDE DFLOAT.
C...
        DFLOAT=DBLE(FLOAT(I))
```

```
        RETURN
        END
```

DFLOAT converts a single precision integer to a double precision floating point number and is included as part of the programming since some Fortran compilers do not have an intrinsic DFLOAT (but generally FLOAT is included as an intrinsic in Fortran compilers).

The output from the preceding programming pertaining to eqs. (H1.18) and (H1.28) is listed below:

```
NP = 100 T =    50.0

A = 0.00   B1 = 1.77   ERF(B1) = 0.9875807   T1 = 27.7943495
A = 0.00   B2 = 7.07   ERF(B2) = 1.0000000   T2 = 25.0000000

ERF FROM HASTINGS APPROXIMATION

B1 =   1.77     ERF(B1) =   0.9875806
B2 =   7.07     ERF(B2) =   1.0000000

NP = 100 T =   100.0

A = 0.00   B1 = 1.25   ERF(B1) = 0.9229001   T1 = 42.3474711
A = 0.00   B2 = 5.00   ERF(B2) = 1.0000000   T2 = 25.0000000

ERF FROM HASTINGS APPROXIMATION

B1 =   1.25     ERF(B1) =   0.9229002
B2 =   5.00     ERF(B2) =   1.0000000

NP = 100 T =   200.0

A = 0.00   B1 = 0.88   ERF(B1) = 0.7887005   T1 = 72.5423981
A = 0.00   B2 = 3.54   ERF(B2) = 0.9999994   T2 = 25.0001290

ERF FROM HASTINGS APPROXIMATION

B1 =   0.88     ERF(B1) =   0.7887003
B2 =   3.54     ERF(B2) =   0.9999994
```

We can note the following points about the output:

(1) The temperatures at $x = 0.25$ are $u(0.25, 50) = 27.7943495$, $u(0.25, 100) = 42.3474711$, and $u(0.25, 200) = 72.5423981$, so that the temperature at $x = 0.25$ increases with t, as expected (the initial temperature is $u_o = 25$, and the surface temperature is $u_s = 250$).

(2) The temperatures at $x = 1$ are $u(1, 50) = 25.0000000, u(1, 100) = 25.0000000$, and $u(1, 200) = 25.0001290$, so that the temperature at $x = 1$ remains essentially constant. In other words, enough time has not passed for significant penetration of the surface temperature, $u_s = 250$, to $x = 1$, which seems reasonable (see Figure H1.1). In fact, eqs. (H1.18) and (H1.28) are based on boundary condition (H1.9), which states that for sufficiently large x, the temperature does not depart from its initial value (e.g., $u_o = 25$).

(3) The integrals computed by Simpson's rule via function SIMP are in good agreement with *erf* computed from the Hastings approximation. For example, at $x = 0.25$ and $t = 200$, from SIMP we obtain ERF(B1) = 0.7887005 and from the Hastings approximation, ERF(B1) = 0.7887003. Thus, 100 intervals appear to be adequate in using SIMP (although this could easily be studied by varying NP in calling SIMP).

In summary, we have applied both analytical methods (a similarity transformation and the Laplace transform) and a numerical method (Simpson's rule) to the analysis of the heat conduction problem. This is an example, then, of how analytical and numerical methods can be used together. Also, the problem consisting of eqs. (H1.1) to (H1.4) is a relatively straightforward problem (i.e., linear, one dimensional with constant coefficients) so that analytical methods can be applied. More complicated problems generally preclude the use of analytical methods so that we must then turn to numerical methods. As an indication of the potential utility of numerical methods, we have observed that SIMP is a general quadrature routine that can be applied to integrals that cannot be done analytically (such as *erf*); all that is required to use SIMP is the programming of the integrand.

References

Abramowitz, M., and I. Stegun, eds. 1964. *Handbook of Mathematical Functions*. National Bureau of Standards Applied Mathematics Series. Washington, DC: U.S. Government Printing Office.

Bird, R. B., W. E. Stewart, and E. N. Lightfoot. 1960. *Transport Phenomena*. New York: John Wiley and Sons.

Forsythe, G. E., M. A. Malcolm, and C. B. Moler. 1977. *Computer Methods for Mathematical Computations*. Englewood Cliffs, NJ: Prentice-Hall.

Greenberg, M. D. 1978. *Foundations of Applied Mathematics*. Englewood Cliffs, NJ: Prentice-Hall.

Hastings, C., Jr. 1955. *Approximations for Digital Computers*. Princeton, NJ: Princeton University Press.

Luke, Y. L. 1975. *Mathematical Functions and Their Approximation*. New York: Academic Press.

Spiegel, M. R. 1991. *Mathematical Handbook of Formulas and Tables*. New York: McGraw-Hill.

H2

One-dimensional heat conduction

The conduction of heat in a one-dimensional solid is modeled by Fourier's second law:

$$\frac{\partial T}{\partial t} = \alpha \left\{ \frac{\partial^2 T}{\partial x^2} + \frac{n}{x} \frac{\partial T}{\partial x} \right\} \tag{H2.1}$$

where

- T solid temperature
- x spatial coordinate in the solid
- t time
- α thermal diffusivity of the solid
- n exponent defining the coordinate system with the values
  - 0–Cartesian coordinates
  - 1–cylindrical coordinates
  - 2–spherical coordinates

Equation (H2.1) is first order in t and second order in x; it therefore requires one initial condition and two boundary conditions, which are taken as

$$T(x, 0) = T_0 \tag{H2.2}$$

$$\frac{\partial T(0, t)}{\partial x} = 0 \tag{H2.3}$$

$$k \frac{\partial T(x_0, t)}{\partial x} = h\{T_s - T(x_0, t)\} \tag{H2.4}$$

where

- T_0 initial solid temperature
- T_s surrounding fluid temperature
- k solid thermal conductivity
- h fluid-to-solid heat transfer cofficient
- x_0 solid half-width

Equation (H2.2) indicates the solid initially is at a uniform temperature, T_0. At the left face, $x = 0$, the solid has a zero temperature gradient, due to symmetry (or insulation), and at the right face, $x = x_0$, the solid exchanges heat with the surrounding fluid at temperature T_s.

Compute a numerical solution to eqs. (H2.1) to (H2.4) for $n = 0, 1,$ and 2 (Cartesian, cylindrical and spherical coordinates, respectively).

Solution

We now develop a method of lines (MOL) solution of eqs. (H2.1) to (H2.4) for $n = 0, 1,$ and 2. First we define some dimensionless variables as

$$T' = \frac{T - T_s}{T_0 - T_s} \tag{H2.5}$$

$$x' = x/x_0 \tag{H2.6}$$

$$t' = t\alpha/x_0^2 \tag{H2.7}$$

Substitution of T', x', and t' from eqs. (H2.5) to (H2.7) in eqs. (H2.1) to (H2.4) gives

$$(T_0 - T_s)(\alpha/x_0^2)\frac{\partial T'}{\partial t'} = (T_0 - T_s)(\alpha/x_0^2)\left\{\frac{\partial^2 T'}{\partial x'^2} + \frac{n}{x'}\frac{\partial T'}{\partial x'}\right\} \tag{H2.8}$$

$$T'(x', 0) = 1 \tag{H2.9}$$

$$\{(T_0 - T_s)/x_0\}\frac{\partial T'(0, t')}{\partial x'} = 0 \tag{H2.10}$$

$$\{(T_0 - T_s)/x_0\}k\frac{\partial T'(1, t')}{\partial x'} = (T_0 - T_s)h\{0 - T'(1, t')\} \tag{H2.11}$$

Equations (H2.8) to (H2.11) are the equations subsequently programmed (after dropping the primes with dimensionless variables understood, cancellation of the common factors, and $B_i = hx_0/k$)

$$\frac{\partial T}{\partial t} = \frac{\partial^2 T}{\partial x^2} + \frac{n}{x}\frac{\partial T}{\partial x} \tag{H2.12}$$

$$T(x, 0) = 1 \tag{H2.13}$$

$$\frac{\partial T(0, t)}{\partial x} = 0 \tag{H2.14}$$

$$\frac{\partial T(1, t)}{\partial x} + B_i T(1, t) = 0 \tag{H2.15}$$

We now consider the MOL solution of eqs. (H2.12) to (H2.15).

Subroutine INITAL for initial condition (H2.13) is

```
      SUBROUTINE INITAL
C...
C...  HEAT CONDUCTION IN ONE DIMENSIONAL SOLIDS
C...
C...  DOUBLE PRECISION CODING IS USED
      IMPLICIT DOUBLE PRECISION (A-H,O-Z)
C...
C...  COMMON AREA WITH THE COORDINATE SYSTEM
      COMMON/I/       NC
C...
C...  SELECT COORDINATE SYSTEM
C...
C...     NC = 0, CARTESIAN COODINATES
C...
C...     NC = 1, CYLINDRICAL COORDINATES
C...
C...     NC = 2, SPHERICAL COORDINATES
C...
      NC=0
C...
C...  INITIAL CONDITIONS
      IF(NC.EQ.0)CALL INIT10
      IF(NC.EQ.1)CALL INIT11
      IF(NC.EQ.2)CALL INIT12
      RETURN
      END
```

We can note the following points about INITAL:

(1) An integer variable, NC, is used to select the coordinate system, which is then passed through COMMON/I/ to the other subroutines that have the MOL code for eqs. (H2.12) to (H2.15)

```
C...
C...  COMMON AREA WITH THE COORDINATE SYSTEM
      COMMON/I/       NC
```

(2) Once NC is set for a particular coordinate system, one of three initialization subroutines, INIT10 (for NC = 0), INIT11 (for NC = 1), or INIT12 (for NC = 2), is called

```
C...
C...  INITIAL CONDITIONS
      IF(NC.EQ.0)CALL INIT10
      IF(NC.EQ.1)CALL INIT11
      IF(NC.EQ.2)CALL INIT12
```

For example, if NC = 0 (Cartesian coordinates), subroutine INIT10 is called

```
      SUBROUTINE INIT10
C...
C...  INITIAL CONDITIONS FOR CARTESIAN COORDINATES
C...
      IMPLICIT DOUBLE PRECISION (A-H,O-Z)
      PARAMETER (NX=21)
      COMMON/T/      TIME,        NSTOP,        NORUN
     +      /Y/      T(NX)
     +      /F/      TT(NX)
     +      /S/      TX(NX),      TXX(NX)
     +      /P/      HXOK
     +      /I/      NC
C...
C...  SELECT BIOT NUMBER, H*XO/K
      IF(NORUN.EQ.1)HXOK=0.0D0
      IF(NORUN.EQ.2)HXOK=1.0D0/5.0D0
      IF(NORUN.EQ.3)HXOK=1.0D0/2.0D0
      IF(NORUN.EQ.4)HXOK=1.0D0/1.0D0
      IF(NORUN.EQ.5)HXOK=1.0D0/0.5D0
      IF(NORUN.EQ.6)HXOK=1.0D0/0.05D0
C...
C...  INITIAL CONDITION
      DO 1 I=1,NX
         T(I)=1.0D0
1     CONTINUE
      RETURN
      END
```

We can note the following points about INIT10:

(1) The COMMON area is for a 21-point grid in x

```
      IMPLICIT DOUBLE PRECISION (A-H,O-Z)
      PARAMETER (NX=21)
      COMMON/T/      TIME,        NSTOP,        NORUN
     +      /Y/      T(NX)
     +      /F/      TT(NX)
     +      /S/      TX(NX),      TXX(NX)
     +      /P/      HXOK
     +      /I/      NC
```

The dependent variable of eq. (H2.12), $T(x, t)$, is in COMMON/Y/ as array T(NX). The temporal derivative of eq. (H2.12), $\partial T/\partial t$, is in COMMON/F/ as array TT(NX), and the spatial derivatives, $\partial T/\partial x$ and $\partial^2 T/\partial x^2$, are in COMMON/S/ as arrays TX(NX) and TXX(NX), respectively.

(2) The Biot number, $B_i = hx_0/k$, in boundary condition (H2.15) is then set to one of six values, $B_i = 0, 0.2, 0.5, 1, 2,$ and 20

```
C...
C...      SELECT BIOT NUMBER, H*XO/K
          IF(NORUN.EQ.1)HXOK=0.0D0
          IF(NORUN.EQ.2)HXOK=1.0D0/5.0D0
          IF(NORUN.EQ.3)HXOK=1.0D0/2.0D0
          IF(NORUN.EQ.4)HXOK=1.0D0/1.0D0
          IF(NORUN.EQ.5)HXOK=1.0D0/0.5D0
          IF(NORUN.EQ.6)HXOK=1.0D0/0.05D0
```

that is, the program is executed six times to produce solutions for the six values of B_i. For the first solution (NORUN $= 1$, set by the main program through COMMON/T/), $B_i = 0$, so that no heat transfer between the solid and surroundings takes place and the solid remains at the initial condition $T(x, 0) = 1$ of eq. (H2.13); this special case serves as a check on the MOL coding. Also, for NORUN $= 6$, $B_i = 20$, which corresponds to negligible resistance to heat transfer at the surface of the solid so that the surface temperature essentially equals the temperature of the surroundings; this special case is programmed for boundary condition (H2.15), that is, $T(1, t) = T_s$, in subroutine DERV10.

(3) Initial condition (H2.13) is then programmed in DO loop 1

```
C...
C...      INITIAL CONDITION
          DO 1 I=1,NX
             T(I)=1.0D0
1         CONTINUE
```

Note that the NX values of $T(x, 0) = 1$ are set in DO loop 1 and passed through COMMON/Y/ to start the numerical integration in t.

Subroutines INIT11 and INIT12 are essentially identical to INIT10. The only difference is the use of R in place of X in the programming of the spatial coordinate, since in cylindrical and spherical coordinates, radial position is the spatial independent variable, that is, in $T(r, t)$.

Subroutine DERV10 computes the temporal derivative, $\partial T/\partial t$, of eq. (H2.12)

```
          SUBROUTINE DERV10
C...
C...      TEMPORAL DERIVATIVES FOR CARTESIAN COORDINATES
C...
          IMPLICIT DOUBLE PRECISION (A-H,O-Z)
          PARAMETER (NX=21)
          COMMON/T/    TIME,      NSTOP,     NORUN
         +       /Y/   T(NX)
```

```
      +      /F/    TT(NX)
      +      /S/    TX(NX),   TXX(NX)
      +      /P/    HXOK
      +      /I/         NC
C...
C...  BOUNDARY CONDITION AT X = 0
      NL=2
      TX(1)=0.0D0
C...
C...  BOUNDARY CONDITION AT X = X0
      IF(NORUN.NE.6)THEN
         NU=2
         TX(NX)=-HXOK*T(NX)
      ELSE IF(NORUN.EQ.6)THEN
         NU=1
         T(NX)=0.0D0
      END IF
C...
C...  TXX
      CALL DSS044(0.0D0,1.0D0,NX,T,TX,TXX,NL,NU)
C...
C...  FOURIER'S SECOND LAW
      DO 1 I=1,NX
         TT(I)=TXX(I)
1     CONTINUE
      RETURN
      END
```

We can note the following points about DERV10:

(1) The COMMON area is the same as in subroutine INIT10.

(2) Boundary condition (H2.14) is implemented first

```
C...
C...  BOUNDARY CONDITION AT X = 0
      NL=2
      TX(1)=0.0D0
```

Note that this is a Neumann boundary condition, so $NL = 2$ (used in the call to the DSS044 for the calculation of $\partial T/\partial x$ in eq. (H2.12)).

(3) Next, boundary condition (H2.15) is implemented

```
C...
C...  BOUNDARY CONDITION AT X = X0
      IF(NORUN.NE.6)THEN
         NU=2
         TX(NX)=-HXOK*T(NX)
```

```
      ELSE IF(NORUN.EQ.6)THEN
         NU=1
         T(NX)=0.0D0
      END IF
```

For the first five values of B_i, the spatial derivative $\partial T(x_0 = 1, t)/\partial x$ is computed, and since this is a Neumann boundary condition, NU = 2. For the sixth value of $B_i (= 20)$, the surface temperature, $T(x_0 = 1, t)$, is essentially the surrounding temperature, T_s, which is a Dirichlet boundary condition, so NU = 1.

(4) With the two boundary conditions defined, the second spatial derivative, $\partial^2 T/\partial x^2$, can be computed by a call to DSS044

```
C...
C...  TXX
      CALL DSS044(0.0D0,1.0D0,NX,T,TX,TXX,NL,NU)
```

In this call to DSS044, T, TX(1), and TX(NX) (or T(NX) for $B_i = 20$) are inputs, and the second derivative, TXX ($= \partial^2 T/\partial x^2$), is the output, which is then used in the programming of eq. (H2.12)

```
C...
C...  FOURIER'S SECOND LAW
      DO 1 I=1,NX
         TT(I)=TXX(I)
1     CONTINUE
```

Note the close resemblance of this coding to eq. (H2.12), which illustrates one of the principal advantages of the MOL.

Finally, the solution from the integration of the temporal derivative in array TT, as computed in DERV10, is written and plotted by subroutine PNT10

```
      SUBROUTINE PRT10(NI,NO)
C...
C...  WRITING AND PLOTTING SOLUTION FOR CARTESIAN
C...  COORDINATES
C...
      IMPLICIT DOUBLE PRECISION (A-H,O-Z)
      PARAMETER (NX=21)
      COMMON/T/     TIME,      NSTOP,     NORUN
     +      /Y/     T(NX)
     +      /F/     TT(NX)
     +      /S/     TX(NX),    TXX(NX)
     +      /P/     HXOK
     +      /I/     NC
C...
C...  MONITOR OUTPUT
```

```fortran
            IF(TIME.LT.0.001D0)WRITE(*,2)HX0K
2           FORMAT(//,' H*X0/K = ',F8.3,'   FROM PRT10')
            WRITE(*,1)TIME,T(1),T(NX)
C...
C...    OPEN A FILE FOR MATLAB PLOTTING
            IF((NORUN.EQ.1).AND.(TIME.LT.0.001D0))THEN
                OPEN(8,FILE='h2.out')
            END IF
C...
C...    HEADING FOR THE NUMERICAL SOLUTION
            IF(TIME.LT.0.001D0)WRITE(NO,3)
3           FORMAT(8X,'t ',3X,' x = 0 ',3X,'x = 0.2',3X,'x = 0.4',
     +                         3X,'x = 0.6',3X,'x = 0.8',3X,' x = 1 ')
C...
C...    WRITE NUMERICAL SOLUTION FOR 21 GRID POINTS
            NXI=4
C...
C...        WRITE SOLUTION ON FILE OUTPUT
            WRITE(NO,1)TIME,(T(I),I=1,NX,NXI)
C...
C...        WRITE SOLUTION ON FILE H2.OUT
            WRITE( 8,1)TIME,(T(I),I=1,NX,NXI)
1           FORMAT(F10.2,6F10.5)
            RETURN
            END
```

We can note the following points about PNT10:

(1) The COMMON area is the same as in INIT10 and DERV10.

(2) A WRITE statement to unit * displays t, $T(0, t)$ and $T(x_0 = 1, t)$ to monitor the progress of the calculations

```fortran
C...
C...    MONITOR OUTPUT
            IF(TIME.LT.0.001D0)WRITE(*,2)HX0K
2           FORMAT(//,' H*X0/K = ',F8.3,'   FROM PRT10')
            WRITE(*,1)TIME,T(1),T(NX)
```

At $t = 0$, the value of $B_i (=$ HX0K$)$ is also displayed.

(3) File h2.out is then opened on unit 8 to store the numerical solution for subsequent plotting by Matlab

```fortran
C...
C...    OPEN A FILE FOR MATLAB PLOTTING
            IF((NORUN.EQ.1).AND.(TIME.LT.0.001D0))THEN
                OPEN(8,FILE='h2.out')
            END IF
```

This procedure creates an output file that can then be sent to any plotting system, for example, a spreadsheet.

(4) At $t = 0$, a heading is written for the numerical solution

```
C...
C...    HEADING FOR THE NUMERICAL SOLUTION
        IF(TIME.LT.0.001D0)WRITE(NO,3)
3       FORMAT(8X,'t ',3X,' x = 0 ',3X,'x = 0.2',3X,'x = 0.4',
     +                  3X,'x = 0.6',3X,'x = 0.8',3X,' x = 1 ')
```

(5) Finally, the numerical solution, $T(x, t)$, is written for the particular t corresponding to the current call to PNT10 (i.e., the value of TIME in COMMON/T/) and for $x = 0, 0.2, 0.4, \ldots, 1$

```
C...
C...    WRITE NUMERICAL SOLUTION FOR 21 GRID POINTS
        NXI=4
C...
C...    WRITE SOLUTION ON FILE OUTPUT
        WRITE(NO,1)TIME,(T(I),I=1,NX,NXI)
C...
C...    WRITE SOLUTION ON FILE H2.OUT
        WRITE( 8,1)TIME,(T(I),I=1,NX,NXI)
1       FORMAT(F10.2,6F10.5)
```

Note that the same solution is also written to file h2.out (unit 8) for plotting.

This completes the MOL programming of eqs. (H2.12) to (H2.15) for NC = 0 (Cartesian coordinates). The integration of the NX = 21 ODEs in t is done by RKF45 (Forsythe, Malcolm, and Moler 1977). The data listed below control the integration (and are read by the main program, which calls RKF45)

```
ONE-DIMENSIONAL HEAT CONDUCTION, HRO/K = 0
0.          5.5         0.1
   21                       0.0001
ONE-DIMENSIONAL HEAT CONDUCTION, HRO/K = 1/5
0.          5.5         0.1
   21                       0.0001
ONE-DIMENSIONAL HEAT CONDUCTION, HRO/K = 1/2
0.          5.5         0.1
   21                       0.0001
ONE-DIMENSIONAL HEAT CONDUCTION, HRO/K = 1/1
0.          5.5         0.1
   21                       0.0001
ONE-DIMENSIONAL HEAT CONDUCTION, HRO/K = 1/0.5
0.          5.5         0.1
   21                       0.0001
```

ONE-DIMENSIONAL HEAT CONDUCTION, HR0/K = 20
0.   2.75   0.05
   21   0.0001
END OF RUNS

As expected, six runs are programmed. In each case, $t = 0$ at the beginning of the solution. The final value is $t = 5.5$ for the first five solutions, and $t = 2.75$ for the sixth solution (for this last case, the solution changes more rapidly in t, so the total time is shorter). In all cases, 56 calls are made to PNT10 by the main program to output the solution (i.e., $5.5/0.1 + 1 = 2.75/0.05 + 1 = 56$). Also, in all cases, 21 ODEs are integrated with an error tolerance of 0.0001.

The output from the preceding routines is summarized below:

```
RUN NO. -    1   ONE-DIMENSIONAL HEAT CONDUCTION, HR0/K = 0

INITIAL T -    .000D+00

   FINAL T -    .550D+01

   PRINT T -    .100D+00

NUMBER OF DIFFERENTIAL EQUATIONS -   21

MAXIMUM INTEGRATION ERROR -    .100D-03

       t      x = 0    x = 0.2   x = 0.4   x = 0.6   x = 0.8    x = 1
      .00   1.00000   1.00000   1.00000   1.00000   1.00000   1.00000
      .10   1.00001   1.00000   1.00000   1.00000   1.00000   1.00001
      .20   1.00000   1.00000   1.00000   1.00000   1.00000   1.00000
      .30   1.00000   1.00000   1.00000   1.00000   1.00000   1.00000
      .40   1.00004   1.00000   1.00000   1.00000   1.00000   1.00004
      .50    .99999   1.00000   1.00000   1.00000   1.00000    .99999
                        .                   .
                        .                   .
                        .                   .
     5.00   1.00000   1.00000   1.00000   1.00000   1.00000   1.00000
     5.10   1.00000   1.00000   1.00000   1.00000   1.00000   1.00000
     5.20    .99996   1.00000   1.00000   1.00000   1.00000    .99996
     5.30   1.00000   1.00000   1.00000   1.00000   1.00000   1.00000
     5.40   1.00000   1.00000   1.00000   1.00000   1.00000   1.00000
     5.50   1.00000   1.00000   1.00000   1.00000   1.00000   1.00000

RUN NO. -    2   ONE-DIMENSIONAL HEAT CONDUCTION, HR0/K = 1/5

   INITIAL T -    .000D+00
```

FINAL T -   .550D+01

PRINT T -   .100D+00

NUMBER OF DIFFERENTIAL EQUATIONS -   21

MAXIMUM INTEGRATION ERROR -   .100D-03

t	x = 0	x = 0.2	x = 0.4	x = 0.6	x = 0.8	x = 1
.00	1.00000	1.00000	1.00000	1.00000	1.00000	1.00000
.10	.99847	.99726	.99277	.98257	.96339	.93242
.20	.98827	.98552	.97692	.96155	.93832	.90650
.30	.97298	.96968	.95966	.94265	.91835	.88655
.40	.95601	.95256	.94216	.92478	.90039	.86904
.50	.93866	.93519	.92478	.90748	.88335	.85252
.						
.						
.						
5.00	.40408	.40257	.39804	.39053	.38010	.36679
5.10	.39658	.39510	.39065	.38328	.37304	.36003
5.20	.38922	.38776	.38340	.37617	.36612	.35331
5.30	.38200	.38057	.37628	.36919	.35932	.34678
5.40	.37491	.37350	.36930	.36233	.35265	.34031
5.50	.36795	.36657	.36245	.35561	.34611	.33403

RUN NO. -   3 ONE-DIMENSIONAL HEAT CONDUCTION, HRO/K = 1/2

INITIAL T -   .000D+00

FINAL T -   .550D+01

PRINT T -   .100D+00

NUMBER OF DIFFERENTIAL EQUATIONS -   21

MAXIMUM INTEGRATION ERROR -   .100D-03

t	x = 0	x = 0.2	x = 0.4	x = 0.6	x = 0.8	x = 1
.00	1.00000	1.00000	1.00000	1.00000	1.00000	1.00000
.10	.99632	.99346	.98284	.95894	.91450	.84394
.20	.97260	.96631	.94671	.91199	.86015	.79015
.30	.93815	.93083	.90872	.87145	.81880	.75101
.40	.90105	.89360	.87128	.83422	.78276	.71756
.50	.86411	.85683	.83506	.79909	.74944	.68688

5.00	.12669	.12561	.12239	.11708	.10978	.10061
5.10	.12139	.12036	.11727	.11219	.10519	.09640
5.20	.11632	.11533	.11237	.10750	.10080	.09237
5.30	.11146	.11051	.10768	.10301	.09658	.08851
5.40	.10681	.10590	.10318	.09871	.09255	.08481
5.50	.10234	.10147	.09887	.09458	.08868	.08127

RUN NO. -   4 ONE-DIMENSIONAL HEAT CONDUCTION, HRO/K = 1/1

INITIAL T -   .000D+00

  FINAL T -   .550D+01

  PRINT T -   .100D+00

NUMBER OF DIFFERENTIAL EQUATIONS -   21

MAXIMUM INTEGRATION ERROR -   .100D-03

t	x = 0	x = 0.2	x = 0.4	x = 0.6	x = 0.8	x = 1
.00	1.00000	1.00000	1.00000	1.00000	1.00000	1.00000
.10	.99311	.98782	.96839	.92517	.84620	.72357
.20	.95064	.93966	.90565	.84613	.75879	.64339
.30	.89180	.87957	.84284	.78159	.69637	.58885
.40	.83095	.81897	.78326	.72453	.64413	.54417
.50	.77253	.76121	.72755	.67244	.59739	.50452
5.00	.02764	.02724	.02602	.02404	.02135	.01800
5.10	.02567	.02529	.02417	.02233	.01983	.01674
5.20	.02384	.02349	.02244	.02073	.01841	.01556
5.30	.02214	.02181	.02084	.01926	.01710	.01444
5.40	.02056	.02026	.01935	.01788	.01588	.01341
5.50	.01909	.01881	.01797	.01661	.01475	.01245

RUN NO. -   5 ONE-DIMENSIONAL HEAT CONDUCTION, HRO/K = 1/0.5

INITIAL T -   .000D+00

  FINAL T -   .550D+01

  PRINT T -   .100D+00

NUMBER OF DIFFERENTIAL EQUATIONS -   21

One-dimensional heat conduction

MAXIMUM INTEGRATION ERROR -   .100D-03

t	x = 0	x = 0.2	x = 0.4	x = 0.6	x = 0.8	x = 1
.00	1.00000	1.00000	1.00000	1.00000	1.00000	1.00000
.10	.98778	.97864	.94549	.87333	.74523	.55360
.20	.91789	.90052	.84726	.75573	.62487	.45764
.30	.82778	.80967	.75567	.66699	.54631	.39840
.40	.73990	.72308	.67324	.59236	.48380	.35235
.50	.65962	.64445	.59961	.52708	.43011	.31310
.	.		.		.	
.	.		.		.	
.	.		.		.	
5.00	.00357	.00349	.00325	.00285	.00233	.00170
5.10	.00318	.00311	.00289	.00254	.00207	.00153
5.20	.00283	.00277	.00258	.00226	.00185	.00134
5.30	.00252	.00247	.00229	.00202	.00164	.00120
5.40	.00225	.00220	.00204	.00179	.00146	.00107
5.50	.00200	.00196	.00182	.00160	.00130	.00095

RUN NO. -   6   ONE-DIMENSIONAL HEAT CONDUCTION, HR0/K = 20

INITIAL T -   .000D+00

FINAL T -   .275D+01

PRINT T -   .500D-01

NUMBER OF DIFFERENTIAL EQUATIONS -   21

MAXIMUM INTEGRATION ERROR -   .100D-03

t	x = 0	x = 0.2	x = 0.4	x = 0.6	x = 0.8	x = 1
.00	1.00000	1.00000	1.00000	1.00000	1.00000	.00000
.05	.99687	.98844	.94221	.79409	.47291	.00000
.10	.94931	.91907	.81854	.62856	.34522	.00000
.15	.86419	.82741	.71609	.53132	.28404	.00000
.20	.77231	.73633	.63040	.46165	.24425	.00000
.25	.68545	.65250	.55638	.40543	.21366	.00000
.	.		.		.	
.	.		.		.	
.	.		.		.	
2.50	.00267	.00254	.00216	.00157	.00082	.00000
2.55	.00235	.00224	.00191	.00139	.00073	.00000
2.60	.00208	.00198	.00169	.00122	.00064	.00000
2.65	.00184	.00175	.00149	.00108	.00057	.00000

2.70	.00163	.00155	.00132	.00096	.00050	.00000
2.75	.00144	.00137	.00116	.00085	.00044	.00000

We can note the following points about this output:

(1) Initial condition (H2.13) gives the starting value of each solution at $t = 0$.

(2) For the case $B_i = 0$ of the first solution, no heat transfer takes place, so solution remains at the initial condition, that is, $T(x, t) = T(x, 0) = 1$.

(3) With increasing values of B_i, that is, increasing heat transfer, the temperature is closer to the surrounding temperature for a given t. This point is illustrated by the following representative values:

B_i	$T(x = 0, t = 0.5)$
0	0.99999
1/5	0.93866
1/2	0.86411
1	0.77253
2	0.65962
20	0.37077

The graphical output produced from file h2.out is given in Figure H2.1.

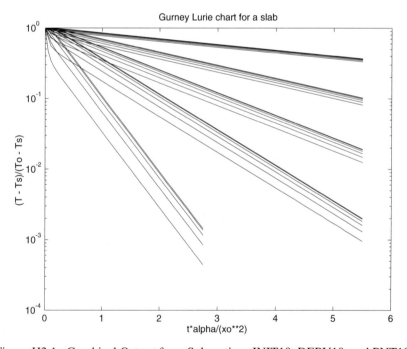

Figure H2.1. Graphical Output from Subroutines INIT10, DERV10, and PNT10

Figure H2.1 is in agreement with the well-known Gurney Lurie chart (Welty, Wicks, and Wilson 1984, 736). The Matlab file used to produce Figure H2.1 is available as explained in the preface. The main program that calls INIT10, PNT10, and RKF45 (the latter calls DERV10) is similar to RKF45M listed in Appendix A and is therefore not listed here.

The programming of the numerical solutions to eqs. (H2.12) to (H2.15) for cylindrical coordinates in subroutines INIT11, DERV11, and PNT11 and spherical coordinates in subroutines INIT12, DERV12, and PNT12 is similar to the preceding subroutines INIT10, DERV10, and PNT10. The only essential difference is in the derivative subroutines. To complete the discussion, we list here DERV11 and DERV12

```
      SUBROUTINE DERV11
C...
C...  TEMPORAL DERIVATIVES FOR CYLINDRICAL COORDINATES
C...
      IMPLICIT DOUBLE PRECISION (A-H,O-Z)
      PARAMETER (NR=21)
      COMMON/T/     TIME,       NSTOP,      NORUN
     +      /Y/     T(NR)
     +      /F/     TT(NR)
     +      /S/     TR(NR),     TRR(NR)
     +      /P/     HROK
     +      /I/         NC
C...
C...  DIRICHLET BOUNDARY CONDITION AT R = R0
      IF(NORUN.EQ.6)THEN
          NU=1
          T(NR)=0.0D0
      END IF
C...
C...  TR
      CALL DSS004(0.0,1.0D0,NR,T,TR)
C...
C...  BOUNDARY CONDITION AT R = 0
      NL=2
      TR(1)=0.0D0
C...
C...  NEUMANN BOUNDARY CONDITION AT R = R0
      IF(NORUN.NE.6)THEN
          NU=2
          TR(NR)=-HROK*T(NR)
      END IF
C...
C...  TRR
```

```
      CALL DSS044(0.0D0,1.0D0,NR,T,TR,TRR,NL,NU)
C...
C... FOURIER'S SECOND LAW
      DO 1 I=1,NR
         IF(I.EQ.1)THEN
            TT(I)=2.0D0*TRR(I)
         ELSE
            R=DFLOAT(I-1)/DFLOAT(NR-1)
            TT(I)=TRR(I)+(1.0D0/R)*TR(I)
         END IF
1     CONTINUE
      RETURN
      END

      SUBROUTINE DERV12
C...
C... TEMPORAL DERIVATIVES FOR SPHERICAL COORDINATES
C...
      IMPLICIT DOUBLE PRECISION (A-H,O-Z)

      PARAMETER (NR=21)

      COMMON/T/     TIME,      NSTOP,      NORUN
     +      /Y/     T(NR)
     +      /F/     TT(NR)
     +      /S/     TR(NR),    TRR(NR)
     +      /P/     HROK
     +      /I/         NC
C...
C... DIRICHLET BOUNDARY CONDITION AT R = R0
      IF(NORUN.EQ.6)THEN
         NU=1
         T(NR)=0.0D0
      END IF
C...
C... TR
      CALL DSS004(0.0,1.0D0,NR,T,TR)
C...
C... BOUNDARY CONDITION AT R = 0
      NL=2
      TR(1)=0.0D0
C...
C... NEUMANN BOUNDARY CONDITION AT R = R0
      IF(NORUN.NE.6)THEN
         NU=2
```

```
            TR(NR)=-HROK*T(NR)
         END IF
C...
C...     TRR
         CALL DSS044(0.0D0,1.0D0,NR,T,TR,TRR,NL,NU)
C...
C...     FOURIER'S SECOND LAW
         DO 1 I=1,NR
            IF(I.EQ.1)THEN
               TT(I)=3.0D0*TRR(I)
            ELSE
               R=DFLOAT(I-1)/DFLOAT(NR-1)
               TT(I)=TRR(I)+(2.0D0/R)*TR(I)
            END IF
1        CONTINUE
         RETURN
         END
```

We can note the following points about DERV11 and DERV12:

(1) In each case, the mathematical boundary conditions are the same as in DERV10 (note that R is used in place of X to denote the radial coordinate), but the programming of the boundary conditions is done in a different order than in DERV10. This reordering is required because eq. (H2.12) with $n = 1$ or 2 has the first derivative, $\partial T/\partial x$, (or $\partial T/\partial r$), that must be programmed in DERV11 and DERV12.

(2) For high B_i (e.g., $B_i = 20$ for NORUN = 6), the surface temperature is $T(1, t) = 0$, which is a Dirichlet boundary condition that is programmed first

```
C...
C...     DIRICHLET BOUNDARY CONDITION AT R = R0
         IF(NORUN.EQ.6)THEN
            NU=1
            T(NR)=0.0D0
         END IF
```

(3) The boundary temperature of (2) must be set so that it can be used in the calculation of the derivative $\partial T/\partial r$ by a call to subroutine DSS004

```
C...
C...     TR
         CALL DSS004(0.0,1.0D0,NR,T,TR)
```

Array T is the input to DSS004, and the derivative $\partial T/\partial r$ is the output (in array TR). This call to DSS004 is the programming in DERV11 and DERV12 that did not appear in DERV10.

(4) The Neumann boundary conditions of eqs. (H2.14) and (H2.15) are programmed next

```
C...
C...  BOUNDARY CONDITION AT R = 0
      NL=2
      TR(1)=0.0D0
C...
C...  NEUMANN BOUNDARY CONDITION AT R = R0
      IF(NORUN.NE.6)THEN
         NU=2
         TR(NR)=-HROK*T(NR)
      END IF
```

so that $\partial T/\partial r$ from DSS004 is reset to the boundary values (at $R = 0$ and at $R = 1$ if the latter is used, for example, NORUN $\leq$ 5).

(5) All of the boundary conditions are now set, so the derivative $\partial^2 T/\partial r^2$ is calculated by a call to DSS044

```
C...
C...  TRR
      CALL DSS044(0.0D0,1.0D0,NR,T,TR,TRR,NL,NU)
```

T is the input to DSS044, along with the boundary derivatives from (4) (in array TR). The derivative $\partial^2 T/\partial r^2$ is the output from DSS044 (in array TRR).

(6) Now all of the partial differential equations (PDEs) spatial derivatives have been calculated, and the coding of the PDE, eq. (H2.12), can be added, for example, from DERV11

```
C...
C...  FOURIER'S SECOND LAW
      DO 1 I=1,NR
         IF(I.EQ.1)THEN
            TT(I)=2.0D0*TRR(I)
         ELSE
            R=DFLOAT(I-1)/DFLOAT(NR-1)
            TT(I)=TRR(I)+(1.0D0/R)*TR(I)
         END IF
1     CONTINUE
```

The IF–THEN–ELSE is used to handle the special case of grid point 1 at $r = 0$. In eq. (H2.12), at $r = 0$, the term

$$\frac{1}{r}\frac{\partial T}{\partial r}$$

is indeterminate since from boundary condition (H2.14)

$$\frac{\partial T(0, t)}{\partial r} = 0$$

This indeterminate form can be removed by an application of l'Hospital's rule

$$\lim_{r \to 0} \frac{1}{r}\frac{\partial T}{\partial r} = \lim_{r \to 0} \frac{1}{1}\frac{\partial^2 T}{\partial r^2}$$

and eq. (H2.12) at $r = 0$ becomes

$$\frac{\partial T}{\partial t} = 2\frac{\partial^2 T}{\partial r^2}$$

This result is used in DO loop 1 of subroutine DERV11 (with I = 1). For I ≠ 1, eq. (H2.12) is programmed directly.

(7) A similar result for eq. (H2.12) with $n = 2$ (spherical coordinates) is used in DERV12

$$\lim_{r \to 0} \frac{2}{r}\frac{\partial T}{\partial r} = \lim_{r \to 0} \frac{2}{1}\frac{\partial^2 T}{\partial r^2}$$

and therefore

$$\frac{\partial T}{\partial t} = 3\frac{\partial^2 T}{\partial r^2}$$

The programming in DERV12 is then

```
C...
C...    FOURIER'S SECOND LAW
        DO 1 I=1,NR
          IF(I.EQ.1)THEN
            TT(I)=3.0D0*TRR(I)
          ELSE
            R=DFLOAT(I-1)/DFLOAT(NR-1)
            TT(I)=TRR(I)+(2.0D0/R)*TR(I)
          END IF
1       CONTINUE
```

Note that the only difference between DERV11 and DERV12 is that the first spatial derivative, $\partial T/\partial r$, in DERV12 is multiplied by a larger constant. But this larger term contributes directly to the temporal derivative, $\partial T/\partial t$, so the temperature changes more rapidly in the case of spherical coordinates than for cylindrical coordinates. This is evident in the two plots of the solutions in Figures H2.2 and Figure H2.3.

126   *Heat transport*

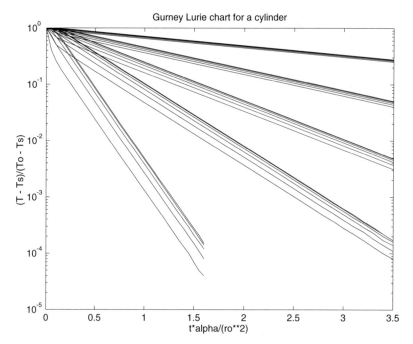

Figure H2.2. Graphical Output from Subroutines INIT11, DERV11, and PNT11

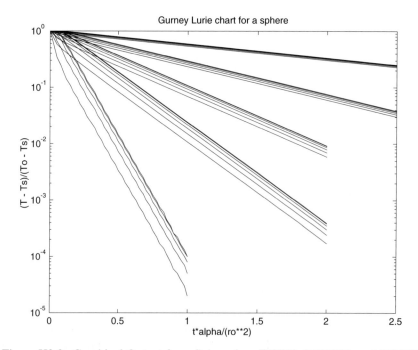

Figure H2.3. Graphical Output from Subroutines INIT12, DERV12, and PNT12

Note the smaller time (horizontal) scale in Figure H2.3 reflecting the faster response in spherical coordinates. This is not surprising since a sphere has a higher surface-to-volume ratio than a cylinder, thereby generally leading to faster heat transfer.

Also, the solution for $B_i = 20$ is wavy, particularly in the case of spherical coordinates (Figure H2.3), which is an indication of the marginal spatial resolution of the 21-point grid (Figures H2.2 and H2.3 are essentially plots of T versus t, but the spatial resolution of the grid in r affects the time response through the PDE, eq. (H2.12)). In other words, for high B_i (with rapid time variations), the sharp spatial variations in T produce inaccuracies in the solution, which suggests that more radial points should be used.

Finally, we observe that eqs. (H2.12) to (H2.15) can be solved analytically as well as numerically. Thus, we could check the preceding numerical solutions with the corresponding analytical solutions. In general, this would be a relatively tedious procedure, and we therefore consider here only one special-case analytical solution, which is relatively straightforward to derive, namely, eq. (H2.12) with $n = 0$ and eq. (H2.15) with $B_i = \infty$ (for the latter, we have $T(1, t) = 0$ as a special case of eq. (H2.15)).

Following the usual separation of variables argument (Schiesser 1994, 400–15), a product solution to eqs. (H2.12) to (H2.15) is assumed

$$T(x, t) = e^{-\lambda_n^2 t} \cos(\lambda_n x) \tag{H2.16}$$

If $\lambda_n = (n - 1/2)\pi, n = 1, 2, 3, \ldots$, boundary conditions (H2.14) and (H2.15) are satisfied (again with $B_i = \infty$). Initial condition (H2.13) can be satisfied by a superposition of the solutions of eq. (H2.16), which is possible since eq. (H2.12) is linear

$$T(x, t) = \sum_{n=1}^{\infty} e^{-\lambda_n^2 t} \cos(\lambda_n x) \tag{H2.17}$$

Then, application of eq. (H2.13) to eq. (H2.17) gives

$$1 = \sum_{n=1}^{\infty} C_n \cos(\lambda_n x) \tag{H2.18}$$

If eq. (H2.18) is multiplied by $\cos(\lambda_m x)$ and integrated from $x = 0$ to $x = 1$

$$\int_0^1 \cos(\lambda_m x) dx = \sum_{n=1}^{\infty} C_n \int_0^1 \cos(\lambda_m x) \cos(\lambda_n x) dx \tag{H2.19}$$

Because of the orthogonality of the cosines, the right hand side (RHS) integral of eq. (H2.19) is zero except when $n = m$. Thus,

$$C_n = \frac{\int_0^1 \cos(\lambda_n x) dx}{\int_0^1 \cos^2(\lambda_n x) dx} = \frac{(1/\lambda_n) \sin(\lambda_n x)|_0^1}{(1/2)\{x + (1/2\lambda_n) \sin(2\lambda_n x)\}|_0^1} = (2/\lambda_n)(-1)^{n+1} \quad \text{(H2.20)}$$

where we have used $\cos^2(nx) = (1/2)(1 + \cos(2nx))$ in the denominator of eq. (H2.20). Thus, the analytical solution is, from eqs. (H2.17) and (H2.20)

$$T(x,t) = 2 \sum_{n=1}^{\infty} \frac{(-1)^{n+1}}{\lambda_n} e^{-\lambda_n^2 t} \cos(\lambda_n x) \quad \text{(H2.21)}$$

Equation (H2.21) is evaluated by the following program for comparison with the numerical solution of eqs. (H2.12) to (H2.15) with $n = 0$ and $B_i = 20$ (which is effectively infinite). Note that for $t > 0$, this series converges rapidly, and therefore only five terms are used in the program

```
      PROGRAM SERIES
C...
C...  DOUBLE PRECISION CODING IS USED
      IMPLICIT DOUBLE PRECISION (A-H,O-Z)
C...
C...  NUMBER OF TIMES, POSITIONS, TERMS IN THE SERIES
      PARAMETER(NT=4,NX=6,NS=5)
C...
C...  DECLARE ARRAYS
      DIMENSION X(NX), SUM(NX)
C...
C...  DEFINE OUTPUT FILE
      OPEN(1,FILE='output')
C...
C...  PI
      PI=4.0D0*DATAN(1.0D0)
C...
C...  HEADING FOR THE COMPUTED SERIES
      WRITE(1,5)
5     FORMAT(8X,'t ',3X,' x = 0 ',3X,'x = 0.2',3X,'x = 0.4',
     +                 3X,'x = 0.6',3X,'x = 0.8',3X,' x = 1 ')
C...
C...  STEP THROUGH A SERIES OF TIMES
      DO 1 IT=1,NT
      T=0.5*DFLOAT(IT)
C...
C...  STEP THROUGH A SERIES OF POSITIONS
      DO 2 IX=1,NX
```

```
              X(IX)=0.2D0*DFLOAT(IX-1)
C...
C...    SUM THE SERIES
              SUM(IX)=0.0D0
              DO 3 N=1,NS
              EN=(DFLOAT(N)-0.5D0)*PI
              TERM=2.0D0*(-1.0D0)**(N+1)/EN*DEXP(-(EN**2)*T)
         +            *COS(EN*X(IX))
              SUM(IX)=SUM(IX)+TERM
3             CONTINUE
C...
C...    NEXT X
2             CONTINUE
C...
C...    PRINT SERIES
              WRITE(1,4)T,(SUM(IX),IX=1,6)
4             FORMAT(F10.2,6F10.5)
C...
C...    NEXT T
1             CONTINUE
C...
C...    END OF CALCULATION
              STOP
              END

              DOUBLE PRECISION FUNCTION DFLOAT(I)
              DFLOAT=DBLE(FLOAT(I))
              RETURN
              END
```

We can note the following points about program SERIES:

(1) Four values of t, six values of x, and five terms in the series are used in eq. (H2.21)

```
C...
C...    NUMBER OF TIMES, POSITIONS, TERMS IN THE SERIES
              PARAMETER(NT=4,NX=6,NS=5)
```

(2) The four values of t are 0.5, 1.0, 1.5, and 2.0

```
C...
C...    STEP THROUGH A SERIES OF TIMES
              DO 1 IT=1,NT
              T=0.5*DFLOAT(IT)
```

(3) The six values of x are 0, 0.2, 0.4, 0.6, 0.8, and 1

```
C...
C...   STEP THROUGH A SERIES OF POSITIONS
       DO 2 IX=1,NX
       X(IX)=0.2D0*DFLOAT(IX-1)
```

(4) The five terms summed in the series are computed in DO loop 3

```
C...
C...   SUM THE SERIES
       SUM(IX)=0.0D0
       DO 3 N=1,NS
       EN=(DFLOAT(N)-0.5D0)*PI
       TERM=2.0D0*(-1.0D0)**(N+1)/EN*DEXP(-(EN**2)*T)
      *    *COS(EN*X(IX))
       SUM(IX)=SUM(IX)+TERM
3      CONTINUE
```

First, the nth eigenvalue is computed, $\lambda_n = (n - 1/2)\pi$ ($=$ EN). Then the nth term ($=$ TERM) in eq. (H2.21) is computed

$$2\frac{(-1)^{n+1}}{\lambda_n}e^{-\lambda_n^2 t}\cos(\lambda_n x)$$

Finally, this term is added to the sum of eq. (H2.21).

(5) The preceding calculations are performed for the six values of x, and the series solution is then printed

```
C...
C...   NEXT X
2      CONTINUE
C...
C...   PRINT SERIES
       WRITE(1,4)T,(SUM(IX),IX=1,6)
4      FORMAT(F10.2,6F10.5)
```

(6) The preceding calculations are performed for the four values of t to complete the calculations

```
C...
C...   NEXT T
1      CONTINUE
C...
C...   END OF CALCULATION
       STOP
       END
```

(7) Function DFLOAT converts a single precision integer to a double precision real (DFLOAT may be included in the Fortran compiler for some computers, but it is provided here in case it is needed).

The output from program SERIES is listed below:

t	x = 0	x = 0.2	x = 0.4	x = 0.6	x = 0.8	x = 1
0.50	0.37078	0.35263	0.29997	0.21795	0.11458	0.00000
1.00	0.10798	0.10269	0.08736	0.06347	0.03337	0.00000
1.50	0.03144	0.02991	0.02544	0.01848	0.00972	0.00000
2.00	0.00916	0.00871	0.00741	0.00538	0.00283	0.00000

Note that boundary condition (H2.15) is satisfied (with $B_i = \infty$). The corresponding output from the MOL program (subroutines INIT10, DERV10, and PRT10) is listed below:

t	x = 0	x = 0.2	x = 0.4	x = 0.6	x = 0.8	x = 1
.50	.37077	.35263	.29997	.21795	.11458	.00000
1.00	.10798	.10269	.08736	.06347	.03337	.00000
1.50	.03144	.02991	.02544	.01848	.00972	.00000
2.00	.00916	.00871	.00741	.00538	.00283	.00000

The difference in the appearance of the preceding output is due to the different compilers used to perform the calculations. The MOL program was executed on a workstation that does not output leading zeros, and program SERIES was executed on a PC that does output leading zeros. In any case, the agreement between the numerical and analytical solutions is to five figures, which generally indicates that the 21-point grid used in the MOL program and the five terms used in program SERIES are adequate.

Finally, we note an important extension of the solution to the one-dimensional problem of eqs. (H2.12) to (H2.15) to two and three dimensions. For example, if we consider the two-dimensional heat conduction problem in Cartesian coordinates

$$\frac{\partial T}{\partial t} = \frac{\partial^2 T}{\partial x^2} + \frac{\partial^2 T}{\partial y^2} \qquad (H2.22)$$

$$T(x, y, 0) = 1 \qquad (H2.23)$$

$$\frac{\partial T(0, y, t)}{\partial x} = 0 \qquad (H2.24)$$

$$\frac{\partial T(x, 0, t)}{\partial y} = 0 \qquad (H2.25)$$

$$\frac{\partial T(1, y, t)}{\partial x} + B_i T(1, y, t) = 0 \qquad (H2.26)$$

$$\frac{\partial T(x, 1, t)}{\partial y} + B_i T(x, 1, t) = 0 \qquad (H2.27)$$

If $T_1(x, t)$ is a solution to eqs. (H2.12) to (H2.15) and $T_2(y, t)$ is a solution to the analogous problem in y (i.e., with x replaced by y in eqs. (H2.12) to (H2.15)), then the solution to eqs. (H2.22) to (H2.27) is

$$T(x, y, t) = T_1(x, t)T_2(y, t) \qquad \text{(H2.28)}$$

$T(x, y, t)$ of eq. (H2.28) can be verified as a solution to eqs. (H2.22) to (H2.27) by substitution in these equations. Thus, eq. (H2.28) substituted in eq. (H2.22) gives

$$T_1 \frac{\partial T_2}{\partial t} + T_2 \frac{\partial T_1}{\partial t} = T_2 \frac{\partial^2 T_1}{\partial x^2} + T_1 \frac{\partial^2 T_2}{\partial y^2}$$

or

$$T_1 \left\{ \frac{\partial T_2}{\partial t} - \frac{\partial^2 T_2}{\partial y^2} \right\} + T_2 \left\{ \frac{\partial T_1}{\partial t} - \frac{\partial^2 T_1}{\partial x^2} \right\} = 0$$

Each of the terms in brackets is zero (since $T_1(x, t)$ and $T_2(y, t)$ satisfy the one-dimensional problem of eq. (H2.12)).

Similarly, substitution of eq. (H2.28) in eq. (H2.23) gives

$$T(x, y, 0) = T_1(x, 0)T_2(y, 0) = 1 \cdot 1 = 1$$

Substitution of eq. (H2.28) in eqs. (H2.24) and (H2.25) gives

$$\frac{\partial T(0, y, t)}{\partial x} = T_2(y, t) \frac{\partial T_1(0, t)}{\partial x} = T_2(y, t) \cdot 0 = 0$$

$$\frac{\partial T(x, 0, t)}{\partial y} = T_1(x, t) \frac{\partial T_2(0, t)}{\partial y} = T_1(x, t) \cdot 0 = 0$$

Substitution of eqs. (H2.28) in eqs. (H2.26) and (H2.27) gives

$$\frac{\partial T(1, y, t)}{\partial x} + B_i T(1, y, t) = T_2(y, t) \frac{\partial T_1(1, t)}{\partial x} + B_i T_2(y, t) T_1(1, t)$$

$$= T_2(y, t) \left\{ \frac{\partial T_1(1, t)}{\partial x} + B_i T_1(1, t) \right\}$$

$$= T_2(y, t) \cdot 0 = 0$$

$$\frac{\partial T(x, 1, t)}{\partial y} + B_i T(x, 1, t) = T_1(x, t) \frac{\partial T_2(1, t)}{\partial y} + B_i T_1(x, t) T_2(1, t)$$

$$= T_1(x, t) \left\{ \frac{\partial T_2(1, t)}{\partial y} + B_i T_2(1, t) \right\}$$

$$= T_1(x, t) \cdot 0 = 0$$

Thus, eq. (H2.28) is a solution to eqs. (H2.22) to (H2.27), that is, the solution to the two-dimensional problem, $T(x, y, t)$, is the product of the solutions to the two one-dimensional problems, $T_1(x, t)$ and $T_2(y, t)$. But we already have the MOL program to compute $T_1(x, t)$ and $T_2(y, t)$, and we can multiply these two numerical solutions (for particular values of x, y, and t) to obtain $T(x, y, t)$. Also, although boundary conditions (H2.26) and (H2.27) are specified at $x = 1$ and $y = 1$, respectively, the half-lengths in x and y, x_0, and y_0, do not have to be equal to apply the product solution of eq. (H2.28).

This same procedure of multiplying the solutions to two one-dimensional problems can be used to obtain the numerical solution to the following two-dimensional problem in cylindrical coordinates

$$\frac{\partial T}{\partial t} = \frac{\partial^2 T}{\partial r^2} + \frac{1}{r}\frac{\partial T}{\partial r} + \frac{\partial^2 T}{\partial z^2} \tag{H2.29}$$

$$T(x, r, 0) = 1 \tag{H2.30}$$

$$\frac{\partial T(0, z, t)}{\partial r} = 0 \tag{H2.31}$$

$$\frac{\partial T(r, 0, t)}{\partial z} = 0 \tag{H2.32}$$

$$\frac{\partial T(1, z, t)}{\partial r} + B_i T(1, z, t) = 0 \tag{H2.33}$$

$$\frac{\partial T(r, 1, t)}{\partial z} + B_i T(r, 1, t) = 0 \tag{H2.34}$$

Thus

$$T(r, z, t) = T_1(r, t) T_2(z, t) \tag{H2.35}$$

where $T_1(r, t)$ and $T_2(z, t)$ can be computed with the MOL program (with NC $= 0$ and 1, respectively).

Finally, a product of the solutions of three one-dimensional problems gives the solution to the following three-dimensional problem in Cartesian coordinates

$$\frac{\partial T}{\partial t} = \frac{\partial^2 T}{\partial x^2} + \frac{\partial^2 T}{\partial y^2} + \frac{\partial^2 T}{\partial z^2} \tag{H2.36}$$

$$T(x, y, z, 0) = 1 \tag{H2.37}$$

$$\frac{\partial T(0, y, z, t)}{\partial x} = 0 \tag{H2.38}$$

$$\frac{\partial T(x, 0, z, t)}{\partial y} = 0 \tag{H2.39}$$

$$\frac{\partial T(x, y, 0, t)}{\partial z} = 0 \tag{H2.40}$$

$$\frac{\partial T(1, y, z, t)}{\partial x} + B_i T(1, y, z, t) = 0 \tag{H2.41}$$

$$\frac{\partial T(x, 1, z, t)}{\partial y} + B_i T(x, 1, z, t) = 0 \tag{H2.42}$$

$$\frac{\partial T(x, y, 1, t)}{\partial z} + B_i T(x, y, 1, t) = 0 \tag{H2.43}$$

The solution to eqs. (H2.37) to (H2.43) is then the product solution

$$T(x, y, z, t) = T_1(x, t)T_2(y, t)T_3(z, t) \qquad (H2.44)$$

where $T_1(x, t)$, $T_2(y, t)$, and $T_3(z, t)$ can be computed with the MOL program (with NC = 0). Again, boundary conditions (H2.41), (H2.42), and (H2.43) can be generalized to apply at x_0, y_0, and z_0 when the three one-dimensional numerical solutions, $T_1(x, t)$, $T_2(y, t)$, and $T_3(z, t)$ are computed with the MOL program.

The preceding MOL coding is applicable only to the linear Fourier's second law, eq. (H2.12), and the extensions to two and three dimensions discussed previously. But experience over many years has indicated these solutions can be applied to a spectrum of problems. Also, a program is available as explained in the preface that computes MOL solutions for more general two- and three-dimensional problems by using two- and three-dimensional finite difference approximations for the spatial derivatives in Fourier's second law.

References

Forsythe, G. E., M. A. Malcolm, and C. B. Moler. 1977. *Computer Methods for Mathematical Computations*. Englewood Cliffs, NJ: Prentice-Hall.

Schiesser, W. E. 1994. *Computational Mathematics in Engineering and Applied Science*. Boca Raton, FL: CRC Press.

Welty, J. R., C. E. Wicks, and R. E. Wilson. 1984. *Fundamentals of Momentum, Heat, and Mass Transfer*. 3d ed. New York: John Wiley and Sons.

H3

Heat transfer in a circular fin

Consider the problem of heat transfer from a circular fin (Bird, Stewart, and Lightfoot (BSL) 1960, 308), as depicted in Figure H3.1.

(a) Derive the steady state energy balance that defines the fin temperature as a function of radial position within the fin.

(b) Obtain an analytical solution and a numerical solution to the energy balance of (a).

(c) From the solutions of (b), compute the total heat transfer from the fin.

Solution

To obtain the numerical solution required in (b), we will numerically integrate the unsteady state energy balance to steady state. Therefore we start with the unsteady

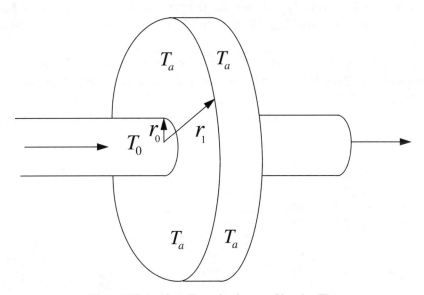

Figure H3.1. Heat Transfer from a Circular Fin

state energy balance for an incremental section of the fin of width Δr and thickness B:

$$2\pi r B \Delta r \rho C_p \frac{\partial T}{\partial t} = -k 2\pi r B \frac{\partial T}{\partial r}\bigg|_r - \left\{ -k 2\pi (r + \Delta r) B \frac{\partial T}{\partial r}\bigg|_{r+\Delta r} \right\}$$
$$- 2\{2\pi r \Delta r h (T - T_a)\} \quad \text{(H3.1)}$$

where we have assumed negligible variation in temperature across the fin (axially) and constant physical properties for the fin. Division of eq. (H3.1) by $2\pi r B \Delta r \rho C_p$, followed by $r \to 0$, gives

$$\frac{\partial T}{\partial t} = \alpha \frac{1}{r} \frac{\partial}{\partial r}\left\{ r \frac{\partial T}{\partial r} \right\} - \frac{2h}{B \rho C_p}(T - T_a) \quad \text{(H3.2)}$$

where $\alpha = k/(\rho C_p)$ (the thermal diffusivity). Also, from the units of eq. (H3.2), we can interpret the group $B\rho C_p/(2h)$ as a time constant for heat transfer, that is, this group has the units of time.

The initial condition for eq. (H3.2) is

$$T(r, 0) = T_i \quad \text{(H3.3)}$$

and the boundary conditions are

$$T(r_0, t) = T_0 \quad \text{(H3.4)}$$
$$\frac{\partial T(r_1, t)}{\partial r} = 0 \quad \text{(H3.5)}$$

Boundary condition (H3.4) indicates the temperature at $r = r_0$ is T_0 (see Figure H3.1), and boundary condition (H3.5) indicates negligible heat loss from the edge of the fin.

We consider first a numerical solution computed by the method of lines (MOL) applied to eqs. (H3.2) to (H3.5) on a 21-point radial grid. The numerical parameters are $r_0 = 0.5, r_1 = 2, B = 0.2, h = 1, \rho C_p = 2, \alpha = 0.01, T_i = 25, T_a = 25, T_0 = 150$. Subroutine INITAL, which implements initial condition (H3.3), is listed first:

```
      SUBROUTINE INITAL
      IMPLICIT DOUBLE PRECISION(A-H,O-Z)
      PARAMETER(NR=21)
      COMMON/T/     T,   NSTOP,   NORUN
     +      /Y/   U(NR)
     +      /F/   UT(NR)
     +      /P/   ALPHA,  BETA,     UA,      U0,    DR,     DRS,
     +            R(NR)
     +      /I/    IP
C...
C...  PROBLEM PARAMETERS
```

```fortran
C...
C...       FIN INSIDE RADIUS
           R0=0.5D0
C...
C...       FIN OUTSIDE RADIUS
           R1=2.0D0
C...
C...       FIN THICKNESS
           B=0.2D0
C...
C...       HEAT TRANSFER COEFFICIENT
           H=0.01D0
C...
C...       RHO*CP
           RHOCP=2.0D0
C...
C...       THERMAL DIFFUSIVITY
           ALPHA=0.01D0
C...
C...       INITIAL TEMPERATURE
           UI=25.0D0
C...
C...       AMBIENT TEMPERATURE
           UA=25.0D0
C...
C...       INSIDE TEMPERATURE
           U0=150.0D0
C...
C...    RECIPROCAL TIME FOR HEAT TRANSFER
        BETA=(2.0D0*H)/(B*RHOCP)
C...
C...    INCREMENT FOR RADIAL GRID, SQUARE OF INCREMENT
        DR=(R1-R0)/DFLOAT(NR-1)
        DRS=DR**2
C...
C...    RADIAL GRID AND INITIAL CONDITION
        DO 1 I=1,NR
           R(I)=R0+DFLOAT(I-1)*DR
           U(I)=UI
1       CONTINUE
C...
C...    INITIAL DERIVATIVES
        CALL DERV
        IP=0
        RETURN
        END
```

We can note the following points about subroutine INITAL:

(1) Double precision coding is used. The radial grid has 21 points. The dependent variable, $T(r, t)$, in eq. (H3.2) is in array U(NR) in COMMON/Y/, and the temporal derivative, $\partial T(r, t)/\partial t$, is in array UT(NR) in COMMON/F/. Thus, we have the usual arrangement of integrating a vector of temporal derivatives in COMMON/F/ to produce a dependent variable vector in COMMON/Y/.

```
      IMPLICIT DOUBLE PRECISION(A-H,O-Z)
      PARAMETER(NR=21)
      COMMON/T/     T,  NSTOP,  NORUN
     +       /Y/  U(NR)
     +       /F/  UT(NR)
     +       /P/  ALPHA,  BETA,    UA,    U0,    DR,    DRS,
     +            R(NR)
     +       /I/    IP
```

Problem parameters are in COMMON/P/, and an integer counter for use in subroutine PRINT is in /I/.

(2) The problem parameters are then set

```
C...
C...  PROBLEM PARAMETERS
C...
C...      FIN INSIDE RADIUS
          R0=0.5D0
              .
              .
              .
C...
C...      INSIDE TEMPERATURE
          U0=150.0D0
C...
C...  RECIPROCAL TIME FOR HEAT TRANSFER
      BETA=(2.0D0*H)/(B*RHOCP)
```

(3) The radial grid spacing is computed next, then used in DO loop 1 to define the radial grid required in the programming of eq. (H3.2) in subroutine DERV. Also, initial condition (H3.3) is programmed in DO loop 1

```
C...
C...  INCREMENT FOR RADIAL GRID, SQUARE OF INCREMENT
      DR=(R1-R0)/DFLOAT(NR-1)
      DRS=DR**2
C...
C...  RADIAL GRID AND INITIAL CONDITION
```

```
      DO 1 I=1,NR
         R(I)=R0+DFLOAT(I-1)*DR
         U(I)=UI
1     CONTINUE
```

(4) Finally, a call to DERV computes the initial derivatives (and thereby generally tests the programming in DERV).

```
C...
C...  INITIAL DERIVATIVES
      CALL DERV
      IP=0
```

An integer counter, IP, is also initialized for use in subroutine PRINT.

The initial conditions for the NR ordinary differential equations (ODEs) approximating eq. (H3.2) are now defined. The integration of the ODEs can then proceed according to the programming in subroutine DERV, listed next

```
      SUBROUTINE DERV
      IMPLICIT DOUBLE PRECISION(A-H,O-Z)
      PARAMETER(NR=21)
      COMMON/T/     T,    NSTOP,   NORUN
     +      /Y/  U(NR)
     +      /F/  UT(NR)
     +      /P/  ALPHA,  BETA,   UA,    UO,   DR,   DRS,
     +           R(NR)
     +      /I/     IP
C...
C...  BOUNDARY CONDITION AT R = R0
         U(1)=UO
         UT(1)=0.0D0
C...
C...  BOUNDARY CONDITION AT R = R1
C...
C...     RADIAL GROUP NOT INCLUDED
         IF(NORUN.EQ.1)THEN
         UT(NR)=-BETA*(U(NR)-UA)
C...
C...     RADIAL GROUP INCLUDED
         ELSE IF(NORUN.EQ.2)THEN
         UT(NR)=ALPHA*2.0D0*(U(NR-1)-U(NR))/DRS
     +          -BETA*(U(NR)-UA)
         END IF
C...
C...  INTERIOR POINTS
      DO 1 I=2,NR-1
         UT(I)=ALPHA*(U(I+1)-2.0D0*U(I)+U(I-1))/DRS
```

```
     +             +ALPHA*(1.0D0/R(I))*(U(I+1)-U(I-1))/(2.0D0*DR)
     +             -BETA*(U(I)-UA)
1          CONTINUE
           RETURN
           END
```

We can note the following points about DERV:

(1) The COMMON area is the same as for INITAL. Boundary condition (H3.4) at $r = r_0$ is programmed as

```
C...
C...   BOUNDARY CONDITION AT R = R0
       U(1)=U0
       UT(1)=0.0D0
```

Since the boundary temperature is constant at T_0 (or U0), the temporal derivative of the temperature is set to zero.

(2) We can consider two methods for programming boundary condition (H3.5). If we consider eq. (H3.2) with $\partial T(r_1, t)/\partial r = 0$ from eq. (H3.5), we might conclude that the entire radial group in eq. (H3.2) is zero and therefore eq. (H3.2) at $r = r_1$ reduces to

$$\frac{\partial T}{\partial t} = -\frac{2h}{B\rho C_p}(T - T_a) \tag{H3.6}$$

The programming of eq. (H3.6) (for the first "run" or solution, with NORUN = 1) is

```
C...
C...   BOUNDARY CONDITION AT R = R1
C...
C...        RADIAL GROUP NOT INCLUDED
       IF(NORUN.EQ.1)THEN
       UT(NR)=-BETA*(U(NR)-UA)
```

(3) However, we can also reason that the radial group in eq. (H3.2) is not zero at $r = r_1$. To see this, if the radial group is differentiated, eq. (H3.2) becomes

$$\frac{\partial T}{\partial t} = \alpha\left\{\frac{\partial^2 T}{\partial r^2} + \frac{1}{r}\frac{\partial T}{\partial r}\right\} - \frac{2h}{B\rho C_p}(T - T_a) \tag{H3.7}$$

Now, we use boundary condition (H3.5) so that eq. (H3.7) (at $r = r_1$) becomes

$$\frac{\partial T}{\partial t} = \alpha\left\{\frac{\partial^2 T}{\partial r^2}\right\} - \frac{2h}{B\rho C_p}(T - T_a) \tag{H3.8}$$

However, $\partial^2 T/\partial r^2$ is not necessarily zero (e.g., if $T(r, t) = (r - r_1)^2$, $\partial T/\partial r = 2(r - r_1) = 0$ at $r = r_1$, but $\partial^2 T/\partial r^2 = 2 \neq 0$). Thus we should retain $\partial^2 T/\partial r^2$

in programming eq. (H3.8)

```
C...
C...       RADIAL GROUP INCLUDED
           ELSE IF(NORUN.EQ.2)THEN
             UT(NR)=ALPHA*2.0D0*(U(NR-1)-U(NR))/DRS
     +              -BETA*(U(NR)-UA)
           END IF
```

Here we have used the centered approximation

$$\frac{\partial^2 T}{\partial r^2} \approx \frac{T(r_1 + \Delta r, t) - 2T(r_1, t) + T(r_1 - \Delta r, t)}{\Delta r^2} \tag{H3.9}$$

The fictitious value, $T(r_1 + \Delta r, t)$, is then replaced with $T(r_1 - \Delta r, t)$ through the approximation of boundary condition (H3.5) as

$$\frac{\partial T}{\partial r} \approx \frac{T(r_1 + \Delta r, t) - T(r_1 - \Delta r, t)}{2\Delta r} = 0 \tag{H3.10}$$

(or, $T(r_1 + \Delta r, t) = T(r_1 - \Delta r, t)$). Then, eq. (H3.9) becomes

$$\frac{\partial^2 T}{\partial r^2} \approx 2 \frac{T(r_1 - \Delta r, t) - T(r_1, t)}{\Delta r^2} \tag{H3.11}$$

as reflected in the preceding programming. The only difference in the solutions for NORUN = 1 and 2 is the inclusion of eq. (H3.11) in the programming for NORUN = 2. Thus, by comparing the two solutions we can observe the contribution of the radial group in eq. (H3.2) at $r = r_1$.

(4) Equation (H3.7) is then programmed at the interior points to complete the programming of all of the temporal derivatives in COMMON/F/

```
C...
C...    INTERIOR POINTS
        DO 1 I=2,NR-1
          UT(I)=ALPHA*(U(I+1)-2.0D0*U(I)+U(I-1))/DRS
     +         +ALPHA*(1.0D0/R(I))*(U(I+1)-U(I-1))/(2.0D0*DR)
     +         -BETA*(U(I)-UA)
1       CONTINUE
```

Subroutine PRINT for printing the solution is listed below:

```
        SUBROUTINE PRINT(NI,NO)
        IMPLICIT DOUBLE PRECISION(A-H,O-Z)
        PARAMETER(NR=21)
        COMMON/T/    T,   NSTOP,  NORUN
     +       /Y/  U(NR)
     +       /F/  UT(NR)
     +       /P/  ALPHA,   BETA,    UA,    U0,    DR,    DRS,
```

```
      +                RO,     R1, R(NR),        B,   RHOCP
      +        /I/      IP
C...
C...  MONITOR OUTPUT
      WRITE(*,*)' NORUN = ',NORUN,'   T = ',T
C...
C...  IF THE SOLUTION HAS A PHYSICALLY UNREALISTIC VALUE,
C...  TERMINATE EXECUTION OF CURRENT RUN
      DO 2 I=1,NR
         IF((U(I).GT.UO).OR.(U(I).LT.UA))THEN
         WRITE(NO,1)NORUN,T
1        FORMAT(//,' RUN NO = ',I2,' AT T = ',F7.2,'
      +            TERMINATED',//)
         NSTOP=1
         END IF
2     CONTINUE
C...
C...  PRINT THE SOLUTION
      WRITE(NO,3)T,(R(I),I=1,NR,4),(U(I),I=1,NR,4)
3     FORMAT(' t =    ',F6.2,/,'        r ',6F9.2,/,' u(r,t)
      +       ',6F9.3,/)
C...
C...  COMPUTE AND PRINT THE TOTAL HEAT TRANSFER TO THE FIN
      DTDR=(-3.0D0*U(1)+4.0D0*U(2)-U(3))/(2.0D0*DR)
      PI=4.0D0*DATAN(1.0D0)
      TOTAL=-2.0D0*PI*RO*B*ALPHA*RHOCP*DTDR
      WRITE(NO,4)TOTAL
4     FORMAT(' Total heat transfer rate = ',F7.3,//)
      RETURN
      END
```

We can note the following points about subroutine PRINT:

(1) The COMMON area is the same as in INITAL and DERV. A test is then made to determine if the temperature moves beyond physically unrealistc limits, that is, whether it exceeds the maximum possible value, T_0, and the minimum possible value, T_a

```
C...
C...  IF THE SOLUTION HAS A PHYSICALLY UNREALISTIC VALUE,
C...  TERMINATE EXECUTION OF CURRENT RUN
      DO 2 I=1,NR
         IF((U(I).GT.UO).OR.(U(I).LT.UA))THEN
         WRITE(NO,1)NORUN,T
1        FORMAT(//,' RUN NO = ',I2,' AT T = ',F7.2,'
      +            TERMINATED',//)
```

```
            NSTOP=1
         END IF
2        CONTINUE
```

We would, of course, expect that these limits would not be violated, but with the numerical integration of an evolutionary partial differential equation (PDE) such as eq. (H3.2), we might observe that the solution moves forward in t for a short duration, during which it assumes an unrealistic value (remember that the temporal integration of eq. (H3.2) proceeds by finite steps in time, in this case selected by the ODE integrator, RKF45 (Forsythe, Malcolm, and Moler 1977), and conceivably the time step might be large enough to move the solution into a physically unrealistic region). Once this happens, the numerical integration might sense that the time step was too large and repeat the integration with a smaller step. This recovery is not guaranteed, however (it depends to some extent on the error estimation algorithm of the ODE integrator and probably, more importantly, on the PDE(s)). Therefore, in programming the solution of an evolutionary PDE, anticipated problems might be covered with some special programming such as in DO loop 2 (at least in this way, a warning message can be printed, and then an assessment can be made of the importance of the error condition).

(2) The solution is then printed as a function of r (at a given value of t when PRINT is called). Every fourth value of $T(r, t)$ is printed radially to provide a reasonable volume of output

```
C...
C...     PRINT THE SOLUTION
         WRITE(NO,3)T,(R(I),I=1,NR,4),(U(I),I=1,NR,4)
3        FORMAT(' t =     ',F6.2,/,'           r ',6F9.2,/,' u(r,t)
        +           ',6F9.3,/)
```

(3) Finally, the total rate of heat transfer to the fin is computed

```
C...
C...     COMPUTE AND PRINT THE TOTAL HEAT TRANSFER TO THE FIN
         DTDR=(-3.0D0*U(1)+4.0D0*U(2)-U(3))/(2.0D0*DR)
         PI=4.0D0*DATAN(1.0D0)
         TOTAL=-2.0D0*PI*R0*B*ALPHA*RHOCP*DTDR
         WRITE(NO,4)TOTAL
4        FORMAT(' Total heat transfer rate = ',F7.3,//)
```

This programming is for the application of Fourier's first law at $r = r_0$

$$q_{\text{total}} = -2\pi r_0 B k \frac{\partial T(r_0, t)}{\partial r} \quad \text{(H3.12)}$$

(note that the thermal conductivity, k, is given by $k = \alpha \rho C_p$). The radial derivative, $\partial T(r_0, t)/\partial r$, is approximated by the three-point, second-order finite difference

(Schiesser 1991, 100)

$$\frac{\partial T(r_0, t)}{\partial r} \approx \frac{-3T(r_0, t) + 4T(r_0 + \Delta r, t) - T(r_0 + 2\Delta r, t)}{2\Delta r} + O(\Delta r^2)$$

since the two-point, first-order finite difference, which was used at first

$$\frac{\partial T(r_0, t)}{\partial r} \approx \frac{T(r_0 + \Delta r, t) - T(r_0, t)}{\Delta r} + O(\Delta r)$$

introduced an error of approximately 10%.

The data file read by main program RKF45M (in Appendix A) that calls RKF45 is listed below:

```
CONDUCTION IN A CIRCULAR FIN
0.          100.0      10.0
   21                           0.00001
CONDUCTION IN A CIRCULAR FIN
0.          100.0      10.0
   21                           0.00001
END OF RUNS
```

Two runs are programmed (as discussed previously and programmed in DERV). In each case time covers the interval $0 \leq t \leq 100$ with output at intervals of 10; 21 ODEs are specified with time integration to an accuracy of 0.00001.

The output from subroutine PRINT for the two runs is summarized below:

```
RUN NO. -   1   CONDUCTION IN A CIRCULAR FIN

  INITIAL T -   0.000D+00

    FINAL T -   0.100D+03

    PRINT T -   0.100D+02

  NUMBER OF DIFFERENTIAL EQUATIONS -   21

  MAXIMUM INTEGRATION ERROR -   0.100D-04

t =       0.00
       r     0.50     0.80     1.10     1.40     1.70     2.00
   u(r,t) 150.000   25.000   25.000   25.000   25.000   25.000

  Total heat transfer rate =   31.416

t =      10.00
       r     0.50     0.80     1.10     1.40     1.70     2.00
   u(r,t) 150.000   66.735   36.488   27.383   25.354   25.000
```

Total heat transfer rate =   5.387

t =   20.00
     r   0.50   0.80   1.10   1.40   1.70   2.00
u(r,t) 150.000   73.592   43.492   31.446   26.865   25.000

Total heat transfer rate =   4.952

.   .
.   .
.   .

t =   90.00
     r   0.50   0.80   1.10   1.40   1.70   2.00
u(r,t) 150.000   76.756   47.577   34.781   28.612   25.000

Total heat transfer rate =   4.768

t =   100.00
     r   0.50   0.80   1.10   1.40   1.70   2.00
u(r,t) 150.000   76.759   47.581   34.785   28.614   25.000

Total heat transfer rate =   4.768

RUN NO. -   2   CONDUCTION IN A CIRCULAR FIN

INITIAL T -   0.000D+00

  FINAL T -   0.100D+03

  PRINT T -   0.100D+02

NUMBER OF DIFFERENTIAL EQUATIONS -   21

MAXIMUM INTEGRATION ERROR -   0.100D-04

t =   0.00
     r   0.50   0.80   1.10   1.40   1.70   2.00
u(r,t) 150.000   25.000   25.000   25.000   25.000   25.000

Total heat transfer rate = 31.416

t =   10.00
     r   0.50   0.80   1.10   1.40   1.70   2.00
u(r,t) 150.000   66.735   36.488   27.383   25.360   25.079

Total heat transfer rate =   5.387

            .   .

```
t =      90.00
      r       0.50      0.80      1.10      1.40      1.70      2.00
u(r,t)    150.000    77.156    48.422    36.340    31.440    30.149

Total heat transfer rate =    4.749

t =     100.00
      r       0.50      0.80      1.10      1.40      1.70      2.00
u(r,t)    150.000    77.175    48.453    36.378    31.482    30.192

Total heat transfer rate =    4.748
```

We can note the following points about this output:

(1) In both runs, the initial condition (H3.3) is satisfied except at $r = r_0 (= 0.5)$, where boundary condition (H3.4) has already been imposed (checking the initial and boundary conditions is, of course, always a good idea). For example, for the first run

```
t =       0.00
      r       0.50      0.80      1.10      1.40      1.70      2.00
u(r,t)    150.000    25.000    25.000    25.000    25.000    25.000

Total heat transfer rate =   20.944
```

(2) In both runs, the solution appears to be reaching a steady state since there is little change with increasing t. For example, for the first run

```
t =      90.00
      r       0.50      0.80      1.10      1.40      1.70      2.00
u(r,t)    150.000    76.756    47.577    34.781    28.612    25.000

Total heat transfer rate =    4.286

t =     100.00
      r       0.50      0.80      1.10      1.40      1.70      2.00
u(r,t)    150.000    76.759    47.581    34.785    28.614    25.000

Total heat transfer rate =    4.286
```

Thus, we have $T(0.8, 90) = 76.756$ and $T(0.8, 100) = 76.759$. The changes in the second run between $t = 90$ and $t = 100$ are about as small.

(3) There is a difference in the two solutions at the same time. For example, $T(0.8, 100) = 76.759$ for NORUN $= 1$, and $T(0.8, 100) = 77.175$ for

NORUN = 2. The differences in the solutions are larger near $r = r_1$, for example, $T(2, 100) = 25.000$ for NORUN = 1, and $T(2, 100) = 30.192$ for NORUN = 2. We should understand the reason for these differences.

If we return to the discussion of whether to keep the radial group in eq. (H3.2), for the case when we did (the second solution), eq. (H3.8) was used as the special case of eq. (H3.2) at $r = r_1$, with boundary condition (H3.5) included. Since eq. (H3.5) is a Neumann boundary condition (in which the derivative of the dependent variable is specified, in this case $\partial T(r_1, t)/\partial r$, the second solution is for a Neumann boundary condition at $r = r_1$. As a consequence, $T(r_1, t)$ can vary with t according to eq. (H3.8), and a variation can be observed in the numerical solution, for example, $T(2, 0) = 25.000$ and $T(2, 100) = 30.192$.

For the first solution, when we did not keep the radial group in eq. (H3.2), we integrated eq. (H3.6) at $r = r_1$ as a special case of eq. (H3.2), that is, eq. (H3.6) was the boundary condition for eq. (H3.2) at $r = r_1$. Since $T(r_1, 0) = T_a$, the heat transfer term in eq. (H3.6) started at zero and therefore $T(r_1, t)$ did not depart from its initial value, since according to eq. (H3.6), $\partial T(r_1, t)/\partial t = 0$. In other words, eq. (H3.6) is

$$0 = -\frac{2h}{B\rho C_p}(T - T_a)$$

from which we conclude

$$T(r_1, t) = T_a \tag{H3.13}$$

Thus, the first solution was for eq. (H3.6) as a boundary condition, or equivalently, eq. (H3.13), and not for eq. (H3.5). In other words, the first solution is for a Dirichlet boundary condition (in which the dependent variable is specified at the boundary (according to eq. (H3.13)) and not a Neumann boundary condition (e.g., eq. (H3.5)). The fact that eq. (H3.6) was the boundary condition at $r = r_1$ and therefore $\partial T(r_1, t)/\partial t = 0$ is apparent in the numerical solution (note that for NORUN = 1, $T(2, t) = 25.000$, i.e., the boundary value remained constant to five figures).

Thus, the two solutions are for two different types of boundary conditions at $r = r_1$, which explains the difference in the solutions (everything else in the two formulations is the same). To complete the discussion, the preceding program should be executed for the case $T(r_1, 0) \neq T_a$ in the first solution. What change would you expect in the first solution for this case?

To conclude this study, we consider the execution of the preceding program for one other case, when the outside radius is $r_1 = 5$ (rather than $r_1 = 2$). Thus, we changed this one parameter in subroutine INITAL and reexecuted the program. The output for the two runs at $t = 100$ is listed below (the solutions again approached a steady state at $t = 100$):

```
RUN NO. -    1   CONDUCTION IN A CIRCULAR FIN
        .
        .
        .
t =    100.00
      r        0.50     1.40     2.30     3.20     4.10     5.00
u(r,t)      150.000   35.773   26.148   25.126   25.013   25.000

Total heat transfer rate =    4.213

RUN NO. -    2   CONDUCTION IN A CIRCULAR FIN
        .
        .
        .
t =    100.00
      r        0.50     1.40     2.30     3.20     4.10     5.00
u(r,t)      150.000   35.773   26.148   25.126   25.013   25.002

Total heat transfer rate =    4.213
```

We can note the following points about this output:

(1) The solutions agree to about five figures.

(2) By increasing r_1 from 2 to 5, the fin has a longer radial dimension over which heat transfer to the surrounding air takes place. As expected, the outer portion of the fin (at larger radial positions) assumes a temperature close to the ambient temperature, 25. In other words, there is little radial variation of the temperature as $r \to r_1$. (This could be considered an indication of a poor fin design since much of the outer portion of the fin does not contribute to the heat transfer, i.e., an outside radius of $r_1 = 2$ is a better design than $r_1 = 5$; with regard to the total heat transfer rate, for $r_1 = 2$, the rate at $t = 100$ for the first solution is 4.768, and for $r_1 = 5$, the rate at $t = 100$ for the first solution is 4.213 so that the rate is actually somewhat higher for $r_1 = 2$.)

(3) Mathematically, the condition of no radial variation of the temperature at $r = r_1$ is expressed by boundary condition (H3.5). Also, since the fin temperature is essentially the ambient temperature at $r = r_1$, boundary condition (H3.13) is also satisfied. In other words, the first solution with the Dirichlet boundary condition (H3.13) also satisfies the Neumann boundary condition of eq. (H3.5). Similarly, the second solution with Neumann boundary condition (H3.5) also satisfies the Dirichlet boundary condition of eq. (H3.13). Thus, the two solutions should be essentially the same, as observed.

Finally, to conclude this study, we consider an analytical solution of the steady state problem, eqs. (H3.2) to (H3.5) with $\partial T(r, t)/\partial t = 0$ in eq. (H3.2), which we can compare with the preceding numerical solutions. If we define a dimensionless temperature as $T(r) = (T(r) - T_a)/(T_0 - T_a)$, then the steady state problem is

$$\frac{d^2 T'}{dr^2} + \frac{1}{r}\frac{dT'}{dr} - (\beta/\alpha)T' = 0 \quad \text{(H3.14)}$$

$$T'(r_0) = 1 \quad \text{(H3.15)}$$

$$T'(r_1) = 0 \quad \text{(H3.16)}$$

$$\frac{\partial T'(r_1)}{\partial r} = 0 \quad \text{(H3.17)}$$

where $\beta = 2h/(B\rho C_p)$ and we have selected the Dirichlet boundary condition at $r = r_1$, eq. (H3.16), for one case, and the Neumann boundary conditon, eq. (H3.17), for a second case. Also, $T'(r)$ has only one argument, that is, t is not an argument since we are now considering the steady state case. In the subsequent analysis, we drop the primes (') in eqs. (H3.14) to (H3.17) with the understanding that we are using the dimensionless temperature.

To obtain an analytical solution to eqs. (H3.14) to (H3.17), we refer to a very useful theorem, and a corollary of the theorem, reported by Wylie and Barrett (1995, 797–9).

Theorem If $(1 - a)^2 \geq 4c$ and if neither d, p, nor q is zero, then except in the obvious special cases when it reduces to the Euler or Cauchy differential equation (stated below), the differential equation

$$x^2\frac{d^2 y}{dx^2} + x(a+2bx^p)\frac{dy}{dx} + [c+dx^{2q}+b(a+p-1)x^p+b^2 x^{2p}]y = 0 \quad \text{(H3.18)}$$

has a complete solution

$$y = x^\alpha e^{-\beta x^p}[c_1 J_\nu(\lambda x^q) + c_2 Y_\nu(\lambda x^q)] \quad \text{(H3.19)}$$

where

$$\alpha = \frac{1-a}{2}, \quad \beta = \frac{b}{p}, \quad \lambda = \frac{\sqrt{|d|}}{q}, \quad \nu = \frac{\sqrt{(1-a)^2 - 4c}}{2q}$$

If $d < 0$, J_ν and Y_ν are to be replaced by I_ν and K_ν, respectively. If ν is not an integer, Y_ν and K_ν can be replaced by $J_{-\nu}$ and $I_{-\nu}$ if desired (J_ν and Y_ν are *Bessel functions* of the first and second kind of order ν, respectively; I_ν and K_ν are *modified Bessel functions* of the first and second kind of order ν, respectively).

One useful special case of this theorem can be stated as a corollary.

Corollary If $(1 - r)^2 \geq 4b$, then except in the special cases (1) $a = 0$ and (2) $r = 2$, $s = b = 0$, when it reduces to the Euler or Cauchy differential equation (stated below), the differential equation

$$\frac{d(x^r dy/dx)}{dx} + (ax^s + bx^{r-2})y = 0 \tag{H3.20}$$

has a complete solution

$$y = x^{\alpha}[c_1 J_{\nu}(\lambda x^{\gamma}) + c_2 Y_{\nu}(\lambda x^{\gamma})] \tag{H3.21}$$

where

$$\alpha = \frac{1-r}{2}, \quad \gamma = \frac{2-r+s}{2}, \quad \lambda = \frac{2\sqrt{|a|}}{2-r+s}, \quad \nu = \frac{\sqrt{(1-r)^2 - 4b}}{2-r+s}$$

If $a < 0$, J_{ν} and Y_{ν} are to be replaced by I_{ν} and K_{ν}, respectively. If ν is not an integer, Y_{ν} and K_{ν} can be replaced by $J_{-\nu}$ and $I_{-\nu}$ if desired.

The Euler or Cauchy differential equation (Spiegel 1991, 105) is

$$x^2 \frac{d^2 y}{dx^2} + ax \frac{dy}{dx} + by = f(x) \tag{H3.22}$$

where a and b are real constants. With $x = e^t$, eq. (H3.22) becomes

$$\frac{d^2 y}{dt^2} + (a-1)\frac{dy}{dt} + by = f(e^t) \tag{H3.23}$$

which is a linear, nonhomogenous, constant-coefficient differential equation that can be solved by standard methods.

We now apply the preceding corollary to eq. (H3.14). A comparison of eqs. (H3.14) and (H3.20) indicates $r = 1$, $s = 1$, $a = -\frac{\beta}{\alpha}$, $b = 0$, for which $\zeta = 0$, $\gamma = 1$, $\lambda = \sqrt{\frac{\beta}{\alpha}}$, $\nu = 0$. The solution to eq. (H3.14) is therefore

$$T(r) = c_1 I_0\left(\sqrt{\frac{\beta}{\alpha}}r\right) + c_2 K_0\left(\sqrt{\frac{\beta}{\alpha}}r\right) \tag{H3.24}$$

(note that the argument $\sqrt{\frac{\beta}{\alpha}}r$ is dimensionless, as expected). The constants c_1 and c_2 are then evaluated from boundary conditions (H3.15) and (H3.16) or from boundary conditions (H3.15) and (H3.17) (two cases). Considering the first case of Dirichlet boundary conditions

$$T(r_0) = 1 = c_1 I_0\left(\sqrt{\frac{\beta}{\alpha}}r_0\right) + c_2 K_0\left(\sqrt{\frac{\beta}{\alpha}}r_0\right) \tag{H3.25}$$

$$T(r_1) = 0 = c_1 I_0\left(\sqrt{\frac{\beta}{\alpha}}r_1\right) + c_2 K_0\left(\sqrt{\frac{\beta}{\alpha}}r_1\right) \tag{H3.26}$$

If we take $I01 = I_0(\sqrt{\frac{\beta}{\alpha}}r_1)$, $K01 = K_0(\sqrt{\frac{\beta}{\alpha}}r_1)$, then from eq. (H3.26) $c_2 = -c_1 I01/K01$ and from eq. (H3.25)

$$c_1 = \frac{K01}{(I00)K01 - (I01)K00} \tag{H3.27}$$

$$c_2 = \frac{-I01}{(I00)K01 - (I01)K00} \tag{H3.28}$$

where $I00 = I_0(\sqrt{\frac{\beta}{\alpha}}r_0)$, $K00 = K_0(\sqrt{\frac{\beta}{\alpha}}r_0)$. Equations (H3.24), (H3.27), and (H3.28) are the solution of eqs. (H3.14), (H3.15), and (H3.16).

For the second case, we first must differentiate eq. (H3.24) to use boundary condition (H3.17). The following identities can be used for this purpose (Watson 1966):

$$\frac{dI_0(x)}{dx} = I_1(x) \tag{H3.29}$$

$$\frac{dK_0(x)}{dx} = -K_1(x) \tag{H3.30}$$

Thus, from eq. (H3.24),

$$\frac{dT(r)}{dr} = \left\{ c_1 I_1\left(\sqrt{\frac{\beta}{\alpha}}r\right) - c_2 K_1\left(\sqrt{\frac{\beta}{\alpha}}r\right) \right\} \sqrt{\frac{\beta}{\alpha}} \tag{H3.31}$$

Equations (H3.15) and (H3.17) are therefore

$$T(r_0) = 1 = c_1 I_0\left(\sqrt{\frac{\beta}{\alpha}}r_0\right) + c_2 K_0\left(\sqrt{\frac{\beta}{\alpha}}r_0\right) \tag{H3.32}$$

$$\frac{dT(r_1)}{dr} = \left\{ c_1 I_1\left(\sqrt{\frac{\beta}{\alpha}}r_1\right) - c_2 K_1\left(\sqrt{\frac{\beta}{\alpha}}r_1\right) \right\} \sqrt{\frac{\beta}{\alpha}} = 0 \tag{H3.33}$$

If we take $I11 = I_1(\sqrt{\frac{\beta}{\alpha}}r_1)$, $K11 = K_1(\sqrt{\frac{\beta}{\alpha}}r_1)$, then from eq. (H3.33) $c_2 = c_1 I11/K11$, and from eq. (H3.32)

$$c_1 = \frac{K11}{(I00)K11 + (I11)K00} \tag{H3.34}$$

$$c_2 = \frac{I11}{(I00)K11 + (I11)K00} \tag{H3.35}$$

Equations (H3.24), (H3.34), and (H3.35) are the solution of eqs. (H3.14), (H3.15), and (H3.17).

We now consider a program for the evaluation of the two preceding solutions, which can be compared with the numerical steady state solutions from eqs. (H3.2), (H3.3), (H3.4), and (H3.6) or (H3.8).

```fortran
      PROGRAM H3SS1
C...
C... H3 - HEAT CONDUCTION IN A CIRCULAR FIN, STEADY STATE
C...      SOLUTION, VERSION 1
C...
C... DOUBLE PRECISION CODING IS USED
      IMPLICIT DOUBLE PRECISION(A-H,O-Z)
C...
C... ADDITIONAL VARIABLES ARE DECLARED DOUBLE PRECISION (TO
C... ACCOMMODATE THE MODIFIED BESSEL FUNCTIONS)
      DOUBLE PRECISION I00, I01, I11,
     +                 K00, K01, K11
C...
C... COMMON AREA WITH PROBLEM PARAMETERS
      PARAMETER(NR=6)
      COMMON/P/  ALPHA,    BETA,      UA,      U0,   U(NR),
     +              DR,      R0,      R1,    R(NR),      B,
     +           RHOCP
C...
C... OPEN AN OUTPUT FILE
      NO=6
      OPEN(NO,FILE='output')
C...
C... CALL SUBROUTINE BFTEST TO CHECK THE PERFORMANCE OF
C... FUNCTIONS BESSI0, BESSI1, BESSK0 AND BESSK1
      CALL BFTEST(NO)
C...
C... PROBLEM PARAMETERS
C...
C...    FIN INSIDE RADIUS
        R0=0.5D0
C...
C...    FIN OUTSIDE RADIUS
        R1=2.0D0
C...
C...    FIN THICKNESS
        B=0.2D0
C...
C...    HEAT TRANSFER COEFFICIENT
        H=0.01D0
C...
C...    RHO*CP
        RHOCP=2.0D0
C...
C...    THERMAL DIFFUSIVITY
        ALPHA=0.01D0
```

```fortran
C...
C...      AMBIENT TEMPERATURE
          UA=25.0D0
C...
C...      INSIDE TEMPERATURE
          U0=150.0D0
C...
C...   RECIPROCAL TIME FOR HEAT TRANSFER
       BETA=(2.0D0*H)/(B*RHOCP)
C...
C...   INCREMENT FOR RADIAL GRID
       DR=(R1-R0)/DFLOAT(NR-1)
C...
C...   RADIAL GRID
       DO 1 I=1,NR
          R(I)=R0+DFLOAT(I-1)*DR
1      CONTINUE
C...
C...   STEADY STATE SOLUTION WITH DIRICHLET BOUNDARY
C...   CONDITIONS
C...
C...      CONSTANTS C1, C2
          ARG0=DSQRT(BETA/ALPHA)*R0
          ARG1=DSQRT(BETA/ALPHA)*R1
          I00=BESSI0(ARG0)
          I01=BESSI0(ARG1)
          K00=BESSK0(ARG0)
          K01=BESSK0(ARG1)
          DEN=I00*K01-I01*K00
          C1= K01/DEN
          C2=-I01/DEN
C...
C...      EVALUATE THE SOLUTION AT A SERIES OF RADIAL
C...      POSITIONS
          DO 2 I=1,NR
          ARG=DSQRT(BETA/ALPHA)*R(I)
          U(I)=C1*BESSI0(ARG)+C2*BESSK0(ARG)
C...
C...      CONVERSION FROM DIMENSIONLESS TO DIMENSIONAL
C...      TEMPERATURE
          U(I)=UA+(U0-UA)*U(I)
2         CONTINUE
C...
C...   PRINT THE SOLUTION
       WRITE(NO,3)(R(I),I=1,NR),(U(I),I=1,NR)
3      FORMAT(/,' Dirichlet boundary conditions',/,
```

```
      +              '         r ',6F9.2,/,'    u(r) ',6F9.3,/)
C...
C...  COMPUTE AND PRINT THE TOTAL HEAT TRANSFER TO THE FIN
      DTDR=(C1*BESSI1(ARG0)-C2*BESSK1(ARG0))
     +     *DSQRT(BETA/ALPHA)*(U0-UA)
      PI=4.0D0*DATAN(1.0D0)
      TOTAL=-2.0D0*PI*R0*B*ALPHA*RHOCP*DTDR
      WRITE(NO,4)TOTAL
4     FORMAT(' Total heat transfer rate = ',F7.3,//)
C...
C...  STEADY STATE SOLUTION WITH DIRICHLET-NEUMANN BOUNDARY
C...  CONDITIONS
C...
C...     CONSTANTS C1, C2
         ARG0=DSQRT(BETA/ALPHA)*R0
         ARG1=DSQRT(BETA/ALPHA)*R1
         I00=BESSI0(ARG0)
         I11=BESSI1(ARG1)
         K00=BESSK0(ARG0)
         K11=BESSK1(ARG1)
         DEN=I00*K11+I11*K00
         C1=K11/DEN
         C2=I11/DEN
C...
C...     EVALUATE THE SOLUTION AT A SERIES OF RADIAL
C...     POSITIONS
         DO 12 I=1,NR
         ARG=DSQRT(BETA/ALPHA)*R(I)
         U(I)=C1*BESSI0(ARG)+C2*BESSK0(ARG)
C...
C...     CONVERSION FROM DIMENSIONLESS TO DIMENSIONAL
C...     TEMPERATURE
         U(I)=UA+(U0-UA)*U(I)
12       CONTINUE
C...
C...  PRINT THE SOLUTION
      WRITE(NO,13)(R(I),I=1,NR),(U(I),I=1,NR)
13    FORMAT(/,' Dirichlet-Neumann boundary conditions',/,
     +              '         r ',6F9.2,/,'    u(r) ',6F9.3,/)
C...
C...  COMPUTE AND PRINT THE TOTAL HEAT TRANSFER TO THE FIN
      DTDR=(C1*BESSI1(ARG0)-C2*BESSK1(ARG0))
     +     *DSQRT(BETA/ALPHA)*(U0-UA)
      PI=4.0D0*DATAN(1.0D0)
      TOTAL=-2.0D0*PI*R0*B*ALPHA*RHOCP*DTDR
      WRITE(NO,14)TOTAL
```

```
14      FORMAT(' Total heat transfer rate = ',F7.3,//)
        STOP
        END
```

We can note the following points about program H3SS1:

(1) Double precision coding is used. Also, the various modified Bessel functions, $I00$ to $K11$, are declared real

```
C...
C...    DOUBLE PRECISION CODING IS USED
        IMPLICIT DOUBLE PRECISION(A-H,O-Z)
C...
C...    ADDITIONAL VARIABLES ARE DECLARED DOUBLE PRECISION (TO
C...    ACCOMMODATE THE MODIFIED BESSEL FUNCTIONS)
        DOUBLE PRECISION I00, I01, I11,
       +                 K00, K01, K11
```

(2) The problem parameters and radial grid are defined; they are the same as in the case of the solution of eqs. (H3.2) to (H3.6) and (H3.8) discussed previously.

```
C...
C...    PROBLEM PARAMETERS
C...
C...        FIN INSIDE RADIUS
            R0=0.5D0
                .
                .
                .
C...
C...    RADIAL GRID
        DO 1 I=1,NR
            R(I)=R0+DFLOAT(I-1)*DR
1       CONTINUE
```

(3) Subroutine BFTEST is then called to test functions BESSI0, BESSI1, BESSK0, and BESSK1, which compute I_0, I_1, K_0, and K_1, respectively.

```
C...
C...    CALL SUBROUTINE BFTEST TO CHECK THE PERFORMANCE OF
C...    FUNCTIONS BESSI0, BESSI1, BESSK0 AND BESSK1
        CALL BFTEST(NO)
```

The output from these four functions can then be compared with tabulated values of I_0, I_1, K_0, and K_1 (some tabulated values are included in the comments in BFTEST, which is listed subsequently).

(4) The steady state solution for eqs. (H3.14), (H3.15), and (H3.16) is then evaluated. First, the constants c_1 and c_2 from eqs. (H3.27) and (H3.28) are computed

```
C...
C...    STEADY STATE SOLUTION WITH DIRICHLET BOUNDARY
C...    CONDITIONS
C...
C...        CONSTANTS C1, C2
            ARG0=DSQRT(BETA/ALPHA)*R0
            ARG1=DSQRT(BETA/ALPHA)*R1
            I00=BESSI0(ARG0)
            I01=BESSI0(ARG1)
            K00=BESSK0(ARG0)
            K01=BESSK0(ARG1)
            DEN=I00*K01-I01*K00
            C1= K01/DEN
            C2=-I01/DEN
```

The Bessel functions I_0 and K_0 are evaluated by calls to functions BESSI0 and BESSK0. Then constants c_1 and c_2 of eqs. (H3.27) and (H3.28) are computed.

(5) The steady state solution of eq. (H3.24) is then computed and the resulting dimensionless temperature is converted to a dimensional temperature (recall again the definition $T'(r) = (T(r) - T_a)/(T_0 - T_a)$)

```
C...
C...    EVALUATE THE SOLUTION AT A SERIES OF RADIAL
C...    POSITIONS
        DO 2 I=1,NR
        ARG=DSQRT(BETA/ALPHA)*R(I)
        U(I)=C1*BESSI0(ARG)+C2*BESSK0(ARG)
C...
C...    CONVERSION FROM DIMENSIONLESS TO DIMENSIONAL
C...    TEMPERATURE
        U(I)=UA+(U0-UA)*U(I)
```

(6) The computed steady state temperatures are then printed

```
C...
C...    PRINT THE SOLUTION
        WRITE(NO,3)(R(I),I=1,NR),(U(I),I=1,NR)
3       FORMAT(/,' Dirichlet boundary conditions',/,
       +         '    r ',6F9.2,/,'   u(r) ',6F9.3,/)
```

(7) Finally, the total heat transfer rate given by eq. (H3.12) (at steady state) is computed

```
C...
C...    COMPUTE AND PRINT THE TOTAL HEAT TRANSFER TO THE FIN
        DTDR=(C1*BESSI1(ARG0)-C2*BESSK1(ARG0))
       +      *DSQRT(BETA/ALPHA)*(U0-UA)
```

```
      PI=4.0D0*DATAN(1.0D0)
      TOTAL=-2.0D0*PI*R0*B*ALPHA*RHOCP*DTDR
      WRITE(NO,4)TOTAL
4     FORMAT(' Total heat transfer rate = ',F7.3,//)
```

Note that the radial derivative, $dT(r_0)/dr$, is computed from eq. (H3.31), with conversion to dimensional temperature included (by multiplication by $(T_0 - T_a)$).

(8) The same series of calculations follows next for the second case of the Dirichlet boundary condition at $r = r_0$, eq. (H3.15) and the Neumann boundary condition at $r = r_1$, eq. (H3.17). Since these calculations are similar to those for the two Dirichlet boundary conditions, eqs. (H3.15) and (H3.16), they will not be described in detail. Note the use of I_1 and K_1, computed by functions BESSI1 and BESSK1, as required in the calculation of c_1 and c_2 of eqs. (H3.34) and (H3.35).

To complete this discussion of the steady state program, subroutine BFTEST and functions BESSI0, BESSI1, BESSK0, BESSK1, and DFLOAT (called in main program H3SS1) are listed below:

```
      SUBROUTINE BFTEST(NO)
C...
C...  PROGRAM BFTEST CALLS FUNCTIONS BESSI0, BESSI1, BESSK0
C...  AND BESSK1 TO CHECK THE CALCULATION OF I0(X), I1(X),
C...  K0(X) AND K1(X).
C...
C...  THE FOLLOWING TABULATED VALUES CAN BE USED TO CHECK
C...  THE PERFORMANCE OF BESSI0 TO BESSK1 [SPIEGEL (1991),
C...  PP. 246-247]
C...
C...      I0(1) = 1.266,   I0(5) = 27.24
C...
C...      I1(1) = 0.5652,  I1(5) = 24.34
C...
C...      K0(1) = 0.4210,  K0(5) = 0.003691
C...
C...      K1(1) = 0.6019,  K1(5) = 0.004045
C...
C...  SPIEGEL, M. R., (1991), MATHEMATICAL HANDBOOK OF
C...  FORMULAS AND TABLES, MCGRAW-HILL, NEW YORK
C...
C...  DOUBLE PRECISION CODING IS USED
      IMPLICIT DOUBLE PRECISION (A-H,O-Z)
C...
C...  STEP THROUGH A SERIES OF VALUES OF X
      DO 1 I=1,10
         X=0.5D0*DFLOAT(I)
         YI0=BESSI0(X)
```

```
          YI1=BESSI1(X)
          YK0=BESSK0(X)
          YK1=BESSK1(X)
C...
C...  WRITE THE COMPUTED BESSEL FUNCTIONS
      WRITE(NO,2)X, YI0, YI1, YK0, YK1
2     FORMAT(' X = ',F5.2,'   YI0 = ',F8.4,'   YI1 = ',F8.4,
     +               '   YK0 = ',F8.4,'   YK1 = ',F8.4,/)
C...
C...  NEXT X
1     CONTINUE
C...
C...  END OF CALCULATION
      END


      DOUBLE PRECISION FUNCTION BESSI0(X)
C...
C...  FUNCTION BESSI0 COMPUTES THE BESSEL FUNCTION I0(X).
C...  BESSI0 IS TAKEN FROM W. H. PRESS, B. P. FLANNERY,
C...  S. A. TEUKOLSKY AND V. T. VETTERLING, (1985),
C...  NUMERICAL RECIPES, CAMBRIDGE UNIVERSITY PRESS,
C...  CAMBRIDGE, CHAPTER 6, WITH CONVERSION TO DOUBLE
C...  PRECISION AND EDITING BY W. E. SCHIESSER.
C...
C...  DOUBLE PRECISION CODING IS USED
      IMPLICIT DOUBLE PRECISION (A-H,O-Z)
      DATA          P1,              P2,                P3,
     +              P4,              P5,                P6,
     +              P7
     +          /1.0D0,        3.5156229D0,       3.0899424D0,
     +       1.2067492D0,      0.2659732D0,       0.360768D-1,
     +       0.45813D-2/
      DATA          Q1,              Q2,                Q3,
     +              Q4,              Q5,                Q6,
     +              Q7,              Q8,                Q9
     +   /0.39894228D0,     0.1328592D-1,       0.225319D-2,
     +      -0.157565D-2,     0.916281D-2,      -0.2057706D-1,
     +       0.2635537D-1,   -0.1647633D-1,       0.392377D-2/
      IF(DABS(X).LT.3.75D0)THEN
          Y=(X/3.75D0)**2
          BESSI0=P1+Y*(P2+Y*(P3+Y*(P4+Y*(P5+Y*(P6+Y*P7)))))
      ELSE
          AX=DABS(X)
          Y=3.75D0/AX
          BESSI0=(DEXP(AX)/DSQRT(AX))*(Q1+Y*(Q2+Y*(Q3+Y*(Q4
     +           +Y*(Q5+Y*(Q6+Y*(Q7+Y*(Q8+Y*Q9))))))))
```

```
      END IF
      RETURN
      END

      DOUBLE PRECISION FUNCTION BESSI1(X)
C...
C...  FUNCTION BESSI1 COMPUTES THE BESSEL FUNCTION I1(X).
C...  BESSI1 IS TAKEN FROM W. H. PRESS, B. P. FLANNERY,
C...  S. A. TEUKOLSKY AND V. T. VETTERLING, (1985),
C...  NUMERICAL RECIPES, CAMBRIDGE UNIVERSITY PRESS,
C...  CAMBRIDGE, CHAPTER 6, WITH CONVERSION TO DOUBLE
C...  PRECISION AND EDITING BY W. E. SCHIESSER.
C...
C...  DOUBLE PRECISION CODING IS USED
      IMPLICIT DOUBLE PRECISION (A-H,O-Z)
      DATA        P1,            P2,              P3,
     +            P4,            P5,              P6,
     +            P7
     +         /0.5D0,     0.87890594D0,     0.51498869D0,
     +    0.15084934D0,    0.2658733D-1,     0.301532D-2,
     +       0.32411D-3/
      DATA        Q1,            Q2,              Q3,
     +            Q4,            Q5,              Q6,
     +            Q7,            Q8,              Q9
     +    /0.39894228D0,  -0.3988024D-1,    -0.362018D-2,
     +       0.163801D-2,  -0.1031555D-1,    0.2282967D-1,
     +     -0.2895312D-1,   0.1787654D-1,   -0.420059D-2/
      IF(DABS(X).LT.3.75D0)THEN
          Y=(X/3.75D0)**2
          BESSI1=X*(P1+Y*(P2+Y*(P3+Y*(P4+Y*(P5+
     +        Y*(P6+Y*P7))))))
      ELSE
          AX=DABS(X)
          Y=3.75D0/AX
          BESSI1=(DEXP(AX)/DSQRT(AX))*(Q1+Y*(Q2+Y*(Q3+Y*(Q4+
     +        Y*(Q5+Y*(Q6+Y*(Q7+Y*(Q8+Y*Q9))))))))
      END IF
      RETURN
      END

      DOUBLE PRECISION FUNCTION BESSK0(X)
C...
C...  FUNCTION BESSK0 COMPUTES THE BESSEL FUNCTION K0(X).
C...  BESSK0 IS TAKEN FROM W. H. PRESS, B. P. FLANNERY,
C...  S. A. TEUKOLSKY AND V. T. VETTERLING, (1985),
C...  NUMERICAL RECIPES, CAMBRIDGE UNIVERSITY PRESS,
```

```
C...   CAMBRIDGE, CHAPTER 6, WITH CONVERSION TO DOUBLE
C...   PRECISION AND EDITING BY W. E. SCHIESSER.
C...
C...   DOUBLE PRECISION CODING IS USED
       IMPLICIT DOUBLE PRECISION (A-H,O-Z)
       DATA        P1,              P2,                  P3,
      +            P4,              P5,                  P6,
      +            P7
      + /-0.57721566D0,      0.42278420D0,        0.23069756D0,
      +   0.3488590D-1,      0.262698D-2,         0.10750D-3,
      +   0.74D-5/
       DATA        Q1,              Q2,                  Q3,
      +            Q4,              Q5,                  Q6,
      +            Q7
      + /1.25331414D0,      -0.7832358D-1,        0.2189568D-1,
      + -0.1062446D-1,       0.587872D-2,        -0.251540D-2,
      +   0.53208D-3/
       IF(X.LE.2.0D0)THEN
         Y=X*X/4.0D0
         BESSK0=(-DLOG(X/2.0D0)*BESSI0(X))+(P1+Y*(P2+Y*(P3+
      +         Y*(P4+Y*(P5+Y*(P6+Y*P7))))))
       ELSE
         Y=(2.0D0/X)
         BESSK0=(DEXP(-X)/DSQRT(X))*(Q1+Y*(Q2+Y*(Q3+
      +         Y*(Q4+Y*(Q5+Y*(Q6+Y*Q7))))))
       END IF
       RETURN
       END

       DOUBLE PRECISION FUNCTION BESSK1(X)
C...
C...   FUNCTION BESSK1 COMPUTES THE BESSEL FUNCTION K1(X).
C...   BESSK1 IS TAKEN FROM W. H. PRESS, B. P. FLANNERY,
C...   S. A. TEUKOLSKY AND V. T. VETTERLING, (1985),
C...   NUMERICAL RECIPES, CAMBRIDGE UNIVERSITY PRESS,
C...   CAMBRIDGE, CHAPTER 6, WITH CONVERSION TO DOUBLE
C...   PRECISION AND EDITING BY W. E. SCHIESSER.
C...
C...   DOUBLE PRECISION CODING IS USED
       IMPLICIT DOUBLE PRECISION (A-H,O-Z)
       DATA        P1,              P2,                  P3,
      +            P4,              P5,                  P6,
      +            P7
      +       /1.0D0,        0.15443144D0,        -0.67278579D0,
      + -0.18156897D0,      -0.1919402D-1,        -0.110404D-2,
      +   -0.4686D-4/
```

```
      DATA          Q1,              Q2,              Q3,
     +              Q4,              Q5,              Q6,
     +              Q7
     +   /1.25331414D0,     0.23498619D0,    -0.3655620D-1,
     +    0.1504268D-1,    -0.780353D-2,     0.325614D-2,
     +    -0.68245D-3/
      IF(X.LE.2.0D0)THEN
         Y=X*X/4.0D0
         BESSK1=(DLOG(X/2.0D0)*BESSI1(X))+(1.0/X)*(P1+Y*(P2+
     +          Y*(P3+Y*(P4+Y*(P5+Y*(P6+Y*P7))))))
      ELSE
         Y=2.0D0/X
         BESSK1=(DEXP(-X)/DSQRT(X))*(Q1+Y*(Q2+Y*(Q3+
     +          Y*(Q4+Y*(Q5+Y*(Q6+Y*Q7))))))
      END IF
      RETURN
      END

      DOUBLE PRECISION FUNCTION DFLOAT(I)
C...
C...  FUNCTION DFLOAT CONVERTS A SINGLE PRECISION INTEGER
C...  INTO A DOUBLE PRECISION FLOATING POINT.  THIS FUNCTION
C...  IS PROVIDED IN CASE THE USER'S FORTRAN COMPILER DOES
C...  NOT INCLUDE DFLOAT.
C...
      DFLOAT=DBLE(FLOAT(I))
      RETURN
      END
```

We can note the following points about routines BFTEST, BESSI0, BESSI1, BESSK0, BESSK1, and DFLOAT:

(1) BFTEST merely evaluates $I_0(x)$, $I_1(x)$, $K_0(x)$, and $K_1(x)$ for a series of values of x by calling BESSI0, BESSI1, BESSK0, and BESSK1, respectively. The output agrees with the tabulated values (Spiegel 1991, 246–7) at the beginning of BFTEST to four figures.

(2) BESSI0, BESSI1, BESSK0, and BESSK1 are taken from Press, et al. (1985, chap. 6), with some minor editing.

(3) DFLOAT converts a single precision integer to a double precision floating point and is included in case the Fortran compiler used to compile and execute this program does not have an intrinsic DFLOAT.

The output from the preceding program is given below:

```
X = 0.50 YI0 = 1.0635 YI1 = 0.2579 YK0 = 0.9244 YK1 = 1.6564
  .                                                        .
```

X = 5.00   YI0 = 27.2399   YI1 = 24.3356   YK0 = 0.0037   YK1 = 0.0040

Dirichlet boundary conditions
```
     r      0.50     0.80     1.10     1.40     1.70     2.00
   u(r)   150.000   76.688   47.530   34.756   28.602   25.000
```

Total heat transfer rate =   4.887

Dirichlet-Neumann boundary conditions
```
     r      0.50     0.80     1.10     1.40     1.70     2.00
   u(r)   150.000   77.123   48.431   36.384   31.503   30.216
```

Total heat transfer rate =   4.865

We can note the following points about this output:

(1) The output from TESTBF agrees with the tabulated values to four figures.

(2) The steady state temperatures agree with the unsteady state temperatures to three figures. For example, for the latter at $t = 100$, for Dirichlet boundary conditions, we obtained

```
t =    100.00
     r      0.50     0.80     1.10     1.40     1.70     2.00
   u(r,t) 150.000  76.759   47.581   34.785   28.614   25.000
```

Total heat transfer rate =   4.768

and, for Dirichlet-Neumann boundary conditions,

```
t =    100.00
     r      0.50     0.80     1.10     1.40     1.70     2.00
   u(r,t) 150.000  77.175   48.453   36.378   31.482   30.192
```

Total heat transfer rate =   4.748

Also, the total heat transfer rates are generally in agreement.

Thus, we can conclude that the unsteady state method of solution for eqs. (H3.2) to (H3.5) and the direct steady state solution (via eqs. (H3.14) to (H3.17)) are in agreement. We conclude this discussion by pointing out one computational problem that occurred when working with the steady state solution of eqs. (H3.14) to (H3.17), that is, eq. (H3.24). Depending on the parameters that were selected, the argument of the Bessel functions, $\sqrt{\frac{\beta}{\alpha}}r$, would become large enough that the functions BESSI0 to BESSK1 would not execute correctly (typically, underflows and overflows were reported). These computational problems can be understood

Heat transfer in a circular fin

by considering the following asymptotic approximations for the Bessel functions for large arguments (or, more specifically, $x \gg n$) (Watson 1966)

$$I_n(x) \approx \frac{1}{\sqrt{2\pi x}} e^x \tag{H3.36}$$

$$K_n(x) \approx \frac{\pi}{\sqrt{2\pi x}} e^{-x} \tag{H3.37}$$

(note that the right hand side (RHS) of these approximations do not involve the order of the Bessel function, n). Thus, for large x, the exponentials e^x and e^{-x} take on large and small values, respectively (at least for a 32-bit computer, performing 64-bit double precision arithmetic).

To try to circumvent these numerical problems, we can substitute eqs. (H3.36) and (H3.37) in the analytical solutions and then consider what might be done with the exponentials to avoid such large and small values. Equation (H3.24) with c_1 and c_2 from eqs. (H3.27) and (H3.28) is

$$T(r) = \frac{1}{I00 - (I01/K01)K00} I_0\left(\sqrt{\frac{\beta}{\alpha}} r\right)$$

$$+ \frac{-I01/K01}{I00 - (I01/K01)K00} K_0\left(\sqrt{\frac{\beta}{\alpha}} r\right)$$

$$= \frac{1}{I00 K01 - I01 K00} \left\{ K01 I_0\left(\sqrt{\frac{\beta}{\alpha}} r\right) - I01 K_0\left(\sqrt{\frac{\beta}{\alpha}} r\right) \right\} \tag{H3.38}$$

If we use as the arguments for the Bessel functions $x_0 = \sqrt{\frac{\beta}{\alpha}} r_0$, $x_1 = \sqrt{\frac{\beta}{\alpha}} r_1$, and $x = \sqrt{\frac{\beta}{\alpha}} r$ and substitute approximations (H3.36) and (H3.37) in eq. (H3.38)

$$T(x) = \frac{1}{\left(\frac{1}{\sqrt{2\pi x_0}} e^{x_0}\right) \frac{\pi}{\sqrt{2\pi x_1}} e^{-x_1} - \left(\frac{1}{\sqrt{2\pi x_1}} e^{x_1}\right) \frac{\pi}{\sqrt{2\pi x_0}} e^{-x_0}}$$

$$\cdot \left\{ \left(\frac{1}{\sqrt{2\pi x_1}} e^{-x_1}\right) \frac{\pi}{\sqrt{2\pi x}} e^{-x} - \left(\frac{1}{\sqrt{2\pi x_1}} e^{x_1}\right) \frac{\pi}{\sqrt{2\pi x}} e^{-x} \right\}$$

$$= \frac{1}{\frac{1}{\sqrt{x_0 x_1}} \{e^{(x_0-x_1)} - e^{-(x_0-x_1)}\}/2} \frac{1}{\sqrt{x_1 x}} \{e^{(x-x_1)} - e^{-(x-x_1)}\}/2$$

$$= \sqrt{\frac{x_0}{x}} \frac{\sinh(x - x_1)}{\sinh(x_0 - x_1)} \tag{H3.39}$$

As expected, eq. (H3.39) satisfies Dirichlet boundary conditions (H3.15) and (H3.16). Also, we note that eq. (H3.39) involves only differences in x in the exponentials of the *sinh* function, which makes the calculation of the steady state solution considerably easier than using eq. (H3.24) for large x (and, of course, eq. (H3.39) is somewhat less complicated than eq. (H3.24)). Equation (H3.39) was then used to compute the steady state solution of eqs. (H3.15) to (H3.17) for large x, using the following program:

```
      PROGRAM H3SS2
C...
C...  H3 - HEAT CONDUCTION IN A CIRCULAR FIN, STEADY STATE
C...       SOLUTION, VERSION 2
C...
C...  DOUBLE PRECISION CODING IS USED
      IMPLICIT DOUBLE PRECISION(A-H,O-Z)
C...
C...  COMMON AREA WITH PROBLEM PARAMETERS
      PARAMETER(NR=6)
      COMMON/P/  ALPHA,    BETA,      UA,       UO,    U(NR),
     +                      DR,       RO,       R1,    R(NR),      B,
     +                    RHOCP
C...
C...  OPEN AN OUTPUT FILE
      NO=6
      OPEN(NO,FILE='output')
C...
C...  PROBLEM PARAMETERS
C...
C...     FIN INSIDE RADIUS
         RO=0.5D0
C...
C...     FIN OUTSIDE RADIUS
         R1=2.0D0
C...
C...     FIN THICKNESS
         B=0.2D0
C...
C...     HEAT TRANSFER COEFFICIENT
         H=0.01D0
C...
C...     RHO*CP
         RHOCP=2.0D0
C...
C...     THERMAL DIFFUSIVITY
         ALPHA=0.01D0
C...
C...     AMBIENT TEMPERATURE
         UA=25.0D0
C...
C...     INSIDE TEMPERATURE
         U0=150.0D0
```

```
C...
C...    RECIPROCAL TIME FOR HEAT TRANSFER
        BETA=(2.0D0*H)/(B*RHOCP)
C...
C...    INCREMENT FOR RADIAL GRID
        DR=(R1-R0)/DFLOAT(NR-1)
C...
C...    RADIAL GRID
        DO 1 I=1,NR
           R(I)=R0+DFLOAT(I-1)*DR
1       CONTINUE
C...
C...    STEADY STATE SOLUTION WITH DIRICHLET BOUNDARY
C...    CONDITION
C...
C...       CONSTANT X0
           X0=DSQRT(BETA/ALPHA)*R0
           X1=DSQRT(BETA/ALPHA)*R1
C...
C...       EVALUATE THE SOLUTION AT A SERIES OF RADIAL
C...       POSITIONS
           DO 2 I=1,NR
           X=DSQRT(BETA/ALPHA)*R(I)
C...
C...       ANALYTICAL SOLUTION
           U(I)=DSQRT(X0/X)*DSINH(X1-X)/DSINH(X1-X0)
C...
C...       CONVERSION FROM DIMENSIONLESS TO DIMENSIONAL
C...       TEMPERATURE
           U(I)=UA+(U0-UA)*U(I)
2          CONTINUE
C...
C...    PRINT THE SOLUTION
        WRITE(NO,3)(R(I),I=1,NR),(U(I),I=1,NR)
3       FORMAT(/,' Dirichlet boundary conditions',/,
       +          '      r ',6F9.2,/,'   u(r) ',6F9.3,/)
        STOP
        END
```

We can note the following points about program H3SS2:

(1) The beginning of the program, up to and including DO loop 1, is the same as for the preceding program H3SS1.

(2) The steady state solution of eq. (H3.39) is then evaluated for a series of values of x.

```
C...
C...    STEADY STATE SOLUTION WITH DIRICHLET BOUNDARY
```

```
C...      CONDITION
C...
C...      CONSTANT X0
          X0=DSQRT(BETA/ALPHA)*R0
          X1=DSQRT(BETA/ALPHA)*R1
C...
C...      EVALUATE THE SOLUTION AT A SERIES OF RADIAL
C...      POSITIONS
          DO 2 I=1,NR
          X=DSQRT(BETA/ALPHA)*R(I)
C...
C...      ANALYTICAL SOLUTION
          U(I)=DSQRT(X0/X)*DSINH(X1-X)/DSINH(X1-X0)
C...
C...      CONVERSION FROM DIMENSIONLESS TO DIMENSIONAL
C...      TEMPERATURE
          U(I)=UA+(U0-UA)*U(I)
2         CONTINUE
```

This programming follows immediately from eq. (H3.39). Since we have not previously considered any cases for large x, we will not report any output from program H3SS2. Briefly, H3SS2 produced steady state solutions that were in agreement with the corresponding numerical solutions for large t (and no difficulties were encountered in executing the MOL program for the numerical solutions, even for large x).

In this study, we have developed the solution of eqs. (H3.2) to (H3.5) numerically and analytically. This study also illustrates the advantageous use of library routines, for example, RKF45, and BESSI0 to BESSK1.

References

Bird, R. B., W. E. Stewart, and E. N. Lightfoot. 1960. *Transport Phenomena.* New York: John Wiley and Sons.

Forsythe, G. E., M. A. Malcolm, and C. B. Moler. 1977. *Computer Methods for Mathematical Computations.* Englewood Cliffs, NJ: Prentice-Hall.

Press, W. H., B. P. Flannery, S. A. Teukolsky, and W. T. Vetterling. 1985. *Numerical Recipes.* Cambridge: Cambridge University Press.

Schiesser, W. E. 1991. *The Numerical Method of Lines Integration of Partial Differential Equations.* San Diego: Academic Press.

Spiegel, M. R. 1991. *Mathematical Handboook of Formulas and Tables.* New York: McGraw-Hill.

Watson, G. N. 1966. *A Treatise on the Theory of Bessel Functions.* 2d ed. Cambridge: Cambridge University Press.

Wylie, C. R., and L. C. Barrett. 1995. *Advanced Engineering Mathematics.* 6th ed. New York: McGraw-Hill.

H4

The Graetz problem with constant wall heat flux

Consider the following elliptic problem, which is usually termed the "Graetz problem" with constant wall heat flux (Eckert and Drake 1972, 333–42):

$$\rho C_p v(r) \frac{\partial T}{\partial z} = k \frac{1}{r} \left\{ r \frac{\partial T}{\partial r} \right\}, \qquad v(r) = 2v_{\text{avg}}(1 - (r/r_0)^2) \qquad \text{(H4.1)}$$

The physical system modeled by eq. (H4.1) is illustrated in Figure H4.1.
Compute a solution to eq. (H4.1) subject to the boundary conditions

$$\frac{\partial T(0, z)}{\partial r} = 0 \qquad \text{(H4.2)}$$

$$-k \frac{\partial T(r_0, z)}{\partial r} = q_w \qquad \text{(H4.3)}$$

$$T(r, 0) = T_1 = q_w r_0 / z \qquad \text{(H4.4)}$$

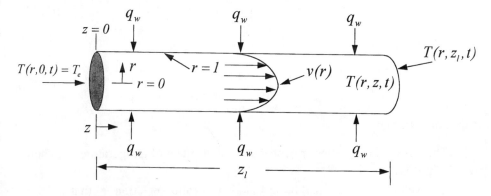

Figure H4.1. Laminar Flow in a Circular Tube with Constant Wall Heat Flux

Compare your solution with the solution for a fully developed thermal field reported by Eckert and Drake (1972)

$$T(r, z) = T(r_0, z) + \frac{2\rho C_p v_{\text{avg}} r_0^2}{k} \frac{\partial T}{\partial z} \left\{ \frac{1}{4}\left(\frac{r}{r_0}\right)^2 - \frac{1}{16}\left(\frac{r}{r_0}\right)^4 - \frac{3}{16} \right\} \quad \text{(H4.5)}$$

where $\partial T/\partial z$ is the axial temperature gradient, which is constant in a fully developed thermal field.

Solution

We start with a derivation of the "dynamic Graetz equation," which is an unsteady state variant of eq. (H4.1). The associated dynamic problem will then be used to calculate a solution to the steady state problem of eqs. (H4.1) to (H4.4).

Consider the laminar flow of a Newtonian fluid through a cylindrical tube. An energy balance on an incremental section of the fluid gives

$$2\pi r \Delta r \Delta z \rho C_p \frac{\partial T}{\partial t} = 2\pi r \Delta r v_z \rho C_p T|_z - 2\pi r \Delta r v_z \rho C_p T|_{z+\Delta z}$$

$$- k 2\pi r \Delta z \frac{\partial T}{\partial r}\bigg|_r - \left\{ -k 2\pi (r+\Delta r) \Delta z \frac{\partial T}{\partial r}\bigg|_{r+\Delta r} \right\} \quad \text{(H4.6)}$$

Here we have assumed

(1) Convection in the z direction only with velocity v_z.
(2) Conduction in the r direction only with conductivity k.

Division of eq. (H4.6) by $2\pi r \Delta r \Delta z$ gives

$$\rho C_p \frac{\partial T}{\partial t} = \frac{v_z \rho C_p T|_z - v_z \rho C_p T|_{z+\Delta z}}{\Delta z}$$
$$+ \frac{k(r+\Delta r)\frac{\partial T}{\partial r}|_{r+\Delta r} - k\frac{\partial T}{\partial r}|_r}{r \Delta r} \quad \text{(H4.7)}$$

In the limit as $\Delta r \to 0$, $\Delta z \to 0$, eq. (H4.7) becomes

$$\rho C_p \frac{\partial T}{\partial t} = -\frac{\partial \{v_z \rho C_p T\}}{\partial z} + \frac{1}{r}\frac{\partial}{\partial r}\left\{ kr \frac{\partial T}{\partial r} \right\} \quad \text{(H4.8)}$$

If we now assume:

(1) The physical properties are constant (independent of temperature and therefore position).
(2) The velocity, v_z, is a parabolic function of radial position, r, that is, $v_z = 2v_{\text{avg}}(1 - (r/r_0)^2)$, where

v_{avg} average velocity defined by $Q = \pi r_0^2 v_{avg}$
r_0 tube radius
Q volumetric flow rate

eq. (H4.8) becomes

$$\frac{\partial T}{\partial t} = -v_z \frac{\partial T}{\partial z} + \alpha \left\{ \frac{\partial^2 T}{\partial r^2} + \frac{1}{r} \frac{\partial T}{\partial r} \right\} \tag{H4.9}$$

where $\alpha = k/(\rho C_p)$ (the thermal diffusivity). Equation (H4.9) is the dynamic form of eq. (H4.1), that is, with $\partial T/\partial t = 0$, eq. (H4.9) reduces to eq. (H4.1).

Next we define temperature $T_1 = T(r, 0, t)$ and dimensionless variables as

$$T' = \frac{T}{q_w r_0/k} \tag{H4.10}$$

$$t' = \alpha t / r_0^2 \tag{H4.11}$$

$$r' = r/r_0 \tag{H4.12}$$

$$z' = z/r_0 \tag{H4.13}$$

Substitution of eqs. (H4.10) to (H4.13) in eq. (H4.9) then gives (with $v' = v_z/v_{avg}$)

$$\left\{ \frac{1}{r_0^2/\alpha} \right\} \frac{\partial T'}{\partial t'} = -v_{avg} v'(r') \left\{ \frac{1}{r_0} \right\} \frac{\partial T'}{\partial z'} + \alpha \left\{ \frac{1}{r_0^2} \right\} \left\{ \frac{\partial^2 T'}{\partial r'^2} + \frac{1}{r'} \frac{\partial T'}{\partial r'} \right\} \tag{H4.14}$$

or after multiplication of eq. (H4.14) by r_0^2/α

$$\frac{\partial T'}{\partial t'} = -P_e v'(r') \frac{\partial T'}{\partial z'} + \alpha \left\{ \frac{\partial^2 T'}{\partial r'^2} + \frac{1}{r'} \frac{\partial T'}{\partial r'} \right\} \tag{H4.15}$$

where $P_e = v_{avg} r_0/\alpha$ (the Peclet number). Finally, we drop the primes (') in eq. (H4.15), with the understanding that we are now using the dimensionless variables defined by eqs. (H4.10) to (H4.13), to arrive at the "dynamic Graetz equation."

Equation (H4.9) requires one initial condition, two boundary conditions in r, and one boundary conditions in z. We take for the initial condition

$$T(r, z, 0) = 1 \tag{H4.16}$$

The dimensional boundary conditions in r are eq. (H4.2) and (H4.3), or in dimensionless form

$$\frac{\partial T(0, z, t)}{\partial r} = 0 \tag{H4.17}$$

$$\frac{\partial T(1, z, t)}{\partial r} = 1 \tag{H4.18}$$

which express symmetry at the centerline, $r = 0$, and the constant heat flux at the wall (for eq. (H4.18), we have used eqs. (H4.10) and (H4.12)). Note that eqs. (H4.17) and (H4.18) are the dynamic counterparts of eqs. (H4.2) and (H4.3).

For the boundary condition in z, we use

$$T(r, 0, t) = \frac{T_1}{q_w r_o / k} \qquad (H4.19)$$

which follows from the definition of the dimensional temperature T_1, and eq. (H4.10).

Finally, in keeping with the usual convention of the numerical analysis literature, we denote the partial differential equation (PDE) dependent variable, the dimensionless temperature of eq. (H4.10), as u rather than T (in the subsequent discussion of the analytical and numerical solutions, we denote the dependent variable as u).

We start the discussion of the numerical solution by again pointing out that eqs. (H4.15) to (H4.19) are programmed. The solution to these equations with $\partial T/\partial t = \partial u/\partial t \approx 0$ in eq. (H4.15) then produces the solution to eqs. (H4.1) to (H4.4). The basic approach to the solution of eqs. (H4.15) to (H4.19) is the numerical method of lines (MOL). Subroutine INITAL, which defines initial condition (H4.16), is listed below:

```
      SUBROUTINE INITAL
      IMPLICIT DOUBLE PRECISION (A-H,O-Z)
      PARAMETER (NR=11,NZ=31)
      COMMON/T/        T,      NSTOP,       NORUN
     +      /Y/     U(NR,NZ)
     +      /F/     UT(NR,NZ)
     +      /G/        R(NR),    Z(NZ),     V(NR)
     +      /C/           PE,       TN,        RO,      ZL,
     +                    DR,      DRS,        DZ,     DZS
     +      /I/           IP,    NCASE
C...
C...  SELECT APPROXIMATION OF SPATIAL DERIVATIVES (SEE THE
C...  COMMENTS IN SUBROUTINE DERV FOR AN EXPLANATION OF THE
C...  TWO CASES)
      NCASE=2
C...
C...  PROBLEM PARAMETERS
C...
C...     PECLET NUMBER
         PE=60.0D0
C...
C...     QW*RO/K = TN (NORMALIZATION TEMPERATURE)
         TN=1.0D0
```

```fortran
C...
C...      OUTSIDE RADIUS
          R0=1.0D0
C...
C...      AXIAL LENGTH
          IF(NORUN.EQ.1)ZL=30.0D0
          IF(NORUN.EQ.2)ZL=90.0D0
C...
C...   RADIAL GRID, VELOCITY PROFILE
          DR=R0/DFLOAT(NR-1)
          DO 1 IR=1,NR
             R(IR)=DFLOAT(IR-1)*DR
             V(IR)=2.0D0*(1.0D0-R(IR)**2)
1         CONTINUE
          DRS=DR**2
C...
C...   AXIAL GRID
          DZ=ZL/DFLOAT(NZ-1)
          DO 2 JZ=1,NZ
             Z(JZ)=DFLOAT(JZ-1)*DZ
2         CONTINUE
          DZS=DZ**2
C...
C...   INITIAL CONDITION
          DO 3 IR=1,NR
          DO 3 JZ=1,NZ
             U(IR,JZ)=1.0D0
3         CONTINUE
C...
C...   GIVE THE TIME DERIVATIVES A VALUE WHICH CAN THEN BE
C...   USED TO TEST FOR UNDEFINED DERIVATIVES BY SETTING
C...   THEM TO PI, THEN CALLING DERV, AND CHECKING IN
C...   SUBROUTINE PRINT IF THEY HAVE ALL BEEN SET TO VALUES
C...   OTHER THAN PI.  THIS IS A SIMPLE PROCEDURE TO ENSURE
C...   THAT ALL NR X NZ DERIVATIVES HAVE BEEN SET IN DERV
          PI=3.1415927D0
          DO 4 IR=1,NR
          DO 4 JZ=1,NZ
             UT(IR,JZ)=PI
4         CONTINUE
C...
C...   INITIAL DERIVATIVES
          CALL DERV
          IP=0
          RETURN
          END
```

We can note the following points about subroutine INITAL:

(1) Double precision coding is used. A two-dimensional grid is defined with 11 points in r and 31 points in z for a total of $11 \times 31 = 341$ grid points.

```
      IMPLICIT DOUBLE PRECISION (A-H,O-Z)
      PARAMETER (NR=11,NZ=31)
      COMMON/T/           T,        NSTOP,       NORUN
     +       /Y/   U(NR,NZ)
     +       /F/   UT(NR,NZ)
     +       /G/        R(NR),      Z(NZ),       V(NR)
     +       /C/           PE,         TN,          RO,      ZL,
     +                     DR,        DRS,          DZ,     DZS
     +       /I/           IP,      NCASE
```

This choice of the number grid points (341) is somewhat arbitrary; it is a balance between acceptable spatial resolution and a reasonable number of ordinary differential equations (ODEs) to be integrated (341). Of course, we could check the adequacy of the number of grid points (i.e., the spatial convergence) by repeating the calculation of the solution with a larger number of grid points and comparing the two solutions.

(2) The case is then selected from a total of two

```
C...
C...  SELECT APPROXIMATION OF SPATIAL DERIVATIVES (SEE THE
C...  COMMENTS IN SUBROUTINE DERV FOR AN EXPLANATION OF THE
C...  TWO CASES)
      NCASE=1
```

The two cases are summarized below and implemented in subroutines DERV1 and DERV2 discussed subsequently:

Case Description

(i) Equations (H4.15) to (H4.19) by explicit finite differences
(ii) Equations (H4.15) to (H4.19) by subroutine DSS034

Explicit finite differences means that second-order finite differences used to approximate the spatial derivatives (in r and z) in eq. (H4.15) are programmed explicitly in the derivative subroutine DERV1 (rather than calling, for example, library routines to compute the finite differences).

Differentiation in Space Subroutine, DSS034, is called from the derivative subroutine DERV2 to compute the spatial derivatives (in r and z) in eq. (H4.15). DSS034 has fourth-order correct finite difference approximations, both centered for derivatives that model diffusion, and five-point biased upwind for derivatives that model convection. DSS034 is described in detail in Schiesser 1991, 262–73, and is available as explained in the preface.

Note that DERV1 and DERV2 are for the same problem, eqs. (H4.15) and (H4.19). The only difference between the two derivative subroutines is the calculation of the spatial derivatives in eq. (H4.15) (second-order finite differences in DERV1 and fourth-order finite differences in DSS034 called from DERV2).

(3) The problem parameters are defined next

```
C...
C...      PROBLEM PARAMETERS
C...
C...         PECLET NUMBER
             PE=60.0D0
C...
C...         QW*RO/K = TN (NORMALIZATION TEMPERATURE)
             TN=1.0D0
C...
C...         OUTSIDE RADIUS
             RO=1.0D0
C...
C...         AXIAL LENGTH
             IF(NORUN.EQ.1)ZL=30.0D0
             IF(NORUN.EQ.2)ZL=90.0D0
```

In this case, the Peclet number, P_e, is 60; the normalization temperature, $q_w r_o/k$, of eq. (H4.10) is 1; the dimensionless radius defined by eq. (H4.12), r, is 1; and the dimensionless length, z, defined by eq. (H4.13), is 30 and 90 for two runs of the program.

(4) The radial grid based on NR = 11 points and the corresponding velocity profile are computed and stored in arrays R and V, respectively, for use in the derivative subroutines DERV1 and DERV2

```
C...
C...      RADIAL GRID, VELOCITY PROFILE
          DR=RO/DFLOAT(NR-1)
          DO 1 IR=1,NR
             R(IR)=DFLOAT(IR-1)*DR
             V(IR)=2.0D0*(1.0D0-R(IR)**2)
1         CONTINUE
          DRS=DR**2
```

Note that the dimensionless velocity, $v(r)$, in array V is computed according to the parabolic velocity profile of eq. (H4.1).

(5) The axial grid is computed next based on NZ = 31 points.

```
C...
C...      AXIAL GRID
          DZ=ZL/DFLOAT(NZ-1)
```

```
      DO 2 JZ=1,NZ
         Z(JZ)=DFLOAT(JZ-1)*DZ
2     CONTINUE
      DZS=DZ**2
```

(6) Initial condition (H4.16) is then set over the NR × NZ grid

```
C...
C...  INITIAL CONDITION
      DO 3 IR=1,NR
      DO 3 JZ=1,NZ
         U(IR,JZ)=1.0D0
3     CONTINUE
```

(7) NR × NZ (= 341) temporal derivatives in the array UT(NR, NZ) (in COMMON /F/) must be computed in each of the two derivative subroutines DERV1 and DERV2, and there is the possibility of errors in the programming of these derivatives (e.g., one or more of the 341 derivatives might be overlooked). Therefore, to check that an error has not been made, an initial value of π (PI) is given to each of the derivatives. This value should not appear for any of the derivatives after they have been computed by a call to DERV (which in turn calls DERV1 or DERV2, depending on the value of NCASE). Of course, this presupposes that the correct value of any of the 341 initial derivatives is not π; also this test does not necessarily guarantee that the computed values are correct.

```
C...
C...  GIVE THE TIME DERIVATIVES A VALUE WHICH CAN THEN BE
C...  USED TO TEST FOR UNDEFINED DERIVATIVES BY SETTING THEM
C...  TO PI, THEN CALLING DERV, AND CHECKING IN SUBROUTINE
C...  PRINT IF THEY HAVE ALL BEEN SET TO VALUES OTHER THAN
C...  PI.  THIS IS A SIMPLE PROCEDURE TO ENSURE THAT ALL
C...  NR x NZ DERIVATIVES HAVE BEEN SET IN DERV
      PI=3.1415927D0
      DO 4 IR=1,NR
      DO 4 JZ=1,NZ
         UT(IR,JZ)=PI
4     CONTINUE
C...
C...  INITIAL DERIVATIVES
      CALL DERV
      IP=0
```

The call to DERV then computes the 341 temporal derivatives in UT(NR, NZ), and therefore all 341 values of π should be overwritten. These initial temporal derivatives are then printed by a call to PRINT (when IP = 0) to check that they have in fact all been computed (that no values of π appear for the initial derivatives). The utility of this check can perhaps best be appreciated when considering the coding in subroutines DERV1 and DERV2, to be discussed subsequently.

Subroutine DERV, which is the derivative subroutine called by the ODE integrator, RKF45 in this case, is listed below:

```
      SUBROUTINE DERV
      IMPLICIT DOUBLE PRECISION (A-H,O-Z)
      COMMON/I/        IP,       NCASE
C...
C...  APPROXIMATIONS OF SPATIAL DERIVATIVES
C...
C...     NCASE = 1 - EXPLICIT FINITE DIFFERENCES (FDM)
         IF(NCASE.EQ.1)THEN
            CALL DERV1
C...
C...     NCASE = 2 - FINITE DIFFERENCES FROM DSS034
         ELSE
     +   IF(NCASE.EQ.2)THEN
            CALL DERV2
         END IF
      RETURN
      END
```

Note that DERV merely calls either DERV1 or DERV2, when NCASE = 1 or 2, respectively.

We now consider subroutines DERV1 and DERV2 to illustrate:

(1) The explicit programming of the finite difference approximations for the spatial derivatives of eq. (H4.15) in DERV1.

(2) The use of a library subroutine, DSS034 in this case, for the calculation of the spatial derivatives of eq. (H4.15) in DERV2.

We start with subroutine DERV1:

```
      SUBROUTINE DERV1
      IMPLICIT DOUBLE PRECISION (A-H,O-Z)
      PARAMETER (NR=11,NZ=31)
      COMMON/T/        T,       NSTOP,       NORUN
     +      /Y/   U(NR,NZ)
     +      /F/   UT(NR,NZ)
     +      /G/        R(NR),   Z(NZ),       V(NR)
     +      /C/        PE,      TN,          RO,       ZL,
     +                 DR,      DRS,         DZ,       DZS
     +      /I/        IP,      NCASE
C...
C...  Z = 0
```

```
              DO 1 IR=1,NR
                 U(IR,1)=1.0D0
                 UT(IR,1)=0.0D0
1             CONTINUE
C...
C...    R = R0, Z NE 0
              DO 2 JZ=2,NZ
                 UNRP1=U(NR-1,JZ)+(2.0D0*DR)
                 UT(NR,JZ)=-PE*0.0D0
     +                    +(UNRP1-2.0D0*U(NR,JZ)+U(NR-1,JZ))/DRS
     +                    +(UNRP1-U(NR-1,JZ))/(2.0D0*DR)
2             CONTINUE
C...
C...    R = 0, Z NE 0
              DO 3 JZ=2,NZ
                 UT(1,JZ)=-PE*V(1)*(U(1,JZ)-U(1,JZ-1))/DZ
     +                    +4.0D0*(U(2,JZ)-U(1,JZ))/DRS
3             CONTINUE
C...
C...    R NE 0, R0, Z NE 0
              DO 4 JZ=2,NZ
              DO 4 IR=2,NR-1
                 UT(IR,JZ)=-PE*V(IR)*(U(IR,JZ)-U(IR,JZ-1))/DZ
     +                    +(U(IR+1,JZ)-2.0D0*U(IR,JZ)
     +                    +U(IR-1,JZ))/DRS+(1.0D0/R(IR))
     +                    *(U(IR+1,JZ)-U(IR-1,JZ))/(2.0D0*DR)
4             CONTINUE
              RETURN
              END
```

We can note the following points about subroutine DERV1:

(1) The COMMON area is the same as in subroutine INITAL.
(2) Boundary condition (H4.19) is implemented first

```
C...
C...    Z = 0
              DO 1 IR=1,NR
                 U(IR,1)=1.0D0
                 UT(IR,1)=0.0D0
1             CONTINUE
```

Note that the second subscript in U(IR, IZ) is one corresponding to $z = 0$. Also, the temporal derivatives, UT(IR, 1), are set to zero to maintain the zero boundary value.

(3) The boundary condition at $r = 1$, eq. (H4.18), is then implemented

```
C...
C...    R = R0, Z NE 0
        DO 2 JZ=2,NZ
            UNRP1=U(NR-1,JZ)+(2.0D0*DR)
            UT(NR,JZ)=-PE*0.0D0
     +                   +(UNRP1-2.0D0*U(NR,JZ)+U(NR-1,JZ))/DRS
     +                   +(UNRP1-U(NR-1,JZ))/(2.0D0*DR)
2       CONTINUE
```

Note that $u(r = 1, z = 0, t) = 1$ (from DO loop 1) since the initial value of the index of DO loop 2 is JZ = 2. The programming within DO loop 2 requires some further explanation:

(3.1) Boundary condition (H4.18) is approximated as

$$\frac{\partial u(1, z, t)}{\partial r} \approx \frac{u(1 + \Delta r, z, t) - u(1 - \Delta r, z, t)}{2\Delta r} = 1$$

which can then be solved for the fictitious value $u(1 + \Delta r, z, t)$ (i.e., the radial position $1 + \Delta r$ is outside the radial grid, which is defined on the interval $0 \leq r \leq 1$)

$$u(1 + \Delta r, z, t) = u(1 - \Delta r, z, t) + 2\Delta r(1) \tag{H4.20}$$

Equation (H4.20) is programmed in DO loop 2 as

`UNRP1=U(NR-1,JZ)+(2.0D0*DR)`

(3.2) $u(1 + \Delta r, z, t)$ is then used in the finite difference approximations of the radial derivatives in eq. (H4.15) at $r = 1$

$$\frac{\partial^2 u}{\partial r^2} + \frac{1}{r}\frac{\partial u}{\partial r} \approx \frac{u(1 + \Delta r, z, t) - 2u(1, z, t) + u(1 - \Delta r, z, t)}{\Delta r^2}$$
$$+ \frac{1}{1}\frac{u(1 + \Delta r, z, t) - u(1 - \Delta r, z, t)}{2\Delta r} \tag{H4.21}$$

The right hand side (RHS) of eq. (H4.21) is programmed in DO loop 2 as

```
     +                   +(UNRP1-2.0D0*U(NR,JZ)+U(NR-1,JZ))/DRS
     +                   +(UNRP1-U(NR-1,JZ))/(2.0D0*DR)
```

(which defines the derivative $\partial u(1, z, t)/\partial t$ = UT(NR, JZ) in eq. (H4.15) with $v(1) = 0$).

(4) Equation (H4.15) is then programmed at $r = 0, z \neq 0$ (which therefore includes boundary condition (H4.17))

```
C...
C...    R = 0, Z NE 0
```

```
      DO 3 JZ=2,NZ
         UT(1,JZ)=-PE*V(1)*(U(1,JZ)-U(1,JZ-1))/DZ
 +               +4.0D0*(U(2,JZ)-U(1,JZ))/DRS
 3    CONTINUE
```

The coding in DO loop 3 requires some further explanation:

(4.1) The convective term $-P_e v(0)\partial u/\partial z$ in eq. (H4.15) is first computed as

```
-PE*V(1)*(U(1,JZ)-U(1,JZ-1))/DZ
```

Note in particular the finite difference approximation of $\partial u/\partial z$ based on the grid spacing DZ, that is,

```
(U(1,JZ)-U(1,JZ-1))/DZ
```

which is an implementation of

$$\frac{\partial u(0,z,t)}{\partial z} \approx \frac{u(0,z,t) - u(0, z - \Delta z, t)}{\Delta z}$$

In general, a centered approximation for a convective derivative such as

$$\frac{\partial u(r,z,t)}{\partial z}$$

should be avoided for a strongly convective problem (see Chapter M3); thus, we use the two-point upwind (backward) difference

$$\frac{\partial u(r,z,t)}{\partial z} \approx \frac{u(r,z,t) - u(r, z - \Delta z, t)}{\Delta z} + O(\Delta z) \qquad (H4.22)$$

and not the centered difference

$$\frac{\partial u(r,z,t)}{\partial z} \approx \frac{u(r, z + \Delta z, t) - u(r, z - \Delta z, t)}{2\Delta z} + O(\Delta z^2)$$

even though the centered approximation is second-order correct, while the upwind approximation is only first-order correct. (This is an example of where the order of the approximation is not the only factor in determining the best approximation to use; in this case, the centered approximation would probably produce unrealistic oscillations in the solution.)

(4.2) The radial conduction group $\partial^2 u/\partial r^2 + (1/r)\partial u/\partial r$ requires special attention. Note that the second term, $(1/r)\partial u/\partial r$, is indeterminate at $r = 0$ as a consequence of boundary condition (H4.17)

$$\lim_{r \to 0} \frac{\partial u(r,z,t)}{\partial r} = \frac{0}{0}$$

Application of l'Hospital's rule to this term (differentiating the numerator, $\partial u/\partial r$, and denominator, r, with respect to r) gives

$$\lim_{r \to 0} \frac{1}{r}\frac{\partial u(r,z,t)}{\partial r} = \lim_{r \to 0} \frac{1}{1}\frac{\partial^2 u(r,z,t)}{\partial r^2} = \frac{\partial^2 u(r,z,t)}{\partial r^2}$$

and therefore the radial group in eq. (H4.15) at $r = 0$ becomes

$$\frac{\partial^2 u}{\partial r^2} + \frac{1}{r}\frac{\partial u}{\partial r} = 2\frac{\partial^2 u}{\partial r^2} \qquad (H4.23)$$

The second derivative is then approximated at $r = 0$ as

$$\frac{\partial^2 u(0, z, t)}{\partial r^2} \approx \frac{u(\Delta r, z, t) - 2u(0, z, t) + u(-\Delta r, z, t)}{\Delta r^2}$$

$$= 2\frac{u(\Delta r, z, t) - u(0, z, t)}{\Delta r^2}$$

The second step follows from the approximation of boundary condition (H4.17) as

$$\frac{\partial u(0, z, t)}{\partial r} \approx \frac{u(\Delta r, z, t) - u(-\Delta r, z, t)}{2\Delta r} = 0$$

or

$$u(-\Delta r, z, t) = u(\Delta r, z, t)$$

Finally, we get for the radial group in eq. (H4.15) at $r = 0$

$$2\frac{\partial^2 u(0, z, t)}{\partial r^2} \approx 4\frac{u(\Delta r, z, t) - u(0, z, t)}{\Delta r^2} \qquad (H4.24)$$

or in terms of the Fortran

```
     +              +4.0D0*(U(2,JZ)-U(1,JZ))/DRS
```

(5) Finally, eq. (H4.15) is programmed at the interior points over $0 < r < r_0$ and $0 < z \le z_l$

```
C...
C...    R NE 0, R0, Z NE 0
        DO 4 JZ=2,NZ
        DO 4 IR=2,NR-1
            UT(IR,JZ)=-PE*V(IR)*(U(IR,JZ)-U(IR,JZ-1))/DZ
     +              +(U(IR+1,JZ)-2.0D0*U(IR,JZ)
     +              +U(IR-1,JZ))/DRS+(1.0D0/R(IR))
     +              *(U(IR+1,JZ)-U(IR-1,JZ))/(2.0D0*DR)
4       CONTINUE
```

which is a straightforward application of the preceding approximations, that is,

(5.1) $-P_e v(r)\partial u/\partial z$ is programmed as

```
-PE*V(IR)*(U(IR,JZ)-U(IR,JZ-1))/DZ
```

(5.2) $\partial^2 u/\partial r^2$ is programmed as

```
+(U(IR+1,JZ)-2.0D0*U(IR,JZ)+U(IR-1,JZ))/DRS
```

(5.3) $(1/r)\partial u/\partial r$ is programmed as

```
      +(1.0D0/R(IR))*(U(IR+1,JZ)-U(IR-1,JZ))/(2.0D0*DR)
```

The required attention to detail in the programming of subroutine DERV1 suggests that the use of a library routine to calculate the various spatial derivatives in eq. (H4.15), with their associated boundary conditions, might facilitate the programming of PDEs. This point will be considered subsequently when subroutine DERV2, which calls the spatial differentiation routine DSS034, is considered.

Subroutine PRINT is listed below:

```
          SUBROUTINE PRINT(NI,NO)
          IMPLICIT DOUBLE PRECISION (A-H,O-Z)
          PARAMETER (NR=11,NZ=31)
          COMMON/T/          T,        NSTOP,       NORUN
         +      /Y/    U(NR,NZ)
         +      /F/    UT(NR,NZ)
         +      /G/       R(NR),       Z(NZ),       V(NR)
         +      /C/          PE,          TN,          RO,      ZL,
         +                   DR,         DRS,          DZ,      DZS
         +      /I/          IP,       NCASE
C...
C...  ARRAYS FOR THE ANALYTICAL SOLUTION
          DIMENSION UZ(NR), UA(NR)
C...
C...  MONITOR SOLUTION
          WRITE(*,*)NCASE,NORUN,T
C...
C...  PRINT THE INITIAL CONDITION AND INITIAL TEMPORAL
C...  DERIVATIVE
          IF(IP.EQ.0)THEN
             WRITE(NO,10)T,((U(IR,JZ),IR=1,NR),JZ=1,NZ)
10           FORMAT(' t = ',F6.2,//,'   u(r,z,t)',/,(11F7.3))
             WRITE(NO,11)(( UT(IR,JZ),IR=1,NR),JZ=1,NZ)
11           FORMAT(//,'  ut(r,z,t)',/,(11F7.1))
C...
C...  CHECK FOR UNDEFINED DERIVATIVES
          DO 12 IR=1,NR
          DO 12 JZ=1,NZ
             IF((UT(IR,JZ).GT.3.14D0).AND.(UT(IR,JZ).LT.3.15D0))
     +       THEN
                WRITE (NO,13)IR,JZ,U(IR,JZ),UT(IR,JZ)
13              FORMAT(' IR = ',I3,'    JZ = ',I3,
         +      '   U(IR,JZ) = ',F7.3,'    UT(IR,JZ) = ',F7.1)
             END IF
12        CONTINUE
          END IF
C...
```

```
C...      PRINT THE NUMERICAL SOLUTION
          WRITE(NO,1)T
1         FORMAT(//,' t = ',F6.2,//,30X,'u(r,z,t)',//,
     +    3X,'z',4X,'r=0.0',4X,'r=0.2',4X,'r=0.4',
     +            4X,'r=0.6',4X,'r=0.8',4X,'r=1.0',/)
          DO 3 JZ=1,NZ
             NRI=2
             WRITE(NO,2)Z(JZ),(U(IR,JZ),IR=1,NR,NRI)
2            FORMAT(F4.1,6F9.4)
3         CONTINUE
C...
C...      AT THE END OF THE RUN, CALCULATE AND PRINT THE
C...      ANALYTICAL, STEADY STATE TEMPERATURE
          IP=IP+1
          IF(IP.EQ.11)THEN
             DO 14 IR=1,NR
             UZ(IR)=(U(IR,NZ)-U(IR,NZ-1))/DZ
             UA(IR)=U(NR,NZ)+2.0D0*PE*TN*UZ(IR)*
     +       ((1.0D0/ 4.0D0)*(R(IR)**2)
     +       -(1.0D0/16.0D0)*(R(IR)**4)-3.0D0/16.0D0)
14           CONTINUE
             WRITE(NO,15)(UZ(IR),IR=1,NR,NRI),(UA(IR),IR=1,NR,
     +       NRI)
15           FORMAT(/,10X,'Axial gradient; analytical, steady
     +       state,',
     +                /,16X,'thermally developed solution',//,
     +                ' uz',6F9.4,/,' ua',6F9.4,/)
          END IF
C...
C...      STORE THE SOLUTION AND PLOT THE SOLUTION AT THE END OF
C...      THE SECOND RUN
C...
C...         PLOT RADIAL PROFILES
             CALL PLOTDR
C...
C...         PLOT CENTERLINE AXIAL PROFILES
             CALL PLOTDZ
          RETURN
          END
```

We can note the following points about subroutine PRINT:

(1) The COMMON area is the same as in subroutines INITAL and DERV1. The initial condition of eq. (H4.16) and the initial temporal derivative vector (computed by the call to DERV in INITAL) are printed at the beginning of the solution (when IP = 0) to confirm the initial condition and observe if the 341 initial derivatives

have reasonable values

```
C...
C...   PRINT THE INITIAL CONDITION AND INITIAL TEMPORAL
C...   DERIVATIVE
       IF(IP.EQ.0)THEN
          WRITE(NO,10)T,((U(IR,JZ),IR=1,NR),JZ=1,NZ)
10        FORMAT(' t = ',F6.2,//,'   u(r,z,t)',/,(11F7.3))
          WRITE(NO,11)(( UT(IR,JZ),IR=1,NR),JZ=1,NZ)
11        FORMAT(//,'  ut(r,z,t)',/,(11F7.1))
```

(2) A check is then performed to ensure that all 341 temporal derivatives have been computed initially by testing for the value of π set in INITAL

```
C...
C...   CHECK FOR UNDEFINED DERIVATIVES
       DO 12 IR=1,NR
       DO 12 JZ=1,NZ
          IF((UT(IR,JZ).GT.3.14D0).AND.(UT(IR,JZ).LT.3.15D0))
          THEN
          WRITE (NO,13)IR,JZ,U(IR,JZ),UT(IR,JZ)
13        FORMAT(' IR = ',I3,'   JZ = ',I3,
      +   '   U(IR,JZ) = ',F7.3,'   UT(IR,JZ) = ',F7.1)
          END IF
12     CONTINUE
       END IF
```

(3) The numerical solution is then printed as the radial temperature profiles, $u(r, z, t)$ for $0 \leq r \leq 1$, starting at $z = 0$ and continuing to $z = z_l$, via DO loop 3.

```
C...
C...   PRINT THE NUMERICAL SOLUTION
       WRITE(NO,1)T
1      FORMAT(//,' t = ',F6.2,//,30X,'u(r,z,t)',//,
      + 3X,'z',4X,'r=0.0',4X,'r=0.2',4X,'r=0.4',
      +     4X,'r=0.6',4X,'r=0.8',4X,'r=1.0',/)
       DO 3 JZ=1,NZ
          NRI=2
          WRITE(NO,2)Z(JZ),(U(IR,JZ),IR=1,NR,NRI)
2         FORMAT(F4.1,6F9.4)
3      CONTINUE
```

Note the increment in the radial index, NRI = 2, so that every second temperature is printed radially.

(4) At the end of the run (IP = 11), the analytical solution of eq. (H4.5) is evaluated and printed for comparison with the steady state numerical solution.

```
C...
C...   AT THE END OF THE RUN, CALCULATE AND PRINT THE
C...   ANALYTICAL, STEADY STATE TEMPERATURE
       IP=IP+1
       IF(IP.EQ.11)THEN
           DO 14 IR=1,NR
           UZ(IR)=(U(IR,NZ)-U(IR,NZ-1))/DZ
           UA(IR)=U(NR,NZ)+2.0D0*PE*TN*UZ(IR)*
     +     ((1.0D0/ 4.0D0)*(R(IR)**2)
     +     -(1.0D0/16.0D0)*(R(IR)**4)-3.0D0/16.0D0)
14         CONTINUE
           WRITE(NO,15)(UZ(IR),IR=1,NR,NRI),(UA(IR),IR=1,NR,
     +     NRI)
15         FORMAT(/,10X,'Axial gradient; analytical, steady
     +     state,',
     +            /,16X,'thermally developed solution',//,
     +            '  uz',6F9.4,/,'  ua',6F9.4,/)
       END IF
```

The coding in DO loop 14 requires some additional explanation:

(4.1) Equation (H4.5) requires the steady state axial derivative, $\partial u/\partial z$. This derivative, which is a constant (i.e., it is independent of r and z as explained by Eckert and Drake (1972)), is first computed numerically with the programming

```
UZ(IR)=(U(IR,NZ)-U(IR,NZ-1))/DZ
```

Here we have used the first-order, backward difference of eq. (H4.22)

$$\partial u(r, z_l, t)/\partial z \approx \frac{T(r, z_l, t) - T(r, z_l - \Delta z, t)}{\Delta z} \quad \text{(H4.25)}$$

(4.2) The analytical solution of eq. (H4.5) is then programmed as

```
       UA(IR)=U(NR,NZ)+2.0D0*PE*TN*UZ(IR)*
     + ((1.0D0/ 4.0D0)*(R(IR)**2)
     + -(1.0D0/16.0D0)*(R(IR)**4)-3.0D0/16.0D0)
```

Note that $\partial T(r, z_l, t)/\partial z$ from eq. (H4.25) is used in this programming.

(5) The radial profiles are then stored by subroutine PLOTDR for subsequent Matlab plotting

```
C...
C...       PLOT RADIAL PROFILES
           CALL PLOTDR
```

Integer counter IP, which is incremented in PRINT, is used in PLOTDR to open file h4r.out (when IP = 1) for subsequent use in Matlab; the radial profiles at $z = z_l$ are then stored in file h4r.out during each call to PLOTDR.

(6) Finally, the centerline axial profiles are stored in file h4z.out opened in subroutine PLOTDZ, which also writes the solution for the axial profiles at $r = 0$

```
C...
C...       PLOT CENTERLINE AXIAL PROFILES
           CALL PLOTDZ
```

The data file read by the main program that calls ODE integrator RKF45 is listed below:

```
Graetz problem, zl = 30
0.            1.0        0.1
   341                            0.00001
Graetz problem, zl = 90
0.            2.0        0.2
   341                            0.00001
END OF RUNS
```

For the first solution, time t in eq. (H4.15) covers the interval $0 \leq t \leq 1$, with the solution printed at intervals of 0.1. In the second run, the interval in t is extended to 2, with an output interval of 0.2, since the longer system ($z_l = 90$ rather than 30) responds more slowly. In each run, $11 \times 31 = 341$ ODEs are integrated with an error tolerance of 0.00001. For the first run (NORUN = 1), $z_l = 30$, and for the second run (NORUN = 2), $z_l = 90$, as set in subroutine INITAL.

The output from the first run of the preceding program (NORUN = 1) is summarized below:

```
RUN NO. -   1  Graetz problem, zl = 30

INITIAL T -  0.000D+00

  FINAL T -  0.100D+01

  PRINT T -  0.100D+00

NUMBER OF DIFFERENTIAL EQUATIONS  -  341

MAXIMUM INTEGRATION ERROR  -   0.100D-04

t =   0.00
                              u(r,z,t)

    z       r=0.0    r=0.2    r=0.4    r=0.6    r=0.8    r=1.0
   0.0     1.0000   1.0000   1.0000   1.0000   1.0000   1.0000
   0.0     1.0000   1.0000   1.0000   1.0000   1.0000   1.0000
   1.0     1.0000   1.0000   1.0000   1.0000   1.0000   1.0000
```

z	r=0.0	r=0.2	r=0.4	r=0.6	r=0.8	r=1.0
2.0	1.0000	1.0000	1.0000	1.0000	1.0000	1.0000
.					.	
.					.	
.					.	
28.0	1.0000	1.0000	1.0000	1.0000	1.0000	1.0000
29.0	1.0000	1.0000	1.0000	1.0000	1.0000	1.0000
30.0	1.0000	1.0000	1.0000	1.0000	1.0000	1.0000

t =   0.10

$u(r,z,t)$

z	r=0.0	r=0.2	r=0.4	r=0.6	r=0.8	r=1.0
0.0	1.0000	1.0000	1.0000	1.0000	1.0000	1.0000
1.0	1.0002	1.0006	1.0034	1.0186	1.0851	1.2636
2.0	1.0012	1.0026	1.0110	1.0456	1.1497	1.3381
.						
.						
.						
15.0	1.0275	1.0368	1.0686	1.1340	1.2467	1.4169
16.0	1.0276	1.0369	1.0687	1.1341	1.2467	1.4169
17.0	1.0277	1.0369	1.0687	1.1341	1.2467	1.4169
.						
.						
.						
28.0	1.0277	1.0369	1.0687	1.1341	1.2467	1.4169
29.0	1.0277	1.0369	1.0687	1.1341	1.2467	1.4169
30.0	1.0277	1.0369	1.0687	1.1341	1.2467	1.4169

.
.
.

t =   0.90

$u(r,z,t)$

z	r=0.0	r=0.2	r=0.4	r=0.6	r=0.8	r=1.0
0.0	1.0000	1.0000	1.0000	1.0000	1.0000	1.0000
1.0	1.0003	1.0007	1.0038	1.0202	1.0899	1.2720
2.0	1.0016	1.0033	1.0132	1.0530	1.1688	1.3680
.						
.						
.						
28.0	1.6467	1.6859	1.7990	1.9721	2.1816	2.3948
29.0	1.6794	1.7186	1.8317	2.0046	2.2137	2.4266
30.0	1.7121	1.7513	1.8642	2.0368	2.2454	2.4580

t =   1.00

u(r,z,t)

z	r=0.0	r=0.2	r=0.4	r=0.6	r=0.8	r=1.0
0.0	1.0000	1.0000	1.0000	1.0000	1.0000	1.0000
1.0	1.0003	1.0007	1.0038	1.0202	1.0899	1.2720
2.0	1.0016	1.0033	1.0132	1.0530	1.1688	1.3680
.	.					.
.	.					.
.	.					.
28.0	1.6473	1.6866	1.8002	1.9741	2.1848	2.3988
29.0	1.6803	1.7197	1.8333	2.0073	2.2179	2.4317
30.0	1.7133	1.7527	1.8664	2.0404	2.2508	2.4645

Axial gradient; analytical, steady state,
thermally developed solution

uz	0.0330	0.0331	0.0331	0.0331	0.0329	0.0328
ua	1.7215	1.7600	1.8722	2.0456	2.2549	2.4645

We can note the following points about the preceding output:

(1) Radial position, r, varies across the top of each output, and axial position, z, varies down the left side of each output.

(2) Boundary condition (H4.19) programmed in DERV1 is clear from the first row at $z = 0$. For example, at $t = 0.1$

t =   0.10

u(r,z,t)

z	r=0.0	r=0.2	r=0.4	r=0.6	r=0.8	r=1.0
0.0	1.0000	1.0000	1.0000	1.0000	1.0000	1.0000
1.0	1.0002	1.0006	1.0034	1.0186	1.0851	1.2636

(3) For short times, the radial profiles attain a constant shape with increasing z. Thus, at $t = 0.1$, $z = 17$, and $z = 30$

17.0	1.0277	1.0369	1.0687	1.1341	1.2467	1.4169
30.0	1.0277	1.0369	1.0687	1.1341	1.2467	1.4169

For short times and large z, $\partial u(r, z, t)/\partial z \approx 0$ in eq. (H4.15) so that the radial term, $\partial^2 u/\partial r^2 + (1/r)\partial u/\partial r$, equals the transient term, $\partial u/\partial t$, in eq. (H4.15), that is, the net radial heat flux equals the heat accumulation.

(4) For long times, the radial profiles have increasing temperatures with increasing z. Thus, at $t = 1$, $z = 28, 29$, and 30

```
28.0    1.6473    1.6866    1.8002    1.9741    2.1848    2.3988
29.0    1.6803    1.7197    1.8333    2.0073    2.2179    2.4317
30.0    1.7133    1.7527    1.8664    2.0404    2.2508    2.4645
```

The increasing temperatures at long times are due to the continuous input of heat at the wall according to boundary condition (H4.18). Also, as we will observe, the axial derivative, $\partial u(r, z, t)/\partial z$, is constant.

(5) The steady state numerical solutions and the analytical steady state solution, eq. (H4.5), are in approximate agreement. Thus, at $t = 1$, $z = 30$

```
30.0    1.7133    1.7527    1.8664    2.0404    2.2508    2.4645
```

```
          Axial gradient; analytical, steady state,
                thermally developed solution

uz       0.0330    0.0331    0.0331    0.0331    0.0329    0.0328

ua       1.7215    1.7600    1.8722    2.0456    2.2549    2.4645
```

Note that the axial derivative, $\partial u(r, z_l, t)/\partial z$, is essentially constant in r at 0.0330 and the numerical and analytical solutions agree to about three figures. The axial derivative $\partial(r, z, t)/\partial z$ is also essentially constant in z, which could be confirmed by comparing its values at $z = 29$ and $z = 30$.

The output for the second solution (NORUN $= 2$, $z_l = 90$) is summarized below:

```
RUN NO. -   2  Graetz problem, zl = 90

INITIAL T -  0.000D+00

  FINAL T -  0.200D+01

  PRINT T -  0.200D+00

MAXIMUM INTEGRATION ERROR -   0.100D-04

t =   0.00
                              u(r,z,t)

  z       r=0.0     r=0.2     r=0.4     r=0.6     r=0.8     r=1.0
```

z						
0.0	1.0000	1.0000	1.0000	1.0000	1.0000	1.0000
3.0	1.0000	1.0000	1.0000	1.0000	1.0000	1.0000
6.0	1.0000	1.0000	1.0000	1.0000	1.0000	1.0000
.						.
.						.
.						.
84.0	1.0000	1.0000	1.0000	1.0000	1.0000	1.0000
87.0	1.0000	1.0000	1.0000	1.0000	1.0000	1.0000
90.0	1.0000	1.0000	1.0000	1.0000	1.0000	1.0000

t =   0.20

u(r,z,t)

z	r=0.0	r=0.2	r=0.4	r=0.6	r=0.8	r=1.0
0.0	1.0000	1.0000	1.0000	1.0000	1.0000	1.0000
3.0	1.0101	1.0147	1.0337	1.0856	1.2025	1.3962
6.0	1.0336	1.0446	1.0848	1.1727	1.3239	1.5212
.						.
.						.
.						.
45.0	1.1679	1.1853	1.2385	1.3304	1.4641	1.6419
48.0	1.1679	1.1853	1.2385	1.3304	1.4641	1.6419
51.0	1.1679	1.1853	1.2385	1.3304	1.4641	1.6419
.						.
.						.
.						.
84.0	1.1679	1.1853	1.2385	1.3304	1.4641	1.6419
87.0	1.1679	1.1853	1.2385	1.3304	1.4641	1.6419
90.0	1.1679	1.1853	1.2385	1.3304	1.4641	1.6419

.
.
.

t =   1.80

u(r,z,t)

z	r=0.0	r=0.2	r=0.4	r=0.6	r=0.8	r=1.0
0.0	1.0000	1.0000	1.0000	1.0000	1.0000	1.0000
3.0	1.0121	1.0172	1.0381	1.0937	1.2162	1.4137
6.0	1.0442	1.0575	1.1051	1.2068	1.3762	1.5848
.						.
.						.
.						.
84.0	3.4642	3.5012	3.6075	3.7702	3.9687	4.1762
87.0	3.5478	3.5841	3.6885	3.8484	4.0441	4.2502

| 90.0 | 3.6278 | 3.6633 | 3.7655 | 3.9224 | 4.1150 | 4.3197 |

t = 2.00

u(r,z,t)

z	r=0.0	r=0.2	r=0.4	r=0.6	r=0.8	r=1.0
0.0	1.0000	1.0000	1.0000	1.0000	1.0000	1.0000
3.0	1.0121	1.0172	1.0381	1.0937	1.2162	1.4137
6.0	1.0442	1.0575	1.1051	1.2068	1.3762	1.5848
.	.					.
.	.					.
.	.					.
84.0	3.4973	3.5359	3.6469	3.8166	4.0222	4.2334
87.0	3.5901	3.6283	3.7382	3.9063	4.1103	4.3206
90.0	3.6807	3.7185	3.8272	3.9933	4.1954	4.4048

Axial gradient; analytical, steady state,
thermally developed solution

| uz | 0.0302 | 0.0301 | 0.0296 | 0.0290 | 0.0284 | 0.0281 |
| ua | 3.7248 | 3.7639 | 3.8744 | 4.0371 | 4.2239 | 4.4048 |

The conclusions from this output are essentially the same as for the first solution, that is, (1) zero axial derivatives for short times, constant axial derivatives for long times, and (2) approximate agreement between the numerical and analytical profiles (although the axial derivative, $\partial u(r, z_l, t)/\partial z$, has not reached a constant value with respect to r as in the first solution and the agreement between the numerical and analytical profiles is not as good as for the first solution, probably because the increase in the final solution time to $t = 2$ was not enough to compensate for the three-fold increase in length from $z_l = 30$ to $z_l = 90$). This conclusion, that a longer time was required to reach steady state, is confirmed by longer runs of the program, discussed subsequently.

The radial profiles plotted by the call to subroutine PLOTDR in PRINT are in Figures H4.2a (NORUN = 1, $z_l = 30$) and H4.2b (NORUN = 2, $z_l = 90$), and the axial profiles from the call to PLOTDZ are in Figures H4.3a (NORUN = 1, $z_l = 30$) and H4.3b (NORUN = 2, $z_l = 90$). These plots essentially confirm the conclusions drawn from the numerical output (note in particular the characteristics of the centerline axial derivative $\partial u(0, z, t)/\partial z$).

The final times of $t = 1$ and $t = 2$ for the two solutions discussed previously were selected so that the radial and axial profiles at the various values of t would be clearly spaced, that is, if longer final times were used, the profiles would tend to bunch together near the steady state solution. However, in order to observe the

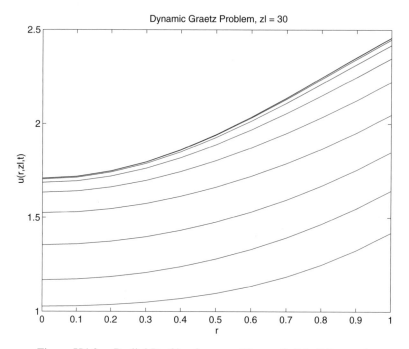

Figure H4.2a. Radial Profiles for $z_l = 30$, $t = 0, 0.1, 0.2, \ldots, 1$

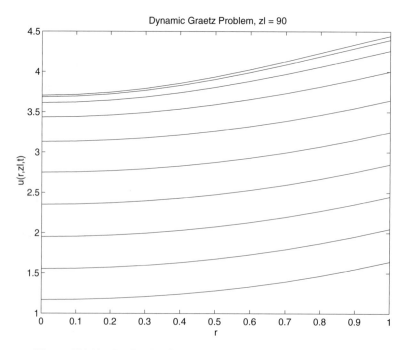

Figure H4.2b. Radial Profiles for $z_l = 90$, $t = 0, 0.2, 0.4, \ldots, 2$

The Graetz problem with constant wall heat flux 191

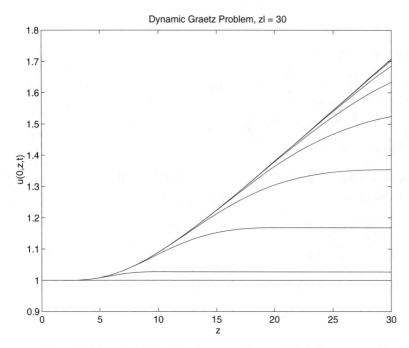

Figure H4.3a. Axial Profiles for $z_l = 30$, $t = 0, 0.1, 0.2, \ldots, 1$

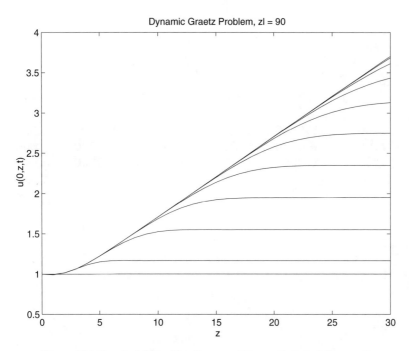

Figure H4.3b. Axial Profiles for $z_l = 90$, $t = 0, 0.1, 0.2, \ldots, 2$

approach to steady state, the program was executed again with the following data file:

```
Graetz problem, zl = 30
0.           5.0          0.5
 341                              0.00001
Graetz problem, zl = 90
0.          10.0          1.0
 341                              0.00001
END OF RUNS
```

Note now that the final times are 5 and 10 for the first and second solutions, respectively. The output from these longer runs is summarized below:

```
RUN NO. -   1  Graetz problem, zl = 30

INITIAL T -  0.000D+00

  FINAL T -  0.500D+01

  PRINT T -  0.500D+00

NUMBER OF DIFFERENTIAL EQUATIONS -  341

MAXIMUM INTEGRATION ERROR -   0.100D-04

t =   0.00
                             u(r,z,t)

   z     r=0.0    r=0.2    r=0.4    r=0.6    r=0.8    r=1.0

  0.0    1.0000   1.0000   1.0000   1.0000   1.0000   1.0000
  1.0    1.0000   1.0000   1.0000   1.0000   1.0000   1.0000
  2.0    1.0000   1.0000   1.0000   1.0000   1.0000   1.0000
          .                                             .
          .                                             .
          .                                             .
 28.0    1.0000   1.0000   1.0000   1.0000   1.0000   1.0000
 29.0    1.0000   1.0000   1.0000   1.0000   1.0000   1.0000
 30.0    1.0000   1.0000   1.0000   1.0000   1.0000   1.0000
    .
    .
    .
t =   4.50
                             u(r,z,t)

   z     r=0.0    r=0.2    r=0.4    r=0.6    r=0.8    r=1.0
```

z	r=0.0	r=0.2	r=0.4	r=0.6	r=0.8	r=1.0
0.0	1.0000	1.0000	1.0000	1.0000	1.0000	1.0000
1.0	1.0003	1.0007	1.0038	1.0202	1.0899	1.2720
2.0	1.0016	1.0033	1.0132	1.0530	1.1688	1.3680
.	.					.
.	.					.
.	.					.
28.0	1.6475	1.6868	1.8006	1.9749	2.1861	2.4004
29.0	1.6806	1.7200	1.8339	2.0084	2.2197	2.4339
30.0	1.7137	1.7532	1.8673	2.0419	2.2532	2.4675

t =   5.00

u(r,z,t)

z	r=0.0	r=0.2	r=0.4	r=0.6	r=0.8	r=1.0
0.0	1.0000	1.0000	1.0000	1.0000	1.0000	1.0000
1.0	1.0003	1.0007	1.0038	1.0202	1.0899	1.2720
2.0	1.0016	1.0033	1.0132	1.0530	1.1688	1.3680
.	.					.
.	.					.
.	.					.
28.0	1.6475	1.6868	1.8006	1.9749	2.1861	2.4004
29.0	1.6806	1.7200	1.8339	2.0084	2.2197	2.4339
30.0	1.7137	1.7532	1.8673	2.0419	2.2532	2.4675

Axial gradient; analytical, steady state, thermally developed solution

uz	0.0332	0.0332	0.0334	0.0335	0.0335	0.0335
ua	1.7216	1.7596	1.8706	2.0432	2.2538	2.4675

RUN NO. -   2 Graetz problem, zl = 90

INITIAL T - 0.000D+00

  FINAL T - 0.100D+02

  PRINT T - 0.100D+01

NUMBER OF DIFFERENTIAL EQUATIONS - 341

MAXIMUM INTEGRATION ERROR - 0.100D-04

t =   0.00

u(r,z,t)

z	r=0.0	r=0.2	r=0.4	r=0.6	r=0.8	r=1.0
0.0	1.0000	1.0000	1.0000	1.0000	1.0000	1.0000
3.0	1.0000	1.0000	1.0000	1.0000	1.0000	1.0000
6.0	1.0000	1.0000	1.0000	1.0000	1.0000	1.0000
.	.					.
.	.					.
.	.					.
84.0	1.0000	1.0000	1.0000	1.0000	1.0000	1.0000
87.0	1.0000	1.0000	1.0000	1.0000	1.0000	1.0000
90.0	1.0000	1.0000	1.0000	1.0000	1.0000	1.0000

.
.
.

t = 9.00

$$u(r,z,t)$$

z	r=0.0	r=0.2	r=0.4	r=0.6	r=0.8	r=1.0
0.0	1.0000	1.0000	1.0000	1.0000	1.0000	1.0000
3.0	1.0121	1.0172	1.0381	1.0937	1.2162	1.4137
6.0	1.0442	1.0575	1.1051	1.2068	1.3762	1.5848
.	.					.
.	.					.
.	.					.
84.0	3.5170	3.5568	3.6715	3.8467	4.0583	4.2726
87.0	3.6172	3.6571	3.7718	3.9470	4.1586	4.3728
90.0	3.7175	3.7573	3.8720	4.0472	4.2588	4.4731

t = 10.00

$$u(r,z,t)$$

z	r=0.0	r=0.2	r=0.4	r=0.6	r=0.8	r=1.0
0.0	1.0000	1.0000	1.0000	1.0000	1.0000	1.0000
3.0	1.0121	1.0172	1.0381	1.0937	1.2162	1.4137
6.0	1.0442	1.0575	1.1051	1.2068	1.3762	1.5848
.	.					.
.	.					.
.	.					.
84.0	3.5170	3.5568	3.6715	3.8467	4.0583	4.2726
87.0	3.6172	3.6571	3.7718	3.9470	4.1586	4.3728
90.0	3.7175	3.7573	3.8720	4.0472	4.2588	4.4731

Axial gradient; analytical, steady state,
thermally developed solution

uz	0.0334	0.0334	0.0334	0.0334	0.0334	0.0334
ua	3.7213	3.7609	3.8752	4.0496	4.2602	4.4731

The numerical output is as expected, that is, for the longer times, the temperature field, $u(r, z, t)$, approaches a steady state, with very little change with respect to t for large t (which is, of course, the condition that would preclude clearly spaced temperature profiles in Figures H4.2a to H4.3b).

The method of lines (MOL) solution in subroutine DERV1 is based on the explicit programming of the finite difference approximations for the spatial derivatives in eq. (H4.15), including the special conditions required at the boundaries. Although conceptually this programming is straightforward, it suggests the possibility for programming errors and the possible benefits from using a library routine that implements the finite difference approximations of derivatives. We now consider the use of such a library spatial differentiator, DSS034, which is called in the following subroutine DERV2 (used in place of the previous DERV1, with the same INITAL and PRINT)

```
      SUBROUTINE DERV2
      IMPLICIT DOUBLE PRECISION (A-H,O-Z)
      PARAMETER (NR=11,NZ=31)
      COMMON/T/          T,       NSTOP,       NORUN
     +       /Y/    U(NR,NZ)
     +       /F/    UT(NR,NZ)
     +       /G/       R(NR),      Z(NZ),      V(NR)
     +       /C/          PE,         TN,         RO,          ZL,
     +                    DR,        DRS,         DZ,          DZS
     +       /I/          IP,      NCASE
C...
C...  ARRAYS FOR UR, URR, UZ, UZZ
      DIMENSION UR(NR,NZ), URR(NR,NZ), UZ(NR,NZ), UZZ(NR,NZ)
C...
C...  Z = 0
      DO 1 IR=1,NR
         U(IR,1)=1.0D0
         UT(IR,1)=0.0D0
1     CONTINUE
C...
C...  UR
```

```
      CALL DSS034(0.0D0,R0,NR,NZ,1,U,UR,0.0D0)
C...
C...  R = R0, Z NE 0
      DO 2 JZ=2,NZ
         UR(NR,JZ)=1.0D0
2     CONTINUE
C...
C...  R = 0, Z NE 0
      DO 3 JZ=2,NZ
         UR(1,JZ)=0.0D0
3     CONTINUE
C...
C...  URR
      CALL DSS034(0.0D0,R0,NR,NZ,1,UR,URR,0.0D0)
C...
C...  UZ (FOR CONVECTION)
      CALL DSS034(0.0D0,ZL,NR,NZ,2,U,UZ,1.0D0)
C...
C...  Z NE 0
      DO 4 JZ=2,NZ
      DO 4 IR=1,NR
C...
C...     R = 0
         IF(IR.EQ.1)THEN
            UT(1,JZ)=-PE*V(1)*UZ(1,JZ)
     +                +2.0D0*URR(1,JZ)
C...
C...     R NE 0
         ELSE
     +   IF(IR.NE.1)THEN
            UT(IR,JZ)=-PE*V(IR)*UZ(IR,JZ)
     +                +URR(IR,JZ)+(1.0D0/R(IR))*UR(IR,JZ)
         END IF
4     CONTINUE
      RETURN
      END
```

We can note the following points about subroutine DERV2:

(1) The COMMON area is the same as in DERV1 (and INITAL and PRINT).

(2) The radial and axial derivatives, $\partial u(r,z,t)/\partial r$, $\partial^2 u(r,z,t)/\partial r^2$, and $\partial u(r,z,t)/\partial z$, in eq. (H4.15) are stored in arrays UR(NR, NZ), URR(NR, NZ), and UZ(NR, NZ), respectively

```
C...
C...  ARRAYS FOR UR, URR, UZ
      DIMENSION UR(NR,NZ), URR(NR,NZ), UZ(NR,NZ)
```

(3) Boundary condition (H4.19) is again implemented as in DERV1

```
C...
C...    Z = 0
        DO 1 IR=1,NR
           U(IR,1)=1.0D0
           UT(IR,1)=0.0D0
1       CONTINUE
```

(4) The first-order radial derivative in eq. (H4.15), $\partial u/\partial r$, is then computed by a call to subroutine DSS034

```
C...
C...    UR
        CALL DSS034(0.0D0,R0,NR,NZ,1,U,UR,0.0D0)
```

DSS034 essentially differentiates a two-dimensional array, for example, U, to produce a two-dimensional array of the first-order derivative, for example, UR. The arguments of DSS034 are briefly explained below in terms of the preceding programming:

0.0D0,R0:

Beginning and end of the interval over which the arrays are defined for the independent variable in the partial derivative, for example, 0 to R0 corresponding to $0 \le r \le r_0$, so the partial derivative is with respect to r (inputs).

NR:

Number of grid points for the independent variable in the partial derivative, for example, NR ($= 11$) (input).

NZ:

Number of grid points in the second independent variable which is not in the partial derivative, for example, NZ ($= 31$), so the partial derivative is not with respect to z (input).

1:

Number of the independent variable with respect to which the partial derivative is computed, for example, $1 \Leftrightarrow r$, $2 \Leftrightarrow z$. A value of 1 corresponds to the first subscript in arrays U and UR, and 2 corresponds to the second subscript; for example, the first subscript of U(NR, NZ) is for r, so the derivative is with respect to r (input).

U:

Two-dimensional array to be differentiated (input)

UR:

Two-dimensional array with the computed first-order partial derivative (output)

0.0D0:

Variable to select the type of finite difference approximation used in the calculation of the derivative. If zero, a five-point, centered approximation is used, which is the appropriate choice in this case since the first-order derivative, $\partial u/\partial r$, in eq. (H4.15) is for radial heat conduction (Schiesser 1991). If positive, five-point, biased upwind approximations are used for convection in the positive direction of the independent variable of the derivative. If negative, five-point, biased upwind approximations are used for convection in the negative direction of the independent variable in the derivative (for a discussion of upwinding, see Chapter M3) (input).

(5) The derivative UR computed by the preceding call to DSS034 in general will not have the correct boundary values. Therefore, boundary conditions (H4.18) and (H4.17) are used in DO loops 2 and 3, respectively

```
C...
C...    R = R0, Z NE 0
        DO 2 JZ=2,NZ
           UR(NR,JZ)=1.0D0
2       CONTINUE
C...
C...    R = 0, Z NE 0
        DO 3 JZ=2,NZ
           UR(1,JZ)=0.0D0
3       CONTINUE
```

(6) The convective derivative, $\partial u/\partial z$, in eq. (H4.15) is then computed by a call to DSS034

```
C...
C...    UZ (FOR CONVECTION)
        CALL DSS034(0.0D0,ZL,NR,NZ,2,U,UZ,1.0D0)
```

Note that the independent variable in the derivative is number 2 (i.e., z). Also, the last argument is positive (i.e., 1.0D0) since the flow is in the positive z direction (again, this specifies that UZ is computed with five-point, biased upwind finite difference approximations, biased in the positive z direction).

(7) Finally, the temporal derivative, $\partial u/\partial t$, is computed in DO loop 4

```
C...
C...    Z NE 0
        DO 4 JZ=2,NZ
        DO 4 IR=1,NR
C...
C...       R = 0
           IF(IR.EQ.1)THEN
              UT(1,JZ)=-PE*V(1)*UZ(1,JZ)
     +                 +2.0D0*URR(1,JZ)
C...
C...       R NE 0
```

```
             ELSE
     +       IF(IR.NE.1)THEN
               UT(IR,JZ)=-PE*V(IR)*UZ(IR,JZ)
     +                   +URR(IR,JZ)+(1.0D0/R(IR))*UR(IR,JZ)
             END IF
4        CONTINUE
```

Note again that the special form of the radial group in eq. (H4.15) is used at $r = 0$ (IR $= 1$), in accordance with eq. (H4.23).

This completes the programming of eq. (H4.15). In particular, all 341 derivatives in UT have been computed in DO loop 4 and sent to the ODE integrator via COMMON/F/. The ODE integrator (again, RKF45) then returns the 341 dependent variables in array U through COMMON/Y/ for programming of UT in DERV2. All of the other programming (the main program that calls RKF45, subroutines INITAL and PRINT, and the two data files for the short and long runs) is the same. Thus, we can now consider the output, starting with the short time runs

```
RUN NO. -    1  Graetz problem, zl = 30

INITIAL T -   0.000D+00

  FINAL T -   0.100D+01

  PRINT T -   0.100D+00

NUMBER OF DIFFERENTIAL EQUATIONS -  341

MAXIMUM INTEGRATION ERROR -   0.100D-04

t =    0.00
                             u(r,z,t)

   z      r=0.0     r=0.2     r=0.4     r=0.6     r=0.8     r=1.0

  0.0    1.0000    1.0000    1.0000    1.0000    1.0000    1.0000
  1.0    1.0000    1.0000    1.0000    1.0000    1.0000    1.0000
  2.0    1.0000    1.0000    1.0000    1.0000    1.0000    1.0000
           .                                                 .
           .                                                 .
           .                                                 .
 28.0    1.0000    1.0000    1.0000    1.0000    1.0000    1.0000
 29.0    1.0000    1.0000    1.0000    1.0000    1.0000    1.0000
 30.0    1.0000    1.0000    1.0000    1.0000    1.0000    1.0000
           .
           .
           .
```

t =   0.90

$u(r,z,t)$

z	r=0.0	r=0.2	r=0.4	r=0.6	r=0.8	r=1.0
0.0	1.0000	1.0000	1.0000	1.0000	1.0000	1.0000
1.0	1.0003	1.0000	0.9996	1.0133	1.0960	1.2775
2.0	1.0000	1.0001	1.0059	1.0473	1.1763	1.3726
.	.					.
.	.					.
.	.					.
28.0	1.6428	1.6820	1.7954	1.9688	2.1783	2.3906
29.0	1.6759	1.7151	1.8286	2.0020	2.2113	2.4235
30.0	1.7090	1.7483	1.8618	2.0350	2.2442	2.4562

t =   1.00

$u(r,z,t)$

z	r=0.0	r=0.2	r=0.4	r=0.6	r=0.8	r=1.0
0.0	1.0000	1.0000	1.0000	1.0000	1.0000	1.0000
1.0	1.0003	1.0000	0.9996	1.0133	1.0960	1.2775
2.0	1.0000	1.0001	1.0059	1.0473	1.1763	1.3726
.	.					.
.	.					.
.	.					.
28.0	1.6428	1.6820	1.7955	1.9690	2.1787	2.3912
29.0	1.6759	1.7152	1.8287	2.0023	2.2121	2.4246
30.0	1.7091	1.7484	1.8620	2.0357	2.2455	2.4579

Axial gradient; analytical, steady state,
thermally developed solution

uz	0.0332	0.0332	0.0333	0.0333	0.0333	0.0333
ua	1.7115	1.7501	1.8623	2.0354	2.2455	2.4579

RUN NO. -   2 Graetz problem, zl = 90

INITIAL T - 0.000D+00

  FINAL T - 0.200D+01

  PRINT T - 0.200D+00

NUMBER OF DIFFERENTIAL EQUATIONS - 341

MAXIMUM INTEGRATION ERROR - 0.100D-04

t =   0.00

u(r,z,t)

z	r=0.0	r=0.2	r=0.4	r=0.6	r=0.8	r=1.0
0.0	1.0000	1.0000	1.0000	1.0000	1.0000	1.0000
3.0	1.0000	1.0000	1.0000	1.0000	1.0000	1.0000
6.0	1.0000	1.0000	1.0000	1.0000	1.0000	1.0000
.						.
.						.
.						.
84.0	1.0000	1.0000	1.0000	1.0000	1.0000	1.0000
87.0	1.0000	1.0000	1.0000	1.0000	1.0000	1.0000
90.0	1.0000	1.0000	1.0000	1.0000	1.0000	1.0000

.
.
.

t =   1.80

u(r,z,t)

z	r=0.0	r=0.2	r=0.4	r=0.6	r=0.8	r=1.0
0.0	1.0000	1.0000	1.0000	1.0000	1.0000	1.0000
3.0	0.9969	1.0007	1.0224	1.0915	1.2359	1.4347
6.0	1.0177	1.0329	1.0893	1.2088	1.3934	1.6019
.						.
.						.
.						.
84.0	3.5004	3.5387	3.6493	3.8178	4.0216	4.2309
87.0	3.5944	3.6322	3.7410	3.9069	4.1079	4.3158
90.0	3.6855	3.7224	3.8291	3.9917	4.1892	4.3955

t =   2.00

u(r,z,t)

z	r=0.0	r=0.2	r=0.4	r=0.6	r=0.8	r=1.0
0.0	1.0000	1.0000	1.0000	1.0000	1.0000	1.0000
3.0	0.9969	1.0007	1.0224	1.0915	1.2359	1.4347
6.0	1.0177	1.0329	1.0893	1.2088	1.3934	1.6019
.						.
.						.
.						.
84.0	3.5076	3.5468	3.6603	3.8333	4.0421	4.2540

87.0	3.6066	3.6457	3.7587	3.9310	4.1389	4.3503
90.0	3.7048	3.7437	3.8560	4.0271	4.2338	4.4446

Axial gradient; analytical, steady state, thermally developed solution

uz	0.0327	0.0327	0.0324	0.0321	0.0316	0.0314
ua	3.7080	3.7486	3.8644	4.0384	4.2430	4.4446

Generally, we can conclude that the two approximations of eq. (H4.15), the second-order approximations in DERV1, and the fourth-order approximations in DERV2, produced solutions that are in approximate agreement. For example, note the following comparison of $u(r, z, 1)$ for the first (short time) solution ($z_l = 30$)

Second-order approximations in DERV1:

$t = 1$, $z_l = 30$

u(r,z,t)

z	r=0.0	r=0.2	r=0.4	r=0.6	r=0.8	r=1.0
28.0	1.6473	1.6866	1.8002	1.9741	2.1848	2.3988
29.0	1.6803	1.7197	1.8333	2.0073	2.2179	2.4317
30.0	1.7133	1.7527	1.8664	2.0404	2.2508	2.4645

Axial gradient; analytical, steady state, thermally developed solution

uz	0.0330	0.0331	0.0331	0.0331	0.0329	0.0328
ua	1.7215	1.7600	1.8722	2.0456	2.2549	2.4645

Fourth-order approximations in DERV2:

$t = 11$, $z_l = 90$

u(r,z,t)

z	r=0.0	r=0.2	r=0.4	r=0.6	r=0.8	r=1.0
28.0	1.6428	1.6820	1.7955	1.9690	2.1787	2.3912
29.0	1.6759	1.7152	1.8287	2.0023	2.2121	2.4246
30.0	1.7091	1.7484	1.8620	2.0357	2.2455	2.4579

Axial gradient; analytical, steady state,
thermally developed solution

uz	0.0332	0.0332	0.0333	0.0333	0.0333	0.0333
ua	1.7115	1.7501	1.8623	2.0354	2.2455	2.4579

indicates that the two solutions are in agreement to about three figures. The differences might be due to the following causes:

(1) We would expect the second-order approximations of DERV1 to be less accurate than the fourth-order approximations of DERV2.

(2) In both DERV1 and DERV2, the 11 × 31 spatial grid may be so coarse as to produce errors in the third figure, particularly for the second-order approximations of DERV1 (for which the errors are $O(\Delta r^2)$ and $O(\Delta z^2)$). Of course, this point could be studied by computing a solution with a finer grid (more grid points in r and z).

(3) The time integration by RKF45 may be inaccurate, although this is unlikely (recall the error tolerance for the time integration was specified as 0.0001 in the data files, and generally our experience has been that solutions computed by RKF45 surpass the specified error tolerance). This point could be studied by computing a solution with a more stringent (smaller) error tolerance.

(4) The differences between the numerical and analytical solutions may be due to the following considerations:

(4.1) There is insufficient time for the numerical solution to reach steady state (we will consider this point subsequently in discussing the long time solutions).

(4.2) Related to (4.1), the numerical derivative, $\partial u/\partial z$, in eq. (H4.5) may not have reached a constant (steady state) value. Notice also that the analytical solution differs for DERV1 and DERV2, which is due to the different values of the numerical derivative, $\partial u/\partial z$, used in the analytical solution.

Finally, we consider briefly the long time solutions (in which the final values of time are $t = 5$ and 10 for $z_l = 30$ and 90, respectively (recall, again, that the preceding short time solutions are for $t = 1$ and 2 for $z_l = 30$ and 90, respectively). Rather than list the output as before, we consider only the solutions for $u(r, 30, 5)$ for $z_l = 30$ and $u(r, z, 10)$ for $z_l = 90$, from DERV1 and DERV2.

Second-order approximations in DERV1:

$t = 5.00$, $z_l = 30$

u(r,z,t)

z	r=0.0	r=0.2	r=0.4	r=0.6	r=0.8	r=1.0

0.0	1.0000	1.0000	1.0000	1.0000	1.0000	1.0000
1.0	1.0003	1.0007	1.0038	1.0202	1.0899	1.2720
2.0	1.0016	1.0033	1.0132	1.0530	1.1688	1.3680
.						.
.						.
.						.
28.0	1.6475	1.6868	1.8006	1.9749	2.1861	2.4004
29.0	1.6806	1.7200	1.8339	2.0084	2.2197	2.4339
30.0	1.7137	1.7532	1.8673	2.0419	2.2532	2.4675

Axial gradient; analytical, steady state,
thermally developed solution

uz	0.0332	0.0332	0.0334	0.0335	0.0335	0.0335
ua	1.7216	1.7596	1.8706	2.0432	2.2538	2.4675

$t = 10.00$, $z_l = 90$

$$u(r,z,t)$$

z	r=0.0	r=0.2	r=0.4	r=0.6	r=0.8	r=1.0
0.0	1.0000	1.0000	1.0000	1.0000	1.0000	1.0000
3.0	1.0121	1.0172	1.0381	1.0937	1.2162	1.4137
6.0	1.0442	1.0575	1.1051	1.2068	1.3762	1.5848
.						.
.						.
.						.
84.0	3.5170	3.5568	3.6715	3.8467	4.0583	4.2726
87.0	3.6172	3.6571	3.7718	3.9470	4.1586	4.3728
90.0	3.7175	3.7573	3.8720	4.0472	4.2588	4.4731

Axial gradient; analytical, steady state,
thermally developed solution

uz	0.0334	0.0334	0.0334	0.0334	0.0334	0.0334
ua	3.7213	3.7609	3.8752	4.0496	4.2602	4.4731

Fourth-order approximations in DERV2:

$t = 5.00$, $z_l = 30$

$$u(r,z,t)$$

z	r=0.0	r=0.2	r=0.4	r=0.6	r=0.8	r=1.0

z	r=0.0	r=0.2	r=0.4	r=0.6	r=0.8	r=1.0
0.0	1.0000	1.0000	1.0000	1.0000	1.0000	1.0000
1.0	1.0003	1.0000	0.9996	1.0133	1.0960	1.2775
2.0	1.0000	1.0001	1.0059	1.0473	1.1763	1.3726
.	.					.
.	.					.
.	.					.
28.0	1.6428	1.6820	1.7955	1.9690	2.1788	2.3913
29.0	1.6759	1.7152	1.8288	2.0024	2.2122	2.4247
30.0	1.7091	1.7484	1.8621	2.0357	2.2456	2.4581

Axial gradient; analytical, steady state, thermally developed solution

uz	0.0332	0.0332	0.0333	0.0334	0.0334	0.0334
ua	1.7116	1.7503	1.8624	2.0352	2.2452	2.4581

$t = 10.00, z_l = 90$

$u(r,z,t)$

z	r=0.0	r=0.2	r=0.4	r=0.6	r=0.8	r=1.0
0.0	1.0000	1.0000	1.0000	1.0000	1.0000	1.0000
3.0	0.9969	1.0007	1.0224	1.0915	1.2359	1.4347
6.0	1.0177	1.0329	1.0893	1.2088	1.3934	1.6019
.	.					.
.	.					.
.	.					.
84.0	3.5084	3.5479	3.6619	3.8359	4.0459	4.2584
87.0	3.6084	3.6479	3.7619	3.9359	4.1459	4.3584
90.0	3.7084	3.7479	3.8619	4.0359	4.2459	4.4584

Axial gradient; analytical, steady state, thermally developed solution

uz	0.0333	0.0333	0.0333	0.0333	0.0333	0.0333
ua	3.7084	3.7480	3.8620	4.0360	4.2460	4.4584

Again, we observe agreement in the solutions from DERV1 and DERV2 to about three figures. To improve the agreement between the DERV1 and DERV2 solutions, an increase in the number of spatial grid points, particularly in the z direction, which has the steeper temperature profiles, would probably be required (of course, the number of ODEs would go above 341).

However, the agreement between the numerical and analytical solutions is better than for the short time solutions because the system has approached the steady state more closely; note, for example, the DERV2 solution for $z_l = 90$, for which the derivative $\partial u/\partial z$ is constant to three figures (0.0333) and the numerical and analytical solutions agree to four figures. Also, the agreement between the numerical and analytical solutions for $t = 10(z_l = 90)$ is better for DERV2 than for DERV1, even though in both cases the derivative $\partial u/\partial z$ has reached a constant value to three figures (0.0334 for DERV1 and 0.0333 for DERV2). This suggests that the fourth-order approximations of DERV2 have performed better than the second-order approximations of DERV1, as expected.

Finally, we can conclude from this example that the use of library spatial differentiation routines has the following advantages:

(1) The programming of the PDE(s) is facilitated and simplified (e.g., compare the coding in DERV1 and DERV2); thus, there is less chance of programming errors in using a library routine.

(2) As a corollary to (1), library routines are presumably thoroughly tested and have been used in a spectrum of applications, so that any programming errors most likely have been found and corrected.

(3) Perhaps most importantly, the approximations in library routines presumably have been selected for their effectiveness, for example, higher-order accuracy, upwinding to accommodate convective systems as illustrated in DSS034. Thus, we can take advantage of the experience of others in selecting approximations (and algorithms in general), and this is going to be a more efficient and effective process than learning from first-hand (trial and error) experience the features and limitations of various approximations (and algorithms).

Therefore, based on experience with a variety of PDE problems, we can recommend the use of library spatial differentiators (and ODE integrators such as RKF45). Of course, library routines should not be used "blindly," with little or no understanding of the underlying principles and approximations. The informed use of library routines is essential to their correct and effective use.

References

Eckert, E. R. G., and R. M. Drake, Jr. 1972. *Analysis of Heat and Mass Transfer*. New York: McGraw-Hill.

Schiesser, W. E. 1991. *The Numerical Method of Lines Integration of Partial Differential Equations*. San Diego: Academic Press.

H5

The Graetz problem with constant wall temperature

Consider the following elliptic problem, which is a variant of the "Graetz problem" (Finlayson 1992):

$$P_e v(r) T_z = T_{rr} + \frac{1}{r} T_r + T_{zz}, \qquad v(r) = 2(1 - r^2) \tag{H5.1}$$

where we have used subscript notation to denote a partial derivative, for example, $\partial T / \partial z \leftrightarrow T_z$. Compute a solution to eq. (H5.1) subject to the boundary conditions

$$T_r(0, z) = 0, \tag{H5.2}$$

$$T(1, z) = 1 \tag{H5.3}$$

$$T(r, 0) = 0, \tag{H5.4}$$

$$T_z(r, z_l) = 0 \tag{H5.5}$$

for $P_e = 60$, $z_l = 30$ and 90 (two cases). Compare your solution with

(1) The series solution originally reported by Graetz in 1885, with some subsequent corrections (Jakob 1949, 456),

$$T(r, z) = 1 - 1.477 e^{-3.658(1/Pe)z} R_0(r) - 0.810 e^{-22.178(1/Pe)z} R_1(r)$$
$$+ 0.385 e^{-53.05(1/Pe)z} R_2(r) + \cdots \tag{H5.6}$$

where $R_0(r)$, $R_1(r)$, $R_2(r)$ are tabulated by Jakob (1949, 455, table 22-1) (e.g., for $r = 0$, $R_0(0) = R_1(0) = R_2(0) = \cdots = 1$). Thus, $T(0, z)$ is readily available from eq. (H5.6). In particular, compare your computed value of $T(0, 30)$ with the value from eq. (H5.6).

Note that eq. (H5.6) was derived for the case that the derivative T_{zz} in eq. (H5.1) is negligible compared to the term $P_e v(r) T_z$ (i.e., axial conduction is small compared to axial convection, which is the case for $P_e = 60$). In other words, the term T_{zz}

did not appear in the original Graetz equation (and therefore a second boundary condition such as eq. (H5.5) was not required). A principal objective of the following analysis is to study the effect of the boundary condition at $z = z_l$, which is required by the derivative T_{zz} in eq. (H5.1). Note also that eq. (H5.6) has the implied boundary conditions $T(r, z) \to 1$ and $T_z(r, z) \to 0$ as $z \to 0$.

Additionally, we can check the boundary conditions of eq. (H5.6) at $z = 0$ and $r = 1$:

(1.1) $z = 0$, that is, $T(r, 0)$. For $r = 0$, we have from eq. (H5.6) $T(0, 0) = 1 - 1.477 - 0.810 + 0.385 + \cdots = 1 - 1.052 \simeq 0$ as required by boundary condition (H5.4) (the error of -0.052 is due to the use of only three terms in the series of eq. (H5.6)). For $r = 0.5$, $T(0.5, 0) = 1 - (1.477)(0.6147) - (0.810)(-0.3399) + (0.385)(-0.204) + \cdots = 1 - 0.9079 + 0.2753 - 0.079 = 1 - 1.1042 = -0.1042$, and again an error results from using only three terms in the series of eq. (H5.6) (here we have used $R_0(0.5) = 0.6147$, $R_1(0.5) = -0.3399$, $R_2(0.5) = -0.204$ taken from Jakob (1949, 455, table 22-1). Note, however, that the series converges rapidly for $z > 0$, for example, for $z = 30$, only the first term in the series is significant (due to the rapidly decreasing exponentials in z).

(1.2) $r = 1$, that is, $T(1, z)$. Since $R_0(1) = R_1(1) = R_2(1) = \cdots = 0$ (Jakob 1949, 455, table 22-1), $T(1, z) = 1$ as required by boundary condition (H5.2).

(2) The numerical solution reported by Finlayson (1992, 302–7), $T(0, 30) = 1$, computed by finite elements.

Presumably, from the discussion by Finlayson (1992), boundary condition (H5.5) was used, and therefore the problem discussed by Finlayson is not the same as the original Graetz problem discussed in (1) above (because of the axial conduction term T_{zz} and the associated boundary condition (H5.5), used by Finlayson, but again, this axial conduction term should have little effect for $P_e = 60$).

Solution

One approach to the problem defined by eqs. (H5.1) to (H5.5) is to convert eq. (H5.1) to a time-dependent partial differential equation (PDE), for example

$$T_t = -P_e v(r) T_z + T_{rr} + \frac{1}{r} T_r + T_{zz}, \qquad v(r) = 2(1 - r^2) \tag{H5.7}$$

then integrate in t until $T_t \to 0$. This formulation requires an *initial condition* (which is selected arbitrarily since it should have no effect on the final steady state solution), for example

$$T(r, z, 0) = 0 \tag{H5.8}$$

Also, boundary conditions (H5.2) to (H5.4) become

$$T_r(0, z, t) = 0, \tag{H5.9}$$

$$T(1, z, t) = 1 \qquad (H5.10)$$

$$T(r, 0, t) = 0 \qquad (H5.11)$$

We also consider a generalization of boundary condition (H5.5)

$$T_t(r, z_l, t) + P_e v(r) T_z(r, z_l, t) - T_{rr}(r, z_l, t) - \frac{1}{r} T_r(r, z_l, t) = 0 \qquad (H5.12)$$

with $z_l = 30$ or 90 in eq. (H5.12). Note that eq. (H5.12) is obtained directly from eq. (H5.7) by setting $T_{zz} = 0$. We shall subsequently investigate the effect of dropping various terms in eq. (H5.12) on the numerical solution of eqs. (H5.7) to (H5.12), for example, if the derivatives in t and r are dropped, eq. (H5.12) reverts to eq. (H5.5).

We start with a derivation of the "dynamic (unsteady) Graetz equation", eq. (H5.7). Consider the laminar flow of a Newtonian fluid through a cylindrical tube. An energy balance on an incremental section of the fluid gives (in cylindrical coordinates r, z and θ)

$$\begin{aligned}
r \Delta\theta \Delta r \Delta z \rho C_p T_t &= r\Delta\theta \Delta r v \rho C_p T|_z - r\Delta\theta \Delta r v \rho C_p T|_{z+\Delta z} \\
&\quad - r\Delta\theta \Delta z k T_r|_r - [-(r+\Delta r)\Delta\theta \Delta z k T_r|_{r+\Delta r}] \\
&\quad - r\Delta\theta \Delta r k T_z|_z - [r\Delta\theta \Delta r k T_z|_{z+\Delta z} \\
&\quad - \Delta r \Delta z (1/r) k T_\theta|_\theta - [-\Delta r \Delta z (1/r) k T_\theta|_{\theta+\Delta\theta}]
\end{aligned}$$
$$(H5.13)$$

Here we have assumed:

(1) Convection in the z direction only with velocity v
(2) Conduction in the r, θ, and z directions with an isotropic (direction-independent) conductivity k.

Division of eq. (H5.13) by $r\Delta\theta \Delta r \Delta z$ gives

$$\begin{aligned}
\rho C_p T_t &= -\frac{v\rho C_p T|_{z+\Delta z} - v\rho C_p T|_z}{\Delta z} \\
&\quad + \frac{(r+\Delta r) k T_r|_{r+\Delta r} - -r z k T_r|_r}{r \Delta r} \\
&\quad + \frac{k T_z|_{z+\Delta z} - k T_z|_z}{\Delta z} \\
&\quad + \frac{(1/r^2) k T_\theta|_{\theta+\Delta\theta} - (1/r^2) k T_\theta|_\theta}{\Delta \theta} \qquad (H5.14)
\end{aligned}$$

If we now assume the physical properties are constant (independent of temperature and therefore position), and take the limits $\Delta r \to 0$, $\Delta \theta \to 0$, $\Delta z \to 0$, equation

(H5.14) becomes

$$T_t = -vT_z + \alpha\left[\frac{1}{r}(rT_r)_r + T_{zz} + \frac{1}{r^2}T_{\theta\theta}\right] \quad (H5.15)$$

where $\alpha = \frac{k}{\rho C_p}$ (the thermal diffusivity). Note that we assumed the velocity, v, is a function of r only (i.e., parabolic velocity profile, $v(r) = v_{\text{avg}}(1 - (r/r_o)^2)$).

If angular variations in the temperature are neglected, eq. (H5.15) becomes

$$T_t = -v(r)T_z + \alpha\left(T_{rr} + \frac{1}{r}T_r + T_{zz}\right) \quad (H5.16)$$

where

- v_{avg} average velocity defined by $Q = \pi r_o^2 v_{\text{avg}}$
- r_0 tube radius
- Q volumetric flow rate

Next, we define temperatures $T_0 = T(r, 0, t)$ and $T_1 = T(r_0, z, t)$, and dimensionless variables as

$$t' = \frac{\alpha}{r_0^2}t, \quad (H5.17)$$

$$r' = \frac{r}{r_0}, \quad (H5.18)$$

$$z' = \frac{z}{r_0} \quad (H5.19)$$

$$T' = \frac{(T - T_0)}{(T_1 - T_0)}, \quad (H5.20)$$

$$v'(r') = \frac{v(r)}{v_{\text{avg}}} \quad (H5.21)$$

Substitution of eqs. (H5.17) to (H5.21) in eq. (H5.16) then gives (after dropping the primes for the dimensionless variables)

$$(\alpha/r_0^2)T_t = -v_{\text{avg}}v(r)(1/r_0)T_z + (\alpha/r_0^2)T_{rr} + (\alpha/r_0^2)\frac{1}{r}T_r + (\alpha/r_0^2)T_{zz} \quad (H5.22)$$

or after multiplication of eq. (H5.22) by r_0^2/α, we arrive at eq. (H5.7)

$$T_t = -P_e v(r)T_z + T_{rr} + \frac{1}{r}T_r + T_{zz}$$

where $P_e = v_{\text{avg}}r_0/\alpha$ (the Peclet number). As noted previously, this is the "dynamic (extended) Graetz equation".

Equation (H5.7) requires one initial condition, two boundary conditions in r, and two boundary conditions in z. We take for the initial condition eq. (H5.8)

$$T(r, z, 0) = 0$$

which in terms of the dimensional temperature is $T(r, z, 0) = T_0$ from eq. (H5.20). The boundary conditions in r are eqs. (H5.9) and (H5.10)

$$T_r(0, z, t) = 0, \qquad T(1, z, t) = 1$$

which express symmetry at the centerline, $r = 0$, and the dimensional temperature $T(r_0, z, t) = T_1$ at $r = r_0$ used in eq. (H5.20). Note that eqs. (H5.9) and (H5.10) are the dynamic counterparts of eqs. (H5.2) and (H5.3).

Finally, for the first boundary condition in z, we use eq. (H5.11)

$$T(r, 0, t) = 0$$

which specifies the dimensional temperature T_0 at $z = 0$ used in eq. (H5.20) and is the dynamic counterpart of eq. (H5.4). For the second boundary condition in z, we could consider the dynamic counterpart of eq. (H5.5)

$$T_z(r, z_l, t) = 0 \tag{H5.23}$$

where z_l is the dimensionless value of z where the boundary condition is applied (e.g., $z_l = 30$). The problem with eq. (H5.23) is that the solution must maintain a zero slope in z for all r and t at $z = z_l$, which causes significant problems in computing a numerical solution (and which is probably not physically realistic). Therefore, we consider the alternative boundary condition suggested by Douglas (1984), eq. (H5.12)

$$T_t(r, z_l, t) + P_e v(r) T_z(r, z_l, t) - T_{rr}(r, z_l, t) - \frac{1}{r} T_r(r, z_l, t) = 0$$

We can note the following points about eq. (H5.12), which can be called an "outflow" boundary condition since it is applied at a downstream value of z:

(1) Equation (H5.12) is eq. (H5.7) with the derivative T_{zz} removed. In other words, eq. (H5.12) is first order in z, and it can therefore be used as a boundary condition (at $z = z_l$) for eq. (H5.7), which is second order in z.

(2) Equation (H5.12) states that at the boundary $z = z_l$ all of the heat transfer processes are taking place that are reflected in eq. (H5.7), except conduction in the z direction.

(3) At steady state, for which $T_t(r, z_l, t) \to 0$, eq. (H5.12) does not reduce to eq. (H5.5), but rather, includes a radial conduction term

$$T_{rr}(r, z_l, t) + \frac{1}{r} T_r(r, z_l, t)$$

as well as the axial convection derivative

$$T_z(r, z_l, t)$$

of eq. (H5.5). Therefore, the use of eq. (H5.12) at steady state does not correspond exactly to the steady state problem of eqs. (H5.1) to (H5.5) (because of the additional radial conduction term). The effect of this additional radial conduction term on the numerical solution can be investigated by setting this term to zero when computing the numerical solution.

(4) If the axial conduction term, T_{zz}, is dropped from eqs. (H5.1) and (H5.7), these equations become first order in z (corresponding to the problem for which the series solution (H5.6) was derived). Then a second boundary condition in z is not required, and eq. (H5.12) is just a statement of the resulting dynamic PDE, eq. (H5.7), at $z = z_l$. If axial convection is much larger than axial conduction (which is the case for $P_e = 60$), this special case problem (with $T_{zz} = 0$) should have a solution that is close to the solution of the original problems (with the T_{zz} term) of eqs. (H5.1) to (H5.5) and eqs. (H5.7) to (H5.12).

Finally, in keeping with the usual convention of the numerical analysis literature, we denote the PDE dependent variable, the dimensionless temperature of eq. (H5.20), as u rather than T, that is,

$$u = T = \frac{T - T_0}{T_1 - T_0}$$

Thus, in the subsequent discussion of the analytical and numerical solutions for eqs. (H5.1) to (H5.5), we denote the dependent variable as u. Also, the subsequent coding is in terms of U (for $u(r, z, t)$).

We start the discussion of the solution of eqs. (H5.7) to (H5.12) with a method of lines (MOL) code, consisting of (a) subroutine INITAL to define initial condition (H5.8); (b) subroutine DERV to define the PDE, eq. (H5.7) and its boundary conditions, eq. (H5.9) to (H5.12); and (c) subroutine PRINT to print and plot the numerical solution. Also, in order to facilitate the analysis of variants on boundary condition (H5.12), subroutine DERV in turn calls subroutines DERV1 to DERV8 (each of which has alternative programming of eqs. (H5.7) and (H5.9) to (H5.12)). Finally, the series solution of eq. (H5.6) is programmed in subroutine SERIES, which is discussed subsequently. We start with subroutine INITAL

```
      SUBROUTINE INITAL
      IMPLICIT DOUBLE PRECISION (A-H,O-Z)
      PARAMETER (NR=11,NZ=31)
      COMMON/T/          T,       NSTOP,      NORUN
     +       /Y/   U(NR,NZ)
     +       /F/   UT(NR,NZ)
     +       /G/        R(NR),    Z(NZ),      V(NR)
     +       /C/          PE,        RO,        ZL,
     +                    DR,       DRS,        DZ,       DZS
     +       /I/          IP,      NCASE
```

```
C...
C...    SELECT APPROXIMATION OF SPATIAL DERIVATIVES (SEE THE
C...    COMMENTS IN SUBROUTINE DERV FOR AN EXPLANATION OF THE
C...    EIGHT CASES)
        NCASE=1
C...
C...    PROBLEM PARAMETERS
        PE=60.0D0
        R0= 1.0D0
        IF(NORUN.EQ.1)ZL=30.0D0
        IF(NORUN.EQ.2)ZL=90.0D0
C...
C...    RADIAL GRID, VELOCITY PROFILE
        DR=R0/DFLOAT(NR-1)
        DO 1 IR=1,NR
           R(IR)=DFLOAT(IR-1)*DR
           V(IR)=2.0D0*(1.0D0-R(IR)**2)
1       CONTINUE
        DRS=DR**2
C...
C...    AXIAL GRID
        DZ=ZL/DFLOAT(NZ-1)
        DO 2 JZ=1,NZ
           Z(JZ)=DFLOAT(JZ-1)*DZ
2       CONTINUE
        DZS=DZ**2
C...
C...    INITIAL CONDITION
        DO 3 IR=1,NR
        DO 3 JZ=1,NZ
           U(IR,JZ)=0.0D0
3       CONTINUE
C...
C...    GIVE THE TIME DERIVATIVES A VALUE WHICH CAN THEN BE
C...    USED TO TEST FOR UNDEFINED DERIVATIVES BY SETTING THEM
C...    TO PI, THEN CALLING DERV, AND CHECKING IN SUBROUTINE
C...    PRINT IF THEY  HAVE ALL BEEN SET TO VALUES OTHER THAN
C...    PI. THIS IS A SIMPLE  PROCEDURE TO ENSURE THAT ALL NR
C...    X NZ DERIVATIVES HAVE BEEN SET IN DERV
        PI=3.1415927D0
        DO 4 IR=1,NR
        DO 4 JZ=1,NZ
           UT(IR,JZ)=PI
4       CONTINUE
C...
C...    INITIAL DERIVATIVES
```

```
      CALL DERV
      IP=0
      RETURN
      END
```

We can note the following points about subroutine INITAL:

(1) Double precision coding is used. A two-dimensional grid is defined with 11 points in r and 31 points in z for a total of $11 \times 31 = 341$ grid points. This choice of the number of 4 grid points is somewhat arbitrary; it is a balance between acceptable spatial resolution and a reasonable number of ordinary differential equations (ODEs) to be integrated (341). Of course, we could check the adequacy of the number of grid points (i.e., the spatial convergence) by repeating the calculation of the solution with a larger number of grid points and comparing the two solutions.

```
      IMPLICIT DOUBLE PRECISION (A-H,O-Z)
      PARAMETER (NR=11,NZ=31)
      COMMON/T/          T,        NSTOP,       NORUN
     +      /Y/    U(NR,NZ)
     +      /F/    UT(NR,NZ)
     +      /G/        R(NR),      Z(NZ),       V(NR)
     +      /C/           PE,         RO,          ZL,
     +                    DR,        DRS,          DZ,        DZS
     +      /I/           IP,      NCASE
```

(2) The case is then selected from a total of eight

```
C...
C... SELECT APPROXIMATION OF SPATIAL DERIVATIVES (SEE THE
C... COMMENTS IN SUBROUTINE DERV FOR AN EXPLANATION OF THE
C... EIGHT CASES)
      NCASE=1
```

The eight cases are summarized below and implemented in subroutines DERV1 to DERV8 discussed subsequently:

Case Description:

  (i) Equations (H5.7), (H5.8), (H5.9), (H5.10), (H5.11), and (H5.23) by explicit finite differences
  (ii) Equations (H5.7), (H5.8), (H5.9), (H5.10), (H5.11), and (H5.12) by explicit finite differences without the radial group $u_{rr} + \frac{1}{r}u_r$ (again, u is used in place of T) in eq. (H5.12)
  (iii) Equations (H5.7), (H5.8), (H5.9), (H5.10), (H5.11), and (H5.12) by explicit finite differences
  (iv) Equations (H5.7), (H5.8), (H5.9), (H5.10), and (H5.11) by explicit finite differences without the u_{zz} term in eq. (H5.7) (so that eq. (H5.12) is not required)

(v) Equations (H5.7), (H5.8), (H5.9), (H5.10), (H5.11), and (H5.23) by subroutine DSS034
(vi) Equations (H5.7), (H5.8), (H5.9), (H5.10), (H5.11), and (H5.12) by subroutine DSS034 without the radial group $u_{rr} + \frac{1}{r}u_r$ in eq. (H5.12)
(vii) Equations (H5.7), (H5.8), (H5.9), (H5.10), (H5.11), and (H5.12) by subroutine DSS034
(viii) Equations (H5.7), (H5.8), (H5.9), (H5.10), and (H5.11) by subroutine DSS034 without the u_{zz} term in eq. (H5.7) (so that eq. (H5.12) is not required)

Explicit finite differences indicates that second-order finite differences to approximate the spatial derivatives (in r and z) in eq. (H5.7) are programmed explicitly in the derivative subroutines, DERV1 to DERV4 (rather than calling, for example, library routines to compute the finite differences).

Differentiation in Space Subroutine DSS034 is called from the derivative subroutines DERV5 to DERV8 to compute the spatial derivatives (in r and z) in eq. (H5.7). DSS034 has fourth-order correct finite difference approximations, both centered for derivatives that model diffusion, and five-point, biased upwind for derivatives that model convection. DSS034 is described in detail in Schiesser (1991, 262–73).

Note that DERV1 to DERV4 are for the same series of four problems as DERV5 to DERV8, respectively. The only difference between the two sets of derivative subroutines is the calculation of the spatial derivatives in eq. (H5.7) (second-order finite differences in DERV1 to DERV4, and fourth-order finite differences in DSS034 called from DERV5 to DERV8). The four problems differ only in the way that the outflow boundary condition at $z = z_l$ is accommodated (which is required by the derivative u_{zz} in eq. (H5.7)).

(3) The problem parameters are defined next

```
C...
C...    PROBLEM PARAMETERS
        PE=60.0D0
        RO= 1.0D0
        IF(NORUN.EQ.1)ZL=30.0D0
        IF(NORUN.EQ.2)ZL=90.0D0
```

In this case, the Peclet number, P_e, is 60; the dimensionless radius defined by eq. (H5.18), r is 1; and the dimensionless length defined by eq. (H5.19), z, is 30 and 90 for two runs of the program.

(4) The radial grid based on NR = 11 points and the corresponding velocity profile are computed and stored in arrays R and V, respectively, for use in the derivative subroutines

```
C...
C...    RADIAL GRID, VELOCITY PROFILE
```

```
      DR=R0/DFLOAT(NR-1)
      DO 1 IR=1,NR
         R(IR)=DFLOAT(IR-1)*DR
         V(IR)=2.0D0*(1.0D0-R(IR)**2)
1     CONTINUE
      DRS=DR**2
```

Note that $v(r)$ in array V is computed according to the parabolic velocity profile of eq. (H5.1).

(5) The axial grid is computed next based on NZ = 31 points

```
C...
C...  AXIAL GRID
      DZ=ZL/DFLOAT(NZ-1)
      DO 2 JZ=1,NZ
         Z(JZ)=DFLOAT(JZ-1)*DZ
2     CONTINUE
      DZS=DZ**2
```

(6) Initial condition (H5.8) is then set over the NR × NZ grid

```
C...
C...  INITIAL CONDITION
      DO 3 IR=1,NR
      DO 3 JZ=1,NZ
         U(IR,JZ)=0.0D0
3     CONTINUE
```

(7) NR × NZ (= 341) temporal derivatives in the array UT(NR, NZ) (in COMMON /F/) must be computed in each of the eight derivative subroutines DERV1 to DERV8, and there is the possibility of errors in the programming of these derivatives (e.g., one or more of the 341 derivatives might be overlooked). Therefore, to check that an error has not been made, an initial value of π (PI) is given to each of the derivatives. This value should not appear for any of the derivatives after they have been computed by a call to DERV (which in turn calls DERV1 to DERV8, depending on the value of NCASE). Of course, this presupposes that the correct value of any of the 341 initial derivatives is not PI; also this test does not necessarily guarantee that the computed values are correct, although unrealistic values resulting from programming errors might be observed

```
C...
C...  GIVE THE TIME DERIVATIVES A VALUE WHICH CAN THEN BE
C...  USED TO TEST FOR UNDEFINED DERIVATIVES BY SETTING THEM
C...  TO PI, THEN CALLING DERV, AND CHECKING IN SUBROUTINE
C...  PRINT IF THEY HAVE ALL BEEN SET TO VALUES OTHER THAN
C...  PI. THIS IS A SIMPLE PROCEDURE TO ENSURE THAT ALL NR X
```

The Graetz problem with constant wall temperature

```
C...    NZ DERIVATIVES HAVE BEEN SET IN DERV
        PI=3.1415927D0
        DO 4 IR=1,NR
        DO 4 JZ=1,NZ
           UT(IR,JZ)=PI
4       CONTINUE
C...
C...    INITIAL DERIVATIVES
        CALL DERV
        IP=0
```

The call to DERV then computes the 341 temporal derivatives in UT(NR, NZ), and therefore all 341 values of the temporal derivative vector should be overwritten. These initial temporal derivatives are then printed by a call to PRINT to check that they have, in fact, all been computed (that no values of π appear for the initial derivatives). The utility of this check can perhaps best be appreciated when considering the coding in subroutines DERV1, to DERV8, to be discussed subsequently.

Subroutine DERV, which is the derivative subroutine called by the ODE integrator, RKF45 in this case, is listed below:

```
        SUBROUTINE DERV
        IMPLICIT DOUBLE PRECISION (A-H,O-Z)
        COMMON/I/       IP,       NCASE
C...
C...    APPROXIMATIONS OF SPATIAL DERIVATIVES
C...
C...       NCASE = 1 - EXPLICIT FINITE DIFFERENCES (FDM) WITH
C...                   UZZ, UZ(R,ZL,T) = 0
           IF(NCASE.EQ.1)THEN
              CALL DERV1
C...
C...       NCASE = 2 - EXPLICIT FINITE DIFFERENCES WITH UZZ,
C...                   UT(R,ZL,T) = -PE*V(R)*UZ(R,ZL,T)
           ELSE
     +     IF(NCASE.EQ.2)THEN
              CALL DERV2
C...
C...       NCASE = 3 - EXPLICIT FINITE DIFFERENCES WITH UZZ,
C...                   UT(R,ZL,T) = -PE*V(R)*UZ(R,ZL,T) +
C...                   URR(R,ZL,T) + (1/R)*UR(R,ZL,T)
           ELSE
     +     IF(NCASE.EQ.3)THEN
              CALL DERV3
C...
C...       NCASE = 4 - EXPLICIT FINITE DIFFERENCES WITHOUT UZZ
```

```
              ELSE
    +         IF(NCASE.EQ.4)THEN
                 CALL DERV4
C...
C...          NCASE = 5 - FINITE DIFFERENCES FROM DSS034 WITH UZZ,
C...                      UZ(R,ZL,T) = 0
              ELSE
    +         IF(NCASE.EQ.5)THEN
                 CALL DERV5
C...
C...          NCASE = 6 - FINITE DIFFERENCES FROM DSS034 WITH UZZ,
C...                      UT(R,ZL,T) = -PE*V(R)*UZ(R,ZL,T)
              ELSE
    +         IF(NCASE.EQ.6)THEN
                 CALL DERV6
C...
C...          NCASE = 7 - FINITE DIFFERENCES FROM DSS034 WITH UZZ,
C...                      UT(R,ZL,T) = -PE*V(R)*UZ(R,ZL,T) +
C...                      URR(R,ZL,T) + (1/R)*UR(R,ZL,T)
              ELSE
    +         IF(NCASE.EQ.7)THEN
                 CALL DERV7
C...
C...          NCASE = 8 - FINITE DIFFERENCES FROM DSS034 WITHOUT UZZ
              ELSE
    +         IF(NCASE.EQ.8)THEN
                 CALL DERV8
              END IF
           RETURN
           END
```

Note that DERV merely calls one of the eight derivative subroutines, DERV1 to DERV8, depending on the value of NCASE (recall again that NCASE is set in INITAL).

We now consider each of the eight derivative subroutines to illustrate

(1) The explicit programming of the finite difference approximations for the spatial derivatives in eqs. (H5.1) and (H5.7).

(2) The implementation of alternative outflow boundary conditions at $z = z_l$.

(3) The use of a library subroutine, DSS034 in this case, for the calculation of the spatial derivatives in eqs. (H5.1) and (H5.7).

We start with subroutine DERV1:

```
           SUBROUTINE DERV1
           IMPLICIT DOUBLE PRECISION (A-H,O-Z)
           PARAMETER (NR=11,NZ=31)
```

```
              COMMON/T/            T,        NSTOP,       NORUN
     +        /Y/       U(NR,NZ)
     +        /F/       UT(NR,NZ)
     +        /G/          R(NR),    Z(NZ),       V(NR)
     +        /C/          PE,       RO,          ZL,
     +                     DR,       DRS,         DZ,       DZS
     +        /I/          IP,       NCASE
C...
C...  Z = 0
      DO 1 IR=1,NR
          U(IR,1)=0.0D0
          UT(IR,1)=0.0D0
1     CONTINUE
C...
C...  R = RO, Z NE 0
      DO 2 JZ=2,NZ
          U(NR,JZ)=1.0D0
          UT(NR,JZ)=0.0D0
2     CONTINUE
C...
C...  R = 0, Z NE 0, ZL
      DO 3 JZ=2,NZ-1
          UT(1,JZ)=-PE*V(1)*(U(1,JZ)-U(1,JZ-1))/DZ
     +             +4.0D0*(U(2,JZ)-U(1,JZ))/DRS
     +             +(U(1,JZ+1)-2.0D0*U(1,JZ)+U(1,JZ-1))/DZS
3     CONTINUE
C...
C...  R = 0, Z = ZL
          UT(1,NZ)=-PE*V(1)*0.0D0
     +             +4.0D0*(U(2,NZ)   -U(1,NZ))/DRS
     +             +2.0D0*(U(1,NZ-1)-U(1,NZ))/DZS
C...
C...  Z = ZL, R NE 0, RO
      DO 4 IR=2,NR-1
          UT(IR,NZ)=-PE*V(IR)*0.0D0
     +              +(U(IR+1,NZ)-2.0D0*U(IR,NZ)+U(IR-1,NZ))/
     +              DRS+(1.0D0/R(IR))*(U(IR+1,NZ)-U(IR-1,NZ))
     +              /(2.0D0*DR)+2.0D0*(U(IR,NZ-1)-U(IR,NZ))
     +              /DZS
4     CONTINUE
C...
C...  R NE 0, RO, Z NE 0, ZL
      DO 5 JZ=2,NZ-1
      DO 5 IR=2,NR-1
          UT(IR,JZ)=-PE*V(IR)*(U(IR,JZ)-U(IR,JZ-1))/DZ
     +              +(U(IR+1,JZ)-2.0D0*U(IR,JZ)+U(IR-1,JZ))/
```

220   Heat transport

```
      +                 DRS+(1.0D0/R(IR))*(U(IR+1,JZ)-U(IR-1,JZ))
      +                 /(2.0D0*DR)+(U(IR,JZ+1)-2.0D0*U(IR,JZ)+U
                        (IR,JZ-1))/DZS
5     CONTINUE
      RETURN
      END
```

We can note the following points about subroutine DERV1:

(1) The COMMON area is the same as in subroutine INITAL.

(2) Boundary condition (H5.11) is implemented first

```
C...
C...  Z = 0
      DO 1 IR=1,NR
         U(IR,1)=0.0D0
         UT(IR,1)=0.0D0
1     CONTINUE
```

Note that the second subscript in U(IR, IZ) corresponds to $z = 0$. Also, the temporal derivatives, UT(IR, 1), are set to zero to maintain the zero boundary value.

(3) The boundary condition at $r = 1$, eq. (H5.10), is then implemented

```
C...
C...  R = R0, Z NE 0
      DO 2 JZ=2,NZ
         U(NR,JZ)=1.0D0
         UT(NR,JZ)=0.0D0
2     CONTINUE
```

Note that $u(r = 1, z = 0, t) = 0$ (from DO loop 1), since the initial value of the index of DO loop 2 is JZ = 2. Alternatively, we could have used $u(r = 1, z = 0, t) = 1$ in accordance with eq. (H5.10) (by starting the index in DO loop 2 at JZ = 1) or even $u(r = 1, z = 0, t) = 0.5$ (the average of the values specified by boundary conditions (H5.10) and (H5.11)). The change in the numerical solution from the use of either of these alternative boundary conditions would probably be negligible (although this was not actually checked). This is the case because parabolic PDEs such as eq. (H5.1) (which is parabolic in r) usually damp out discontinuities at boundaries (see Schiesser 1991, 49–54 for further discussion of this point). This point also illustrates how easily a grid point can be inadvertently skipped, which is some of the motivation for the check using π in INITAL, as discussed previously.

(4) Equation (H5.7) is then programmed at $r = 0$, $z \neq 0$ or $z \neq z_l$ (which therefore includes boundary condition (H5.9))

```
C...
C...  R = 0, Z NE 0, ZL
```

```
      DO 3 JZ=2,NZ-1
         UT(1,JZ)=-PE*V(1)*(U(1,JZ)-U(1,JZ-1))/DZ
     +              +4.0D0*(U(2,JZ)-U(1,JZ))/DRS
     +              +(U(1,JZ+1)-2.0D0*U(1,JZ)+U(1,JZ-1))/DZS
3     CONTINUE
```

The coding in DO loop 3 requires some further explanation:
(4.1) The convective term

$$-P_e v(0) u_z$$

in eq. (H5.7) is first computed as

```
-PE*V(1)*(U(1,JZ)-U(1,JZ-1))/DZ
```

Note in particular the finite difference approximation of u_z based on the grid spacing DZ, that is

```
(U(1,JZ)-U(1,JZ-1))/DZ
```

which is an implementation of

$$u_z(0, z, t) \approx \frac{u(0, z, t) - u(0, z - \Delta z, t)}{\Delta z}$$

In general, a centered approximation for a convective derivative such as

$$u_z(r, z, t)$$

should be avoided for a strongly convective problem (Schiesser 1991, 122–41); thus, we use the two-point upwind (backward) difference of u

$$u_z(r, z, t) \approx \frac{u(r, z, t) - u(r, z - \Delta z, t)}{\Delta z} + O(\Delta z)$$

and not the centered difference

$$u_z(r, z, t) \approx \frac{u(r, z + \Delta z, t) - u(r, z - \Delta z, t)}{\Delta z} + O(\Delta z^2)$$

even though the centered approximation is second-order correct, while the upwind approximation is only first-order correct. (This is an example of where the order of the approximation is not the only factor in determining the best approximation to use; in this case, the centered approximation would probably produce unrealistic oscillations in the solution.)

(4.2) The radial conduction group

$$\frac{1}{r}(r u_r)_r$$

is approximated at $r = 0$ as

```
+                +4.0D0*(U(2,JZ)-U(1,JZ))/DRS
```

By differentiation of the product, this group expands to

$$u_{rr} + \frac{1}{r}u_r$$

The second term is indeterminate at $r = 0$ as a consequence of boundary condition (H5.9)

$$\lim_{r \to 0} \frac{1}{r} u_r(r, z, t) = \frac{0}{0}$$

Application of l'Hospital's rule to this term (differentiating the numerator, u_r, and denominator, r, with respect to r) gives

$$\lim_{r \to 0} \frac{1}{r} u_r(0, z, t) = \lim_{r \to 0} u_{rr}(r, z, t) = u_{rr}(0, z, t)$$

and therefore the radial group in eq. (H5.7) at $r = 0$ becomes

$$u_{rr} + \frac{1}{r}u_r = 2u_{rr}$$

The second derivative is then approximated at $r = 0$ as

$$2u_{rr} \approx 2 \frac{u(\Delta r, z, t) - 2u(0, z, t) + u(-\Delta r, z, t)}{\Delta r^2} = 4 \frac{u(\Delta r, z, t) - u(0, z, t)}{\Delta r^2}$$

The second step follows from the approximation of boundary condition (H5.9) as

$$u_r(0, z, t) \approx \frac{u(\Delta r, z, t) - u(-\Delta r, z, t)}{2\Delta r} = 0$$

or

$$u(-\Delta r, z, t) = u(\Delta r, z, t)$$

Finally, from the preceding result, the Fortran programming for the radial group at $r = 0$ is

```
+                +4.0D0*(U(2,JZ)-U(1,JZ))/DRS
```

(4.3) The axial conduction derivative in eq. (H5.7), u_{zz}, is approximated at $r = 0$ as

```
+                +(U(1,JZ+1)-2.0D0*U(1,JZ)+U(1,JZ-1))/DZS
```

which is an implementation of

$$u_{zz}(0, z, t) \approx \frac{u(0, z + \Delta z, t) - 2u(0, r, t) + u(0, z - \Delta z, t)}{\Delta r^2}$$

(5) Equation (H5.7) is programmed next at $r = 0, z = z_l$, which therefore includes boundary conditions (H5.9) and (H5.23))

```
C...
C...    R = 0, Z = ZL
        UT(1,NZ)=-PE*V(1)*0.0D0
     +              +4.0D0*(U(2,NZ)   -U(1,NZ))/DRS
     +              +2.0D0*(U(1,NZ-1)-U(1,NZ))/DZS
```

Note the temporal derivative (UT(1, NZ)) is computed at $r = 0, z = z_l$ (and thus the indices of 1 and NZ for r and z, respectively).

(5.1) The convective term

$$-P_e v(0) u_z$$

is programmed according to boundary condition eq. (H5.23) as

```
        -PE*V(1)*0.0D0
```

(5.2) The radial conduction group

$$u_{rr} + \frac{1}{r} u_r$$

is approximated at $r = 0$ and $z = z_l$ as

```
     +              +4.0D0*(U(2,NZ)   -U(1,NZ))/DRS
```

(note the indices for r and z), which follows directly from the preceding discussion of (4.2) (and includes boundary condition eq. (H5.9)).

(5.3) The axial conduction term

$$u_{zz}$$

is approximated at $r = 0$ and $z = z_l$ as

```
     +              +2.0D0*(U(1,NZ-1)-U(1,NZ))/DZS
```

which follows from the approximations

$$u_{zz}(0, z_l, t) \approx \frac{u(0, z_l + \Delta z, t) - 2u(0, z_l, t) + u(0, z_l - \Delta z, t)}{\Delta z^2}$$

$$\approx 2 \frac{u(0, z_l - \Delta z, t) - u(0, z_l, t)}{\Delta z^2}$$

This last step follows from the approximation of boundary condition (H5.23) as

$$u_z(0, z_l, t) \approx \frac{u(0, z_l + \Delta z, t) - u(0, z_l - \Delta z, t)}{2\Delta z} = 0$$

or

$$u(0, z_l + \Delta z, t) = u(0, z_l - \Delta z, t)$$

(6) Equation (H5.7) is programmed next at $r \neq 0$ or 1, $z = z_l$ (which therefore includes boundary condition (H5.23))

```
C...
C...    Z = ZL, R NE 0, R0
        DO 4 IR=2,NR-1
          UT(IR,NZ)=-PE*V(IR)*0.0D0
     +              +(U(IR+1,NZ)-2.0D0*U(IR,NZ)
     +              +U(IR-1,NZ))/DRS+(1.0D0/R(IR))*(U(IR+1,NZ)
     +              -U(IR-1,NZ))/(2.0D0*DR)+2.0D0*(U(IR,NZ-1)
     +              -U(IR,NZ))/DZS
4       CONTINUE
```

(6.1) The convective term

$$-P_e v(r) u_r$$

is programmed according to boundary condition eq. (H5.23) as

```
          -PE*V(IR)*0.0D0
```

(6.2) The radial conduction group

$$u_{rr} + \frac{1}{r} u_r$$

is approximated at $z = z_l$ as

```
     +        +(U(IR+1,NZ)-2.0D0*U(IR,NZ)+U(IR-1,NZ))/DRS
     +        +(1.0D0/R(IR))*(U(IR+1,NZ)-U(IR-1,NZ))/(2.0D0*DR)
```

which is an implementation of

$$u_{rr} + \frac{1}{r} u_r \approx \frac{u(r + \Delta r, z_l, t) - 2u(r, z_l, t) + u(r - \Delta r, z_l, t)}{\Delta r^2}$$
$$+ \frac{1}{r} \frac{u(r + \Delta r, z_l, t) - u(r - \Delta r, z_l, t)}{2\Delta r}$$

(6.3) The axial conduction term

$$u_{zz}$$

is approximated at $z = z_l$ as

```
     +             +2.0D0*(U(IR,NZ-1)-U(IR,NZ))/DZS
```

which is an implementation of

$$u_{zz}(r, z_l, t) \approx \frac{u(r, z_l + \Delta z, t) - 2u(r, z_l, t) + u(r, z_l - \Delta z, t)}{\Delta z^2}$$
$$\approx 2 \frac{u(r, z_l - \Delta z, t) - u(r, z_l, t)}{\Delta z^2}$$

This last step follows from the approximation of boundary condition (H5.23) as

$$u_z(r, z_l, t) \approx \frac{u(r, z_l + \Delta z, t) - u(r, z_l - \Delta z, t)}{2\Delta z} = 0$$

or

$$u(r, z_l + \Delta z, t) = u(r, z_l - \Delta z, t)$$

(7) Finally, eq. (H5.7) is programmed at the interior points (so that boundary conditions do not have to be included)

```
C...
C...    R NE 0, R0, Z NE 0, ZL
        DO 5 JZ=2,NZ-1
        DO 5 IR=2,NR-1
          UT(IR,JZ)=-PE*V(IR)*(U(IR,JZ)-U(IR,JZ-1))/DZ
     +              +(U(IR+1,JZ)-2.0D0*U(IR,JZ)
     +              +U(IR-1,JZ))/DRS+(1.0D0/R(IR))
     +              *(U(IR+1,JZ)-U(IR-1,JZ))/(2.0D0*DR)
     +              +(U(IR,JZ+1)-2.0D0*U(IR,JZ)
     +              +U(IR,JZ-1))/DZS
5       CONTINUE
```

(7.1) The convective term

$$-P_e v(r) u_z$$

is programmed as

```
-PE*V(IR)*(U(IR,JZ)-U(IR,JZ-1))/DZ
```

which is an implementation of

$$-P_e v(r) u_z \approx -P_e v(r) \frac{u(r, z, t) - u(r, z - \Delta z, t)}{\Delta z}$$

that is, a two-point, upwind approximation is used rather than a centered difference, as discussed previously in (4.1).

(7.2) The radial conduction group

$$\frac{1}{r}(r u_r)_r = u_{rr} + \frac{1}{r} u_r$$

is approximated as

```
     +    +(U(IR+1,JZ)-2.0D0*U(IR,JZ)+U(IR-1,JZ))/DRS
     +    +(1.0D0/R(IR))*(U(IR+1,JZ)-U(IR-1,JZ))/(2.0D0*DR)
```

which is an implementation of

$$u_{rr} + \frac{1}{r} u_r \approx \frac{u(r + \Delta r, z, t) - 2u(r, z, t) + u(r - \Delta r, z, t)}{\Delta r^2}$$
$$+ \frac{1}{r} \frac{u(r + \Delta r, z, t) - u(r - \Delta r, z, t)}{2\Delta r}$$

(7.3) The axial conduction term

$$u_{zz}$$

is approximated at $z \neq 0$ and $z \neq z_l$ as

```
    +                    +(U(IR,JZ+1)-2.0D0*U(IR,JZ)+U(IR,JZ-1))/DZS
```

which is an implementation of

$$u_{zz}(r, z, t) \approx \frac{u(r, z + \Delta z, t) - 2u(r, z, t) + u(r, z - \Delta z, t)}{\Delta z^2}$$

The required attention to detail in the programming of subroutine DERV1 suggests that the use of a library routine to calculate the various spatial derivatives in eq. (H5.7), with their associated boundary conditions, might facilitate the programming of PDEs. This point will be considered subsequently when subroutines DERV5 to DERV8, are considered.

Subroutine PRINT is listed below:

```
      SUBROUTINE PRINT(NI,NO)
      IMPLICIT DOUBLE PRECISION (A-H,O-Z)
      PARAMETER (NR=11,NZ=31)
      COMMON/T/          T,       NSTOP,      NORUN
     +      /Y/    U(NR,NZ)
     +      /F/    UT(NR,NZ)
     +      /G/        R(NR),     Z(NZ),      V(NR)
     +      /C/          PE,        RO,         ZL,
     +                   DR,       DRS,         DZ,        DZS
     +      /I/          IP,     NCASE
C...
C...  MONITOR SOLUTION
      WRITE(*,*)NCASE,NORUN,T
C...
C...  PRINT THE INITIAL CONDITION AND INITIAL TEMPORAL
C...  DERIVATIVE
      IF(IP.EQ.0)THEN
         WRITE(NO,10)T,((U(IR,JZ),IR=1,NR),JZ=1,NZ)
10       FORMAT(' t = ',F6.2,//,'   u(r,z,t)',/,(11F6.3))
         WRITE(NO,11)(( UT(IR,JZ),IR=1,NR),JZ=1,NZ)
11       FORMAT(//,'  ut(r,z,t)',/,(11F6.1))
C...
C...  CHECK FOR UNDEFINED DERIVATIVES
      DO 12 IR=1,NR
      DO 12 JZ=1,NZ
      IF((UT(IR,JZ).GT.3.14D0).AND.(UT(IR,JZ).LT.3.15D0))
     +   THEN
         WRITE (NO,13)IR,JZ,U(IR,JZ),UT(IR,JZ)
13       FORMAT(' IR = ',I3,'   JZ = ',I3,
```

The Graetz problem with constant wall temperature 227

```
      +    '   U(IR,JZ) = ',F7.3,'    UT(IR,JZ) = ',F7.1)
           END IF
12      CONTINUE
        END IF
C...
C...    PRINT THE NUMERICAL SOLUTION EVERY FIFTH CALL TO PRINT
        IP=IP+1
        IF(((IP-1)/5*5).EQ.(IP-1))THEN
           WRITE(NO,1)T
1          FORMAT(//,' t = ',F6.2,//,5X,'u(r,z,t)',//,
      +      4X,'z',4X,'r = 0.0',3X,'r = 0.2',3X,'r = 0.4',
      +            3X,'r = 0.6',3X,'r = 0.8',3X,'r = 1.0')
           DO 3 JZ=1,NZ
              NRI=2
              WRITE(NO,2)Z(JZ),(U(IR,JZ),IR=1,NR,NRI)
2             FORMAT(F5.1,6F10.4)
3          CONTINUE
        END IF
C...
C...    OPEN FILE FOR MATLAB PLOTTING
        IF((NORUN.EQ.1).AND.(T.LT.0.001D0))THEN
           OPEN(1,FILE='h5a.out')
           OPEN(2,FILE='h5b.out')
        END IF
C...
C...    WRITE NUMERICAL SOLUTION AT R = 0 FOR MATLAB PLOTTING
        DO 4 JZ=1,NZ
           IF(NORUN.EQ.1)WRITE(1,5)Z(JZ),U(1,JZ)
           IF(NORUN.EQ.2)WRITE(2,5)Z(JZ),U(1,JZ)
5          FORMAT(2F10.4)
4       CONTINUE
        RETURN
        END
```

We can note the following points about subroutine PRINT:

(1) The COMMON area is the same as in subroutines INITAL and DERV1. First, the progress of the solution is monitored by writing NCASE, NORUN, and T to the screen (via unit *). Then, the initial condition of eq. (H5.8) and the initial temporal derivative vector (computed by the call to DERV in INITAL) are printed at the beginning of the solution to confirm the initial condition and observe if the 341 initial derivatives have reasonable values

```
C...
C...    MONITOR SOLUTION
        WRITE(*,*)NCASE,NORUN,T
C...
```

```
C...   PRINT THE INITIAL CONDITION AND INITIAL TEMPORAL
C...   DERIVATIVE
       IF(IP.EQ.0)THEN
          WRITE(NO,10)T,((U(IR,JZ),IR=1,NR),JZ=1,NZ)
10        FORMAT(' t = ',F6.2,//,'    u(r,z,t)',/,(11F6.3))
          WRITE(NO,11)(( UT(IR,JZ),IR=1,NR),JZ=1,NZ)
11        FORMAT(//,'   ut(r,z,t)',/,(11F6.1))
```

(2) A check is then performed to ensure that all 341 temporal derivatives have been computed initially by testing for the value of π set in INITAL

```
C...
C...   CHECK FOR UNDEFINED DERIVATIVES
       DO 12 IR=1,NR
       DO 12 JZ=1,NZ
          IF((UT(IR,JZ).GT.3.14D0).AND.(UT(IR,JZ).LT.3.15D0))
      +   THEN
          WRITE (NO,13)IR,JZ,U(IR,JZ),UT(IR,JZ)
13        FORMAT(' IR = ',I3,'    JZ = ',I3,
      +   '   U(IR,JZ) = ',F7.3,'    UT(IR,JZ) = ',F7.1)
          END IF
12     CONTINUE
       END IF
```

(3) The numerical solution is then printed as the radial temperature profiles, u(r, z, t) for $0 \leq r \leq 1$, starting at $z = 0$, and continuing to $z = z_l$, via DO loop 3. The WRITE statement is executed every fifth call to PRINT to keep the total output to a reasonable level

```
C...
C...   PRINT THE NUMERICAL SOLUTION EVERY FIFTH CALL TO PRINT
       IP=IP+1
       IF(((IP-1)/5*5).EQ.(IP-1))THEN
          WRITE(NO,1)T
1         FORMAT(//,' t = ',F6.2,//,5X,'u(r,z,t)',//,
      +      4X,'z',4X,'r = 0.0',3X,'r = 0.2',3X,'r = 0.4',
      +           3X,'r = 0.6',3X,'r = 0.8',3X,'r = 1.0')
          DO 3 JZ=1,NZ
             NRI=2
             WRITE(NO,2)Z(JZ),(U(IR,JZ),IR=1,NR,NRI)
2            FORMAT(F5.1,6F10.4)
3         CONTINUE
       END IF
```

Note the increment in the radial index, NRI = 2, so that every second temperature is printed radially.

(4) Two files are then opened at the beginning of the first solution for Matlab plotting of the centerline temperature, $u(0, z, t)$, versus z with t as a parameter, corresponding to the two case $z_l = 30$ and 90 (the length in boundary condition (H5.12))

```
C...
C...   OPEN FILE FOR MATLAB PLOTTING
       IF((NORUN.EQ.1).AND.(T.LT.0.001D0))THEN
          OPEN(1,FILE='h5a.out')
          OPEN(2,FILE='h5b.out')
       END IF
```

(5) The centerline temperature, $u(0, z, t)$ is then stored versus z for plotting of the solution

```
C...
C...   WRITE NUMERICAL SOLUTION AT R = 0 FOR MATLAB PLOTTING
       DO 4 JZ=1,NZ
          IF(NORUN.EQ.1)WRITE(1,5)Z(JZ),U(1,JZ)
          IF(NORUN.EQ.2)WRITE(2,5)Z(JZ),U(1,JZ)
5         FORMAT(2F10.4)
4      CONTINUE
```

This writing of the numerical solution to files h5a.out ($z_l = 30$) and h5b.out ($z_l = 90$) during each call to PRINT (and therefore at a succession of values of t) then provides plots with t as a parameter.

The data file read by the main program that calls ODE integrator RKF45 is listed below:

```
Graetz problem, zl = 30
0.            1.0           0.05
   341                                 0.00001
Graetz problem, zl = 90
0.            1.0           0.05
   341                                 0.00001
END OF RUNS
```

Two runs are programmed for $z_l = 30$ and 90. In each run, 341 ODEs = (11 × 31) are programmed over the interval $0 \leq t \leq 1$, with output at intervals of 0.05 (21 calls to PRINT, including the initial condition at $t = 0$). The integration of the 341 ODES is performed with an error tolerance of 0.00001.

The execution of INITAL, DERV, DERV1 (selected by using NCASE = 1 in INITAL), PRINT, and the data, plus the main program in Appendix A and integrator RKF45, produced the following (abbreviated) output:

```
RUN NO. -    1   Graetz problem, zl = 30
```

```
INITIAL T -    .000D+00

  FINAL T -    .100D+01

  PRINT T -    .500D-01

NUMBER OF DIFFERENTIAL EQUATIONS - 341

MAXIMUM INTEGRATION ERROR -    .100D-04

t =      .00
          .
          .
          .
t =     1.00

   u(r,z,t)
```

z	r = 0.0	r = 0.2	r = 0.4	r = 0.6	r = 0.8	r = 1.0
.0	.0000	.0000	.0000	.0000	.0000	.0000
1.0	.0012	.0028	.0139	.0737	.3282	1.0000
2.0	.0054	.0112	.0436	.1680	.5019	1.0000
3.0	.0147	.0268	.0844	.2553	.5965	1.0000
4.0	.0299	.0494	.1306	.3293	.6557	1.0000
5.0	.0511	.0778	.1783	.3910	.6970	1.0000
6.0	.0777	.1107	.2253	.4427	.7281	1.0000
7.0	.1086	.1464	.2704	.4867	.7529	1.0000
8.0	.1426	.1839	.3131	.5249	.7734	1.0000
9.0	.1787	.2221	.3534	.5584	.7909	1.0000
10.0	.2158	.2603	.3911	.5883	.8060	1.0000
11.0	.2533	.2980	.4266	.6152	.8193	1.0000
12.0	.2904	.3346	.4599	.6396	.8313	1.0000
13.0	.3268	.3701	.4911	.6620	.8421	1.0000
14.0	.3622	.4042	.5205	.6826	.8520	1.0000
15.0	.3964	.4368	.5481	.7017	.8611	1.0000
16.0	.4292	.4680	.5741	.7195	.8695	1.0000
17.0	.4605	.4976	.5985	.7360	.8773	1.0000
18.0	.4904	.5257	.6215	.7514	.8845	1.0000
19.0	.5188	.5524	.6432	.7659	.8913	1.0000
20.0	.5458	.5776	.6636	.7795	.8977	1.0000
21.0	.5713	.6015	.6828	.7922	.9036	1.0000
22.0	.5955	.6241	.7010	.8042	.9092	1.0000
23.0	.6184	.6454	.7180	.8154	.9144	1.0000
24.0	.6400	.6655	.7341	.8260	.9193	1.0000
25.0	.6605	.6846	.7493	.8360	.9240	1.0000

26.0	.6798	.7025	.7636	.8454	.9283	1.0000
27.0	.6980	.7194	.7771	.8542	.9324	1.0000
28.0	.7152	.7354	.7899	.8626	.9363	1.0000
29.0	.7329	.7519	.8031	.8715	.9406	1.0000
30.0	.9336	.9377	.9489	.9650	.9830	1.0000

RUN NO. -   2 Graetz problem, zl = 90

INITIAL T -   .000D+00

  FINAL T -   .100D+01

  PRINT T -   .500D-01

NUMBER OF DIFFERENTIAL EQUATIONS - 341

MAXIMUM INTEGRATION ERROR -   .100D-04

t =   .00
         .
         .
         .
t =   1.00

   u(r,z,t)

z	r = 0.0	r = 0.2	r = 0.4	r = 0.6	r = 0.8	r = 1.0
.0	.0000	.0000	.0000	.0000	.0000	.0000
3.0	.0293	.0416	.0919	.2260	.5218	1.0000
6.0	.0948	.1218	.2158	.4056	.6927	1.0000
9.0	.1844	.2206	.3346	.5285	.7707	1.0000
12.0	.2827	.3220	.4379	.6159	.8177	1.0000
15.0	.3789	.4170	.5254	.6820	.8508	1.0000
18.0	.4672	.5021	.5989	.7342	.8760	1.0000
21.0	.5456	.5765	.6609	.7766	.8961	1.0000
24.0	.6137	.6405	.7132	.8117	.9125	1.0000
27.0	.6722	.6952	.7574	.8410	.9262	1.0000
30.0	.7222	.7418	.7947	.8656	.9377	1.0000
33.0	.7647	.7814	.8262	.8863	.9473	1.0000
36.0	.8007	.8149	.8530	.9038	.9554	1.0000
39.0	.8313	.8433	.8755	.9186	.9623	1.0000
42.0	.8572	.8674	.8947	.9311	.9681	1.0000
45.0	.8791	.8877	.9109	.9417	.9730	1.0000
48.0	.8977	.9050	.9245	.9507	.9771	1.0000
51.0	.9134	.9196	.9361	.9582	.9806	1.0000
54.0	.9267	.9319	.9459	.9647	.9836	1.0000

57.0	.9379	.9424	.9542	.9701	.9861	1.0000
60.0	.9475	.9512	.9613	.9747	.9883	1.0000
63.0	.9555	.9587	.9672	.9785	.9901	1.0000
66.0	.9623	.9650	.9722	.9818	.9916	1.0000
69.0	.9680	.9703	.9764	.9846	.9928	1.0000
72.0	.9729	.9748	.9800	.9869	.9939	1.0000
75.0	.9769	.9786	.9830	.9888	.9948	1.0000
78.0	.9803	.9817	.9855	.9905	.9956	1.0000
81.0	.9832	.9844	.9875	.9918	.9962	1.0000
84.0	.9855	.9865	.9893	.9929	.9967	1.0000
87.0	.9875	.9884	.9907	.9939	.9971	1.0000
90.0	.9948	.9951	.9959	.9972	.9986	1.0000

We can note the following points about the preceding output:

(1) Radial position, r, varies across the top of each output, and axial position, z, varies down each output.

(2) Boundary condition (H5.11) programmed in DERV1 is clear from the first row at $z = 0$. For example, at $t = 1$

```
t =    1.00

   u(r,z,t)

   z    r = 0.0  r = 0.2  r = 0.4  r = 0.6  r = 0.8  r = 1.0
   .0    .0000    .0000    .0000    .0000    .0000    .0000
  1.0    .0012    .0028    .0139    .0737    .3282   1.0000
```

(3) Similarly, boundary condition (H5.10) is clear from the right-most column at $r = 1$ (i.e., $u(1.0, z, t) = 1$).

(4) The centerline temperature, $u(0, z, t)$, is not a smooth function of z for the first case of $z = z_l$ (the first run). For example, at $t = 1$, $25 \leq z \leq 30$

```
t =    1.00

   u(r,z,t)

   z    r = 0.0  r = 0.2  r = 0.4  r = 0.6  r = 0.8  r = 1.0
                    .                          .
                    .                          .
                    .                          .
  25.0   .6605    .6846    .7493    .8360    .9240   1.0000
  26.0   .6798    .7025    .7636    .8454    .9283   1.0000
  27.0   .6980    .7194    .7771    .8542    .9324   1.0000
  28.0   .7152    .7354    .7899    .8626    .9363   1.0000
  29.0   .7329    .7519    .8031    .8715    .9406   1.0000
  30.0   .9336    .9377    .9489    .9650    .9830   1.0000
```

Note in particular the successive values $u(0, 28, 1) = .7152$, $u(0, 29, 1) = .7329$, $u(0, 30, 1) = .9336$. This unrealistic behavior (a sudden increase in u at $z = 30$) is due to the effect of boundary condition (H5.23), that is, u near $z = 30$ cannot undergo the transition to a zero slope required by boundary condition (H5.23). This anomaly is clear from the plot of the solution in Figure H5.1a; note in particular the behavior of the solution near $z = 30$.

However, if the solution has reached a nearly constant value in the neighborhood of $z = z_l$, then the application of boundary condition (H5.23) does not present a problem. This is illustrated by the solution for $z = z_l$ (second run). Thus, near $z = 90$, the solution is

```
t =     1.00

    u(r,z,t)

   z     r = 0.0   r = 0.2   r = 0.4   r = 0.6   r = 0.8   r = 1.0
                      .
                      .
                      .
                      .
 75.0     .9769     .9786     .9830     .9888     .9948    1.0000
 78.0     .9803     .9817     .9855     .9905     .9956    1.0000
 81.0     .9832     .9844     .9875     .9918     .9962    1.0000
 84.0     .9855     .9865     .9893     .9929     .9967    1.0000
 87.0     .9875     .9884     .9907     .9939     .9971    1.0000
 90.0     .9948     .9951     .9959     .9972     .9986    1.0000
```

Note that the wall temperature $u(1, z, t) = 1$ dominates the solution (the entering temperature $u(r, 0, t) = 0$) has little effect at $z = 90$, and therefore the application of boundary condition (H5.23) produces a smooth solution as reflected in Figure (H5.1b).

In conclusion, the suitability of boundary condition (H5.23) is dependent on the circumstances (parameters) of the problem. Clearly if a moving front travels from the entrance at $z = 0$ to the end of the system at $z = z_l$, boundary condition (H5.23) can cause problems (and it is physically unrealistic since it requires the front to leave the system at $z = z_l$ with zero slope in z for all r and t). In other words, we conclude that boundary condition (H5.23) causes numerical problems when used as an outflow condition for this strongly convective problem (with $P_e = 60$), and by implication, it cannot be used as an outflow boundary condition for strongly convective PDEs in general.

We therefore will consider some alternate boundary conditions at $z = z_l$, which will circumvent this problem. Specifically, we now turn to boundary condition (H5.12). First, we consider boundary condition (H5.12) without the radial group as

234 *Heat transport*

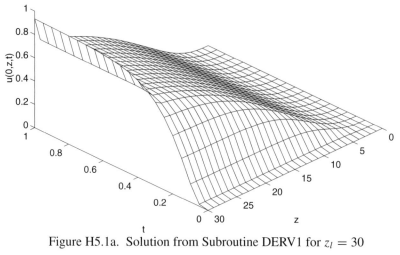

Figure H5.1a. Solution from Subroutine DERV1 for $z_l = 30$

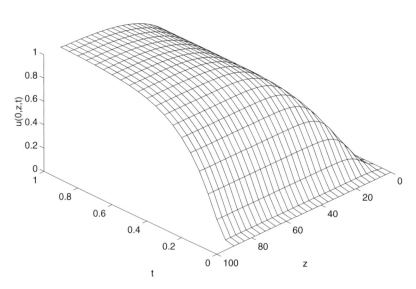

Figure H5.1b. Solution from Subroutine DERV1 for $z_l = 90$

implemented in DERV2 (which is called by setting NCASE = 2 in INITAL)

$$u_t(r, z_l, t) + P_e v(r) u_z(r, z_l, t) = 0 \qquad (H5.24)$$

```
      SUBROUTINE DERV2
      IMPLICIT DOUBLE PRECISION (A-H,O-Z)
      PARAMETER (NR=11,NZ=31)
      COMMON/T/           T,       NSTOP,      NORUN
     +      /Y/     U(NR,NZ)
     +      /F/     UT(NR,NZ)
     +      /G/         R(NR),     Z(NZ),      V(NR)
     +      /C/           PE,       RO,        ZL,
     +                    DR,       DRS,       DZ,       DZS
     +      /I/           IP,       NCASE
C...
C...  Z = 0
      DO 1 IR=1,NR
         U(IR,1)=0.0D0
         UT(IR,1)=0.0D0
1     CONTINUE
C...
C...  R = RO, Z NE 0
      DO 2 JZ=2,NZ
         U(NR,JZ)=1.0D0
         UT(NR,JZ)=0.0D0
2     CONTINUE
C...
C...  R = 0, Z NE 0, ZL
      DO 3 JZ=2,NZ-1
         UT(1,JZ)=-PE*V(1)*(U(1,JZ)-U(1,JZ-1))/DZ
     +            +4.0D0*(U(2,JZ)-U(1,JZ))/DRS
     +            +(U(1,JZ+1)-2.0D0*U(1,JZ)+U(1,JZ-1))/DZS
3     CONTINUE
C...
C...  Z = ZL, R NE RO
      DO 4 IR=1,NR-1
         UT(IR,NZ)=-PE*V(IR)*(U(IR,NZ)-U(IR,NZ-1))/DZ
4     CONTINUE
C...
C...  R NE 0, RO, Z NE 0, ZL
      DO 5 JZ=2,NZ-1
      DO 5 IR=2,NR-1
         UT(IR,JZ)=-PE*V(IR)*(U(IR,JZ)-U(IR,JZ-1))/DZ
     +            +(U(IR+1,JZ)-2.0D0*U(IR,JZ)+U(IR-1,JZ))/
     +            DRS+(1.0D0/R(IR))*(U(IR+1,JZ)-U(IR-1,JZ))
     +            /(2.0D0*DR)+(U(IR,JZ+1)-2.0D0*U(IR,JZ)
     +            +U(IR,JZ-1))/DZS
```

```
5       CONTINUE
        RETURN
        END
```

We can note the following points about DERV2:

(1) The COMMON area is the same as for DERV1 (and will remain the same for DERV3 to DERV8). Also, the programming of boundary conditions (H5.9), (H5.10), and (H5.11) is the same as in DERV1.

(2) The programming of boundary condition (H5.24) as a special case of (H5.12) (with no radial group) is in DO loop 4

```
C...
C...    Z = ZL, R NE RO
        DO 4 IR=1,NR-1
            UT(IR,NZ)=-PE*V(IR)*(U(IR,NZ)-U(IR,NZ-1))/DZ
4       CONTINUE
```

This coding follows directly from eq. (H5.24). Note in particular:

(2.1) The radial group in eq. (H5.12) has been dropped.
(2.2) The axial derivative

$$u_z(r, z_l, t)$$

in eq. (H5.24) is approximated by a two-point, backward difference, that is,

$$u_z(r, z_l, t) \approx \frac{u(r, z_l, t) - u(r, z_l - \Delta z, t)}{\Delta z}$$

All of the other coding (in INITAL, with NCASE = 2), DERV, and PRINT, and the data file, remain the same.

Abbreviated output (at $t = 1$ for $z_l = 30$ and 90) from execution of the program is given below:

```
t =     1.00

        u(r,z,t)

    z      r = 0.0   r = 0.2   r = 0.4   r = 0.6   r = 0.8   r = 1.0
    .0      .0000     .0000     .0000     .0000     .0000     .0000
   1.0      .0012     .0028     .0139     .0737     .3282    1.0000
                .
                .
                .
   25.0     .6605     .6846     .7493     .8360     .9240    1.0000
   26.0     .6798     .7025     .7636     .8454     .9283    1.0000
   27.0     .6980     .7194     .7771     .8542     .9324    1.0000
```

28.0	.7152	.7354	.7898	.8626	.9363	1.0000
29.0	.7313	.7504	.8017	.8703	.9399	1.0000
30.0	.7313	.7504	.8017	.8703	.9399	1.0000

t =   1.00

u(r,z,t)

z	r = 0.0	r = 0.2	r = 0.4	r = 0.6	r = 0.8	r = 1.0
.0	.0000	.0000	.0000	.0000	.0000	.0000
3.0	.0293	.0416	.0919	.2260	.5218	1.0000
.		.		.		
.		.		.		
.		.		.		
72.0	.9729	.9748	.9800	.9869	.9939	1.0000
75.0	.9769	.9786	.9830	.9888	.9948	1.0000
78.0	.9803	.9817	.9855	.9905	.9956	1.0000
81.0	.9832	.9844	.9875	.9918	.9962	1.0000
84.0	.9855	.9865	.9893	.9929	.9967	1.0000
87.0	.9875	.9883	.9907	.9938	.9971	1.0000
90.0	.9871	.9880	.9903	.9934	.9965	1.0000

We can note the following points about this output:

(1) The solution near $z = 30$ has a constant value (to four figures) with respect to z which is due to

(1.1) The removal of the radial conduction in boundary condition (H5.23), that is, heat from the wall at temperature $u(1, z, t) = 1$ (according to boundary condition (H5.10)) cannot flow radially by conduction at $z = z_l$, so the temperature at $z = z_l$ does not vary with z.

(1.2) The use of the two-point approximations for $u_z(r, z_l, t)$ as discussed previously in (2.2) p. 236; this approximation smooths the axial profile in z.

This effect is apparent in the following output:

28.0	.7152	.7354	.7898	.8626	.9363	1.0000
29.0	.7313	.7504	.8017	.8703	.9399	1.0000
30.0	.7313	.7504	.8017	.8703	.9399	1.0000

This unrealistic effect (no variation of $u(z, r, t)$ near $z = 30$) is also apparent in Figure H5.2a.

(2) As in the case of using DERV1, the solution at $z = 90$ is free of large, unrealistic numerical effects (but note the small drop in u, i.e., $u(0, 87, 1) = .9875$, $u(0, 90, 1) = .9871$)

| 84.0 | .9855 | .9865 | .9893 | .9929 | .9967 | 1.0000 |

87.0	.9875	.9883	.9907	.9938	.9971	1.0000
90.0	.9871	.9880	.9903	.9934	.9965	1.0000

The smoothness of this solution is apparent in Figure H5.2b.

(3) Returning to (1) above, the absence of radial heat conduction at $z = z_l$ can be investigated by writing the value of the term

$$u_{rr} + \frac{1}{r}u_r$$

in eq. (H5.7) as a function of r, z, and t. Also, the relative contribution of convection and conduction in the axial direction can be observed by observing the values of the terms

$$-P_e v(r) u_z$$

and

$$u_{zz}$$

respectively, as a function of r, z, and t. Finally, the temporal change of the temperature is elucidated by observing the value of the derivative

$$u_t$$

defined by eq. (H5.7) as a function of r, z, and t; the relative contributions of the right hand side (RHS) spatial derivatives of eq. (H5.7) to this temporal derivative can then be observed to better understand the time-dependent behavior of the solution. All of these derivatives are readily available for writing and/or plotting, for example, by a call to DERV2 via DERV where they can be written into two-dimensional arrays and passed to PRINT through COMMON. This procedure of writing all of the terms in a PDE, as well as the dependent variable, as a function of the independent variables is an important procedure for (a) understanding the behavior of the solution, particularly with respect to the contributions of the individual terms, which generally reflect physical phenomena, for example, conduction, convection, accumulation; (b) finding errors in the programming, and perhaps the mathematical model, especially when a computer code is first developed; (c) checking that the solution conforms to all of the mathematical conditions, for example, initial and boundary conditions and symmetry; and (d) checking that the solution makes sense physically. Also, subsidiary calculations involving the dependent variable and its various derivatives can often be performed to further check the solution, for example, to test for conservation of energy. We recommend this procedure of producing comprehensive output for checking the computed solution and enhancing understanding of the solution.

The Graetz problem with constant wall temperature

Graetz problem with constant wall temperature, zl = 30, DERV2

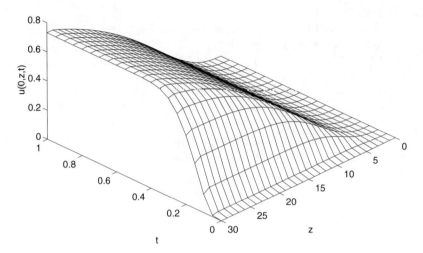

Figure H5.2a. Solution from Subroutine DERV2 for $z_l = 30$

Graetz problem with constant wall temperature, zl = 90, DERV2

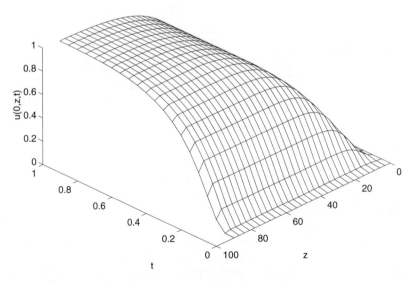

Figure H5.2b. Solution from Subroutine DERV2 for $z_l = 90$

Finally, we conclude from the preceding results produced by DERV2 that eq. (H5.24) is not a suitable boundary condition for eq. (H5.7), and we therefore go on to the use of eq. (H5.12) as a boundary condition, as programmed in DERV3 (and selected with NCASE = 3 in INITAL)

```fortran
      SUBROUTINE DERV3
      IMPLICIT DOUBLE PRECISION (A-H,O-Z)
      PARAMETER (NR=11,NZ=31)
      COMMON/T/          T,        NSTOP,      NORUN
     +      /Y/   U(NR,NZ)
     +      /F/   UT(NR,NZ)
     +      /G/        R(NR),      Z(NZ),      V(NR)
     +      /C/        PE,         RO,         ZL,
     +                 DR,         DRS,        DZ,        DZS
     +      /I/        IP,         NCASE
C...
C...  Z = 0
      DO 1 IR=1,NR
         U(IR,1)=0.0D0
         UT(IR,1)=0.0D0
1     CONTINUE
C...
C...  R = RO, Z NE 0
      DO 2 JZ=2,NZ
         U(NR,JZ)=1.0D0
         UT(NR,JZ)=0.0D0
2     CONTINUE
C...
C...  R = 0, Z NE 0, ZL
      DO 3 JZ=2,NZ-1
         UT(1,JZ)=-PE*V(1)*(U(1,JZ)-U(1,JZ-1))/DZ
     +            +4.0D0*(U(2,JZ)-U(1,JZ))/DRS
     +            +(U(1,JZ+1)-2.0D0*U(1,JZ)+U(1,JZ-1))/DZS
3     CONTINUE
C...
C...  R = 0, Z = ZL
      UT(1,NZ)=-PE*V(1)*(U(1,NZ)-U(1,NZ-1))/DZ
     +         +4.0D0*(U(2,NZ)   -U(1,NZ))/DRS
C...
C...  Z = ZL, R NE 0, RO
      DO 4 IR=2,NR-1
         UT(IR,NZ)=-PE*V(IR)*(U(IR,NZ)-U(IR,NZ-1))/DZ
     +             +(U(IR+1,NZ)-2.0D0*U(IR,NZ)
     +             +U(IR-1,NZ))/DRS+(1.0D0/R(IR))
     +             *(U(IR+1,NZ)-U(IR-1,NZ))/(2.0D0*DR)
4     CONTINUE
```

```
C...
C...    R NE 0, R0, Z NE 0, ZL
        DO 5 JZ=2,NZ-1
        DO 5 IR=2,NR-1
          UT(IR,JZ)=-PE*V(IR)*(U(IR,JZ)-U(IR,JZ-1))/DZ
     +                +(U(IR+1,JZ)-2.0D0*U(IR,JZ)
     +                +U(IR-1,JZ))/DRS+(1.0D0/R(IR))*(U(IR+1,JZ)
     +                -U(IR-1,JZ))/(2.0D0*DR)+(U(IR,JZ+1)
     +                -2.0D0*U(IR,JZ)+U(IR,JZ-1))/DZS
5       CONTINUE
        RETURN
        END
```

We can note the following points about DERV3:

(1) The COMMON area is the same as in DERV1 and DERV2. Also, the programming of boundary conditions (H5.9), (H5.10), and (H5.11) is the same as in DERV2.

(2) The programming of boundary condition (H5.12) is in two parts corresponding to $r = 0$ and $r > 0$. For $r = 0$, we also use boundary condition (H5.9)

```
C...
C...    R = 0, Z = ZL
          UT(1,NZ)=-PE*V(1)*(U(1,NZ)-U(1,NZ-1))/DZ
     +              +4.0D0*(U(2,NZ)   -U(1,NZ))/DRS
```

where we have used the approximation

$$u_{rr} + \frac{1}{r}u_r = 2u_{rr} \approx 4\frac{u(\Delta r, z, t) - u(0, z, t)}{\Delta r^2}$$

For $r > 0$, the programming of eq. (H5.12) is in DO loop 4; note the use of the subscript NZ for $z = z_l$.

```
C...
C...    Z = ZL, R NE 0, R0
        DO 4 IR=2,NR-1
          UT(IR,NZ)=-PE*V(IR)*(U(IR,NZ)-U(IR,NZ-1))/DZ
     +                +(U(IR+1,NZ)-2.0D0*U(IR,NZ)
     +                +U(IR-1,NZ))/DRS+(1.0D0/R(IR))
     +                *(U(IR+1,NZ)-U(IR-1,NZ))/(2.0D0*DR)
4       CONTINUE
```

As expected, this is the same as the programming of eq. (H5.7) without the derivative u_{zz}, which is not in eq. (H5.12); note the use of the subscript NZ for $z = z_l$. All of the other coding (in INITAL, with NCASE = 3), DERV, and PRINT, and the data file, remain the same.

Abbreviated output (at $t = 1$ for $z_l = 30$ and 90) from execution of the program is given below:

```
t =    1.00

    u(r,z,t)

   z    r = 0.0   r = 0.2   r = 0.4   r = 0.6   r = 0.8   r = 1.0
  .0     .0000     .0000     .0000     .0000     .0000     .0000
 1.0     .0012     .0028     .0139     .0737     .3282    1.0000
           .                                        .
           .
           .
25.0     .6605     .6846     .7493     .8360     .9240    1.0000
26.0     .6798     .7025     .7636     .8454     .9283    1.0000
27.0     .6980     .7194     .7771     .8542     .9324    1.0000
28.0     .7152     .7354     .7898     .8626     .9363    1.0000
29.0     .7314     .7505     .8018     .8704     .9399    1.0000
30.0     .7467     .7647     .8132     .8778     .9434    1.0000

t =    1.00

    u(r,z,t)

   z    r = 0.0   r = 0.2   r = 0.4   r = 0.6   r = 0.8   r = 1.0
  .0     .0000     .0000     .0000     .0000     .0000     .0000
 3.0     .0293     .0416     .0919     .2260     .5218    1.0000
           .                                        .
           .
           .
75.0     .9769     .9786     .9830     .9888     .9948    1.0000
78.0     .9803     .9817     .9855     .9905     .9956    1.0000
81.0     .9832     .9844     .9875     .9918     .9962    1.0000
84.0     .9855     .9865     .9893     .9929     .9967    1.0000
87.0     .9875     .9884     .9907     .9938     .9971    1.0000
90.0     .9891     .9898     .9919     .9946     .9975    1.0000
```

We can note the following points about this output:

(1) The solution near $z = 30$ appears smooth with respect to z

```
28.0     .7152     .7354     .7898     .8626     .9363    1.0000
29.0     .7314     .7505     .8018     .8704     .9399    1.0000
30.0     .7467     .7647     .8132     .8778     .9434    1.0000
```

This conclusion is supported by Figure 5.3a.

(2) The solution at $z = z_l = 90$ is smooth

The Graetz problem with constant wall temperature

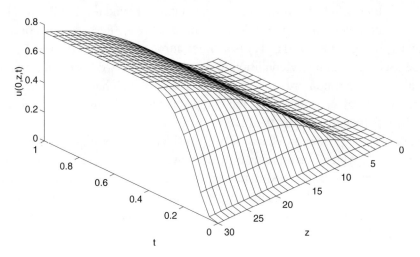

Figure H5.3a. Solution from Subroutine DERV3 for $z_l = 30$

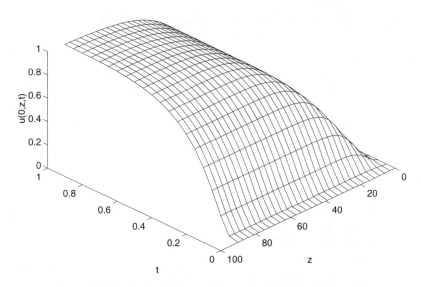

Figure H5.3b. Solution from Subroutine DERV3 for $z_l = 90$

84.0	.9855	.9865	.9893	.9929	.9967	1.0000
87.0	.9875	.9884	.9907	.9938	.9971	1.0000
90.0	.9891	.9898	.9919	.9946	.9975	1.0000

The smoothness of this solution is apparent in Figure H5.3b.

Thus, we conclude from this result that outflow boundary condition (H5.12) gives a smooth solution in the neighborhood of $z = z_l$. Perhaps this conclusion is not surprising since eq. (H5.12) is a "minimum reduction" of eq. (H5.7). Yet it can be used as a boundary condition in z since it is of lower order in z (with the derivative of maximum order, u_z) than eq. (H5.7) (which has the derivative u_{zz}). In other words, given a choice, we recommend the use of boundary conditions that have the least departure from the PDE.

Finally, to complete this discussion, we consider the programming of eq. (H5.7) with $u_{zz} = 0$, which corresponds only to convection in z (but no conduction or diffusion in z). We might predict that the solution should be similar to that from DERV3 since this problem is strongly convective (with $P_e = 60$). This turns out to be the case, as demonstrated by the solution produced by subroutine DERV4 (called by setting NCASE = 4 in INITAL)

```
      SUBROUTINE DERV4
      IMPLICIT DOUBLE PRECISION (A-H,O-Z)
      PARAMETER (NR=11,NZ=31)
      COMMON/T/            T,       NSTOP,      NORUN
     +      /Y/    U(NR,NZ)
     +      /F/    UT(NR,NZ)
     +      /G/        R(NR),       Z(NZ),      V(NR)
     +      /C/           PE,          RO,         ZL,
     +                    DR,         DRS,         DZ,      DZS
     +      /I/           IP,       NCASE
C...
C...  Z = 0
      DO 1 IR=1,NR
          U(IR,1)=0.0D0
          UT(IR,1)=0.0D0
1     CONTINUE
C...
C...  R = RO, Z NE 0
      DO 2 JZ=2,NZ
          U(NR,JZ)=1.0D0
          UT(NR,JZ)=0.0D0
2     CONTINUE
C...
C...  R = 0, Z NE 0
      DO 3 JZ=2,NZ
          UT(1,JZ)=-PE*V(1)*(U(1,JZ)-U(1,JZ-1))/DZ
```

```
     +                  +4.0D0*(U(2,JZ)-U(1,JZ))/DRS
3         CONTINUE
C...
C...      R NE 0, R0, Z NE 0
          DO 5 JZ=2,NZ
          DO 5 IR=2,NR-1
             UT(IR,JZ)=-PE*V(IR)*(U(IR,JZ)-U(IR,JZ-1))/DZ
     +                  +(U(IR+1,JZ)-2.0D0*U(IR,JZ)
     +                  +U(IR-1,JZ))/DRS+(1.0D0/R(IR))
     +                  *(U(IR+1,JZ)-U(IR-1,JZ))/(2.0D0*DR)
5         CONTINUE
          RETURN
          END
```

We can note the following points about DERV4:

(1) The COMMON area is the same as in DERV1 to DERV3. Also, the programming of boundary conditions (H5.9), (H5.10), and (H5.11) is the same as in DERV1 to DERV3.

(2) The programming of boundary condition (H5.12) is not included (since it is not required with $u_{zz} = 0$). In other words, the programming of eq. (H5.7) (with $u_{zz} = 0$) can proceed all the way to $z = z_l$ with no special case at $z = z_l$. The programming for $r = 0, 0 < z \leq z_l$, is

```
C...
C...      R = 0, Z NE 0
          DO 3 JZ=2,NZ
             UT(1,JZ)=-PE*V(1)*(U(1,JZ)-U(1,JZ-1))/DZ
     +                  +4.0D0*(U(2,JZ)-U(1,JZ))/DRS
3         CONTINUE
```

where we have again used the approximation

$$u_{rr} + \frac{1}{r}u_r = 2u_{rr} \approx 4\frac{u(\Delta r, z, t) - u(0, z, t)}{\Delta r^2}$$

For $r > 0, 0 < z \leq z_l$, the programming of eq. (H5.7) is in DO loop 5

```
C...
C...      R NE 0, R0, Z NE 0
          DO 5 JZ=2,NZ
          DO 5 IR=2,NR-1
             UT(IR,JZ)=-PE*V(IR)*(U(IR,JZ)-U(IR,JZ-1))/DZ
     +                  +(U(IR+1,JZ)-2.0D0*U(IR,JZ)
     +                  +U(IR-1,JZ))/DRS+(1.0D0/R(IR))
     +                  *(U(IR+1,JZ)-U(IR-1,JZ))/(2.0D0*DR)
5         CONTINUE
```

This programming follows directly from eq. (H5.7) (with $u_{zz} = 0$) and the finite difference approximations that were discussed previously. All of the other coding (in INITAL, with NCASE = 4), DERV, and PRINT, and the data file, remain the same.

Abbreviated output (at $t = 1$ for $z_l = 30$ and 90) from execution of the program is given below:

```
t =    1.00

    u(r,z,t)

    z     r = 0.0   r = 0.2   r = 0.4   r = 0.6   r = 0.8   r = 1.0
   .0      .0000     .0000     .0000     .0000     .0000     .0000
   1.0     .0011     .0027     .0139     .0741     .3307    1.0000
             .         .         .         .         .
             .         .         .         .         .
             .         .         .         .         .
  25.0     .6610     .6850     .7497     .8362     .9241    1.0000
  26.0     .6802     .7029     .7640     .8456     .9284    1.0000
  27.0     .6984     .7199     .7775     .8545     .9325    1.0000
  28.0     .7156     .7358     .7902     .8628     .9364    1.0000
  29.0     .7318     .7509     .8022     .8706     .9400    1.0000
  30.0     .7471     .7651     .8135     .8780     .9435    1.0000

t =    1.00

    u(r,z,t)

    z     r = 0.0   r = 0.2   r = 0.4   r = 0.6   r = 0.8   r = 1.0
   .0      .0000     .0000     .0000     .0000     .0000     .0000
   3.0     .0292     .0416     .0920     .2264     .5226    1.0000
             .         .         .         .         .
             .         .         .         .         .
             .         .         .         .         .
  75.0     .9770     .9786     .9830     .9889     .9948    1.0000
  78.0     .9804     .9818     .9855     .9905     .9956    1.0000
  81.0     .9832     .9844     .9876     .9918     .9962    1.0000
  84.0     .9856     .9866     .9893     .9929     .9967    1.0000
  87.0     .9875     .9884     .9907     .9939     .9971    1.0000
  90.0     .9891     .9899     .9919     .9946     .9975    1.0000
```

We can note the following points about this output:

(1) The solution near $z = 30$ is smooth with respect to z

```
  28.0      .7156     .7358     .7902     .8628     .9364    1.0000
```

| 29.0 | .7318 | .7509 | .8022 | .8706 | .9400 | 1.0000 |
| 30.0 | .7471 | .7651 | .8135 | .8780 | .9435 | 1.0000 |

This conclusion is supported by Figure H5.4a.

(2) The solution at $z = 90$ is smooth

84.0	.9856	.9866	.9893	.9929	.9967	1.0000
87.0	.9875	.9884	.9907	.9939	.9971	1.0000
90.0	.9891	.9899	.9919	.9946	.9975	1.0000

The smoothness of this solution is apparent in Figure H5.4b.

The contribution of the axial conduction term, u_{zz}, in eq. (H5.7) can be observed by comparing the solutions from DERV3 (with u_{zz}) and DERV4 (without u_{zz}):

DERV3

t =   1.00

   u(r,z,t)

z	r = 0.0	r = 0.2	r = 0.4	r = 0.6	r = 0.8	r = 1.0
.0	.0000	.0000	.0000	.0000	.0000	.0000
.						
.						
.						
28.0	.7152	.7354	.7898	.8626	.9363	1.0000
29.0	.7314	.7505	.8018	.8704	.9399	1.0000
30.0	.7467	.7647	.8132	.8778	.9434	1.0000

t =   1.00

   u(r,z,t)

z	r = 0.0	r = 0.2	r = 0.4	r = 0.6	r = 0.8	r = 1.0
.0	.0000	.0000	.0000	.0000	.0000	.0000
.						
.						
.						
84.0	.9855	.9865	.9893	.9929	.9967	1.0000
87.0	.9875	.9884	.9907	.9938	.9971	1.0000
90.0	.9891	.9898	.9919	.9946	.9975	1.0000

DERV4

t =   1.00

Heat transport

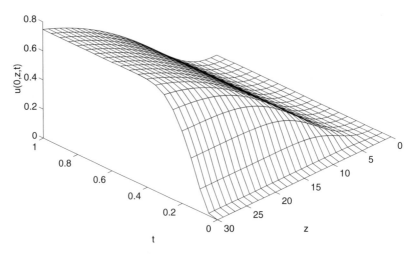

Figure H5.4a. Solution from Subroutine DERV4 for $z_l = 30$

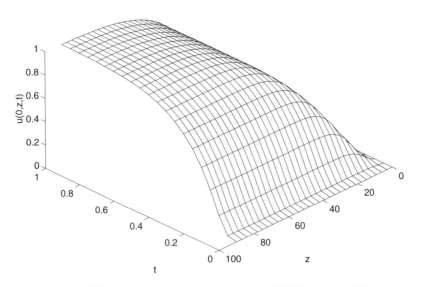

Figure H5.4b. Solution from Subroutine DERV4 for $z_l = 90$

```
u(r,z,t)

  z     r = 0.0   r = 0.2   r = 0.4   r = 0.6   r = 0.8   r = 1.0
 .0      .0000     .0000     .0000     .0000     .0000     .0000
          .         .         .         .         .
          .         .         .         .         .
          .         .         .         .         .
28.0     .7156     .7358     .7902     .8628     .9364    1.0000
29.0     .7318     .7509     .8022     .8706     .9400    1.0000
30.0     .7471     .7651     .8135     .8780     .9435    1.0000

t =    1.00

    u(r,z,t)

  z     r = 0.0   r = 0.2   r = 0.4   r = 0.6   r = 0.8   r = 1.0
 .0      .0000     .0000     .0000     .0000     .0000     .0000
          .         .         .         .         .
          .         .         .         .         .
          .         .         .         .         .
84.0     .9856     .9866     .9893     .9929     .9967    1.0000
87.0     .9875     .9884     .9907     .9939     .9971    1.0000
90.0     .9891     .9899     .9919     .9946     .9975    1.0000
```

These results indicate that the two solutions are in close agreement (and therefore the derivative u_{zz} in eq. (H5.7) has a small effect). For example,

T(0, 30, 1)	T(0, 90, 1)	
.7467	.9891	DERV3
.7471	.9891	DERV4

Thus, somewhat ironically, the preceding discussion of implementing an outflow boundary condition, as programmed in DERV1, DERV2, and DERV3, was basically unnecessary since axial conduction for this system is small compared with axial convection because of the large value of P_e ($= 60$). However, for a problem that is not predominantly convective (e.g., when axial conduction is important), the use of boundary condition (H5.12) is recommended as the best choice of the three boundary conditions considered (eqs. (H5.12), (H5.23), and (H5.24)).

The programming in subroutines DERV1 to DERV4 is rather detailed and prone to error, that is, simplification of the programming would be desirable. To this end, we now consider the use of library routines for the calculation of the spatial derivatives in eqs. (H5.7) to (H5.12) (these routines have DSS in their names, standing for Differentiation in Space Subroutine). The same four problems in DERV1 to DERV4 are programmed in DERV5 to DERV8 using these library routines. DERV5 is the counterpart of DERV1

```fortran
      SUBROUTINE DERV5
      IMPLICIT DOUBLE PRECISION (A-H,O-Z)
      PARAMETER (NR=11,NZ=31)
      COMMON/T/          T,        NSTOP,      NORUN
     +      /Y/     U(NR,NZ)
     +      /F/     UT(NR,NZ)
     +      /G/       R(NR),        Z(NZ),     V(NR)
     +      /C/          PE,           RO,        ZL,
     +                   DR,          DRS,        DZ,       DZS
     +      /I/          IP,        NCASE
C...
C...  ARRAYS FOR UR, URR, UZ, UZZ
      DIMENSION UR(NR,NZ), URR(NR,NZ), UZ(NR,NZ), UZZ(NR,NZ)
C...
C...  Z = 0
      DO 1 IR=1,NR
         U(IR,1)=0.0D0
         UT(IR,1)=0.0D0
1     CONTINUE
C...
C...  R = RO, Z NE 0
      DO 2 JZ=2,NZ
         U(NR,JZ)=1.0D0
         UT(NR,JZ)=0.0D0
2     CONTINUE
C...
C...  UR
      CALL DSS034(0.0D0,RO,NR,NZ,1,U,UR,0.0D0)
C...
C...  R = 0
      DO 3 JZ=1,NZ
         UR(1,JZ)=0.0D0
3     CONTINUE
C...
C...  URR
      CALL DSS034(0.0D0,RO,NR,NZ,1,UR,URR,0.0D0)
C...
C...  UZ (FOR CONDUCTION)
      CALL DSS034(0.0D0,ZL,NR,NZ,2,U,UZ,0.0D0)
C...
C...  Z = ZL
      DO 4 IR=1,NR
         UZ(IR,NZ)=0.0D0
4     CONTINUE
```

```
C...
C...    UZZ (FOR CONDUCTION)
        CALL DSS034(0.0D0,ZL,NR,NZ,2,UZ,UZZ,0.0D0)
C...
C...    UZ (FOR CONVECTION)
        CALL DSS034(0.0D0,ZL,NR,NZ,2,U,UZ,1.0D0)
C...
C...    Z = ZL
        DO 5 IR=1,NR
           UZ(IR,NZ)=0.0D0
5       CONTINUE
C...
C...    R NE R0, Z NE 0
        DO 6 JZ=2,NZ
        DO 6 IR=1,NR-1
C...
C...       R = 0
           IF(IR.EQ.1)THEN
              UT(1,JZ)=-PE*V(1)*UZ(1,JZ)
     +                 +2.0D0*URR(1,JZ)
     +                 +UZZ(1,JZ)
C...
C...       R NE 0
           ELSE
     +     IF(IR.NE.1)THEN
              UT(IR,JZ)=-PE*V(IR)*UZ(IR,JZ)
     +                  +URR(IR,JZ)+(1.0D0/R(IR))*UR(IR,JZ)
     +                  +UZZ(IR,JZ)
           END IF
6       CONTINUE
        RETURN
        END
```

We can note the following points about DERV5:

(1) The COMMON area is the same as in DERV1 to DERV4 (which is to be expected since the reduction of eq. (H5.7) to a system of ODEs will proceed basically the same way in all cases). A set of additional arrays is then defined for the various spatial derivatives in the right hand side (RHS) of eq. (H5.7)

```
C...
C...    ARRAYS FOR UR, URR, UZ, UZZ
        DIMENSION UR(NR,NZ), URR(NR,NZ), UZ(NR,NZ), UZZ(NR,NZ)
```

In addition to providing the necessary storage of the four derivatives, u_r (= UR (NR, NZ)), u_{rr} (= URR(NR, NZ)), u_z (= UZ(NR, NZ)), and u_{zz} (= UZZ(NR,

NZ)), these arrays could be passed through COMMON to subroutine PRINT for printing and/or plotting of the various RHS terms in eq. (H5.7). This procedure was discussed previously (after DERV2) and is recommended for insight into the properties of the numerical solution.

(2) Boundary condition (H5.11) is implemented in DO loop 1 (as in DERV1 to DERV4).

```
C...
C...    Z = 0
        DO 1 IR=1,NR
            U(IR,1)=0.0D0
            UT(IR,1)=0.0D0
1       CONTINUE
```

(3) Boundary condition (H5.10) is then implemented in DO loop 2 (as in DERV1 to DERV4)

```
C...
C...    R = R0, Z NE 0
        DO 2 JZ=2,NZ
            U(NR,JZ)=1.0D0
            UT(NR,JZ)=0.0D0
2       CONTINUE
```

(4) With boundary condition (H5.10) programmed, the derivative u_r can now be computed from U by a call to spatial differentiation routine DSS034

```
C...
C...    UR
        CALL DSS034(0.0D0,R0,NR,NZ,1,U,UR,0.0D0)
```

Subroutine DSS034 is described by Schiesser (1991, 262–73), and therefore only the arguments will be discussed briefly:

(4.1) Since a derivative in r is being computed, u_r, the first two arguments defined the range in r, that is, $0 \leq r \leq r_0$. The third and fourth arguments are the number of grid points in the r and z directions, respectively (NR $= 11$ and NZ $= 31$).

(4.2) DSS034 computes a first-order partial derivative over a two-dimensional domain. Thus, since two independent spatial variables are involved, the fifth argument specifies which independent variable will be used for the derivative ($r \Rightarrow 1, z \Rightarrow 2$).

(4.3) The sixth argument is the two-dimensional array containing the dependent variable to be differentiated, in this case U (an input to DSS034).

(4.4) The seventh argument is the two-dimensional array containing the first partial derivative over the two-dimensional $r - z$ domain, in this case UR (the output from DSS034).

(4.5) The eighth argument specifies whether five-point, centered finite difference approximations are used to calculate the first-order derivative for parabolic or conductive terms in a PDE (= 0), or five-point, biased upwind, finite difference approximations are used to calculate the first-order derivative for hyperbolic or convective terms in a PDE (= 1). In this case, since the computed derivative, u_r, is for conduction, the eighth argument specifies a centered approximation.

(5) The centerline derivative, $u_r(0, z, t)$ (= UR(1, JZ)), from the preceding call to DSS034, is then reset to the value specified by boundary condition (H5.9) in DO loop 3

```
C...
C...    R = 0
        DO 3 JZ=1,NZ
            UR(1,JZ)=0.0D0
3       CONTINUE
```

Note that boundary condition (H5.10), which is Dirichlet (it specifies the dependent variable, $u(1, z, t)$), is set in DO loop 2 before the call to DSS034, so the correct boundary value of u at $r = 1$ is an input to DSS034. Boundary condition (H5.9), which is Neumann (it specifies the derivative of the dependent variable, $u_r(0, z, t)$), is set in DO loop 3 after the call to DSS034, so that this derivative has the correct value from boundary condition (H5.9) (it will not necessarily be $u_r(0, z, t) = 0$ as an output from DSS034).

(6) Now that both boundary conditions in r are defined (eqs. (H5.9) and (H5.10)), $u_{rr}(r, z, t)$ can be computed from $u_r(r, z, t)$ by a second call to DSS034

```
C...
C...    URR
        CALL DSS034(0.0D0,R0,NR,NZ,1,UR,URR,0.0D0)
```

Here we have used the idea that a second derivative (URR) is the derivative of the first derivative (UR). Note that again the eighth argument is zero since the computed derivative, u_{rr}, is for conduction (so that centered approximations are used).

(7) Now we can proceed to an equivalent set of calculations for conduction in the axial direction (based on the derivative u_{zz} in eq. (H5.7)). Since boundary condition (H5.11) (a Dirichlet boundary condition) has already been set in DO loop 1, we can proceed directly to the calculation of u_z by a call to DSS034

```
C...
C...    UZ (FOR CONDUCTION)
        CALL DSS034(0.0D0,ZL,NR,NZ,2,U,UZ,0.0D0)
```

Note that the arguments now apply to z rather than r (but the grid is still NR × NZ = 11 × 31); also, the eighth argument is zero since we are considering conduction in the z direction.

(8) Boundary condition (H5.23) is then implemented in DO loop 4 (note the difference from DERV1)

```
C...
C...    Z = ZL
        DO 4 IR=1,NR
            UZ(IR,NZ)=0.0D0
4       CONTINUE
```

(9) Since both boundary conditions in z (eqs. (H5.11) and (H5.23)) are now defined, we can compute u_{zz} from u_z by a second call to DSS034

```
C...
C...    UZZ (FOR CONDUCTION)
        CALL DSS034(0.0D0,ZL,NR,NZ,2,UZ,UZZ,0.0D0)
```

(10) We then compute u_z for the convective term of eq. (H5.7), $-P_e v(r) u_z$

```
C...
C...    UZ (FOR CONVECTION)
        CALL DSS034(0.0D0,ZL,NR,NZ,2,U,UZ,1.0D0)
```

Note that the boundary condition for this derivative has already been defined in DO loop 1. Also, the eighth argument of DSS034 is now one (for the five-point, biased upwind, finite difference approximation of convective terms).

(11) Again, we reset $u_z(r, zl, t)$ in accordance with boundary condition (H5.23) (this derivative will not necessarily have a zero value as an output from the preceding call to DSS034)

```
C...
C...    Z = ZL
        DO 5 IR=1,NR
            UZ(IR,NZ)=0.0D0
5       CONTINUE
```

(12) At this point, we have all of the spatial derivatives in the RHS of eq. (H5.7), and we next proceed to the programming of eq. (H5.7).

```
C...
C...    R NE R0, Z NE 0
        DO 6 JZ=2,NZ
        DO 6 IR=1,NR-1
C...
C...        R = 0
            IF(IR.EQ.1)THEN
                UT(1,JZ)=-PE*V(1)*UZ(1,JZ)
     +                   +2.0D0*URR(1,JZ)
```

```
     +                    +UZZ(1,JZ)
C...
C...         R NE 0
             ELSE
     +       IF(IR.NE.1)THEN
                UT(IR,JZ)=-PE*V(IR)*UZ(IR,JZ)
     +                   +URR(IR,JZ)+(1.0D0/R(IR))*UR(IR,JZ)
     +                   +UZZ(IR,JZ)
             END IF
6        CONTINUE
```

The programming in DO loop 6 is in two parts:

(12.1) For $r = 0$, we include boundary condition (H5.9) by using

$$u_{rr} + \frac{1}{r}u_r = 2u_{rr}$$

(12.2) For $r > 0$, the programming is a straightforward implementation of the RHS of eq. (H5.7) (in fact, one of the important features of this approach is the close resemblance of the programming to the PDE).

This completes the programming of eq. (H5.7) and its associated boundary conditions in DERV5. Initial condition (H5.8) is again set in INITAL (with NCASE = 5); PRINT and the data file remain the same as for DERV1 to DERV4. The output from this programming is summarized below:

```
t =      1.00

     u(r,z,t)

    z     r = 0.0   r = 0.2   r = 0.4   r = 0.6   r = 0.8   r = 1.0
   .0      .0000     .0000     .0000     .0000     .0000     .0000
  1.0     -.0002    -.0010     .0013     .0752     .4036    1.0000
              .         .         .         .         .
              .         .         .         .         .
              .         .         .         .         .
 25.0      .6775     .7004     .7619     .8441     .9276    1.0000
 26.0      .6970     .7185     .7763     .8536     .9320    1.0000
 27.0      .7127     .7331     .7879     .8612     .9355    1.0000
 28.0      .7396     .7581     .8077     .8741     .9415    1.0000
 29.0      .7057     .7267     .7830     .8582     .9343    1.0000
 30.0      .9667     .9687     .9743     .9823     .9914    1.0000

t =      1.00

     u(r,z,t)

    z     r = 0.0   r = 0.2   r = 0.4   r = 0.6   r = 0.8   r = 1.0
```

| .0 | .0000 | .0000 | .0000 | .0000 | .0000 | .0000 |
| 3.0 | -.0025 | .0144 | .0889 | .2778 | .6066 | 1.0000 |

. . .
. . .
. . .

75.0	.9846	.9857	.9887	.9926	.9966	1.0000
78.0	.9872	.9881	.9906	.9938	.9971	1.0000
81.0	.9893	.9901	.9921	.9948	.9976	1.0000
84.0	.9911	.9917	.9934	.9957	.9980	1.0000
87.0	.9922	.9928	.9942	.9962	.9982	1.0000
90.0	.9950	.9953	.9961	.9973	.9987	1.0000

We can note the following points about this output:

(1) Near $t = 0$, the temperatures are slightly negative due to small oscillations introduced by the five-point, biased upwind approximations used in the calculation of the convective derivative u_z in eq. (H5.7)

$z = 30$

z	r = 0.0	r = 0.2	r = 0.4	r = 0.6	r = 0.8	r = 1.0
.0	.0000	.0000	.0000	.0000	.0000	.0000
1.0	-.0002	-.0010	.0013	.0752	.4036	1.0000

$z = 90$

| .0 | .0000 | .0000 | .0000 | .0000 | .0000 | .0000 |
| 3.0 | -.0025 | .0144 | .0889 | .2778 | .6066 | 1.0000 |

These negative values result from the discontinuous change between initial condition (H5.8) and boundary condition (H5.10) (Schiesser 1991, 136). The oscillations and associated negative values are relatively small (their most pronounced effect is in the automatic scaling of the vertical axis (for $u(0, z, t)$) in the plots of Figures H5.5a and H5.5b). Negative values did not appear in the output from DERV1 because the two-point upwind approximations used to approximate the convective derivative u_z in eq. (H5.7) preclude oscillations, but they introduce numerical diffusion (Schiesser 1991, 133).

(2) Again, we note the numerical distortion in the solution $u(r, 30, t)$, as with DERV1 (which is, of course, to be expected since the problem is boundary condition (H5.23), and not the calculation of the spatial derivatives, either by explicit programming of the finite differences, as in DERV1, or the calculation of the spatial derivatives by DSS034, as in DERV5). This feature of the solution is clear in Figure H5.5a, although again the solution is smooth for $z_l = 90$, as reflected in Figure H5.5b.

Next we consider DERV6, which is the counterpart of DERV2. The essential feature of DERV6 is again the use of boundary condition (H5.24) in place of (H5.23)

The Graetz problem with constant wall temperature 257

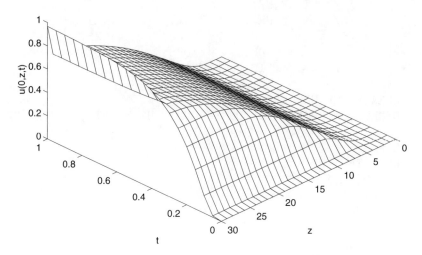

Figure H5.5a. Solution from Subroutine DERV5 for $z_l = 30$

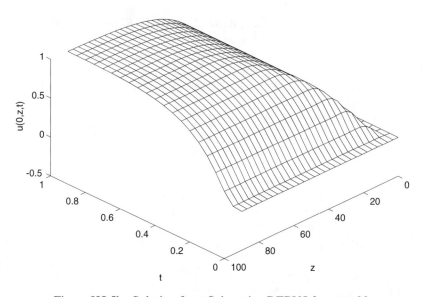

Figure H5.5b. Solution from Subroutine DERV5 for $z_l = 90$

```
      SUBROUTINE DERV6
      IMPLICIT DOUBLE PRECISION (A-H,O-Z)
      PARAMETER (NR=11,NZ=31)
      COMMON/T/           T,       NSTOP,       NORUN
     +      /Y/    U(NR,NZ)
     +      /F/    UT(NR,NZ)
     +      /G/       R(NR),       Z(NZ),       V(NR)
     +      /C/          PE,          RO,          ZL,
     +                   DR,         DRS,          DZ,       DZS
     +      /I/          IP,       NCASE
C...
C...  ARRAYS FOR UR, URR, UZ, UZZ
      DIMENSION UR(NR,NZ), URR(NR,NZ), UZ(NR,NZ), UZZ(NR,NZ)
C...
C...  Z = 0
      DO 1 IR=1,NR
         U(IR,1)=0.0D0
         UT(IR,1)=0.0D0
1     CONTINUE
C...
C...  R = RO, Z NE 0
      DO 2 JZ=2,NZ
         U(NR,JZ)=1.0D0
         UT(NR,JZ)=0.0D0
2     CONTINUE
C...
C...  UR
      CALL DSS034(0.0D0,RO,NR,NZ,1,U,UR,0.0D0)
C...
C...  R = 0
      DO 3 JZ=1,NZ
         UR(1,JZ)=0.0D0
3     CONTINUE
C...
C...  URR
      CALL DSS034(0.0D0,RO,NR,NZ,1,UR,URR,0.0D0)
C...
C...  UZ (FOR CONDUCTION)
      CALL DSS034(0.0D0,ZL,NR,NZ,2,U,UZ,0.0D0)
C...
C...  UZZ (FOR CONDUCTION)
      CALL DSS034(0.0D0,ZL,NR,NZ,2,UZ,UZZ,0.0D0)
C...
C...  UZ (FOR CONVECTION)
      CALL DSS034(0.0D0,ZL,NR,NZ,2,U,UZ,1.0D0)
C...
```

```
C...      R NE RO, Z NE 0, ZL
          DO 4 JZ=2,NZ-1
          DO 4 IR=1,NR-1
C...
C...          R = 0
              IF(IR.EQ.1)THEN
                  UT(1,JZ)=-PE*V(1)*UZ(1,JZ)
     +                    +2.0D0*URR(1,JZ)
     +                    +UZZ(1,JZ)
C...
C...          R NE 0
              ELSE
     +        IF(IR.NE.1)THEN
                  UT(IR,JZ)=-PE*V(IR)*UZ(IR,JZ)
     +                     +URR(IR,JZ)+(1.0D0/R(IR))*UR(IR,JZ)
     +                     +UZZ(IR,JZ)
              END IF
4         CONTINUE
C...
C...      R NE RO, Z = ZL
          DO 5 IR=1,NR-1
              UT(IR,NZ)=-PE*V(IR)*UZ(IR,NZ)
5         CONTINUE
          RETURN
          END
```

We can note the following points about DERV6:

(1) The COMMON area is the same as in DERV1 to DERV5.

(2) The coding closely parallels that in DERV5. The only essential differences are

(2.1) The coding for boundary condition (H5.23) has been removed.

(2.2) The programming for eq. (H5.7) in outer DO loop 4 does not include the last grid point in z (the upper limit of the outer DO loop 4 is $NZ - 1$ rather than NZ as in DERV5).

(2.3) Boundary condition (H5.24) is programmed in DO loop 5

```
C...
C...      R NE RO, Z = ZL
          DO 5 IR=1,NR-1
              UT(IR,NZ)=-PE*V(IR)*UZ(IR,NZ)
5         CONTINUE
```

All of the other programming and the data file remain the same (in INITAL, NCASE = 6).

The output from DERV6 is abbreviated below:

```
t =    1.00

   u(r,z,t)

  z      r = 0.0    r = 0.2    r = 0.4    r = 0.6    r = 0.8    r = 1.0
 .0       .0000      .0000      .0000      .0000      .0000      .0000
1.0      -.0002     -.0010      .0013      .0752      .4036     1.0000
            .
            .
            .
25.0      .6776      .7004      .7619      .8442      .9276     1.0000
26.0      .6966      .7181      .7760      .8534      .9319     1.0000
27.0      .7145      .7348      .7893      .8621      .9360     1.0000
28.0      .7312      .7503      .8016      .8701      .9397     1.0000
29.0      .7482      .7661      .8141      .8783      .9435     1.0000
30.0      .7573      .7746      .8209      .8828      .9456     1.0000

t =    1.00

   u(r,z,t)

  z      r = 0.0    r = 0.2    r = 0.4    r = 0.6    r = 0.8    r = 1.0
 .0       .0000      .0000      .0000      .0000      .0000      .0000
3.0      -.0025      .0144      .0889      .2778      .6066     1.0000
            .
            .
            .
75.0      .9846      .9858      .9887      .9926      .9966     1.0000
78.0      .9872      .9881      .9906      .9938      .9971     1.0000
81.0      .9893      .9901      .9921      .9948      .9976     1.0000
84.0      .9910      .9917      .9934      .9956      .9980     1.0000
87.0      .9925      .9930      .9944      .9963      .9983     1.0000
90.0      .9931      .9936      .9948      .9965      .9983     1.0000
```

We can note the following points about this output:

(1) Near $t = 0$, the temperatures are again slightly negative due to small oscillations introduced by the five-point, biased upwind approximations used in the calculation of the convective derivative u_z in eq. (H5.7).

$z_l = 30$

```
 .0       .0000      .0000      .0000      .0000      .0000      .0000
1.0      -.0002     -.0010      .0013      .0752      .4036     1.0000
```

$z_l = 90$

.0	.0000	.0000	.0000	.0000	.0000	.0000
3.0	-.0025	.0144	.0889	.2778	.6066	1.0000

As noted previously (for DERV5), these negative values result from the discontinuous change between initial condition (H5.8) and boundary condition (H5.10) (Schiesser 1991, 136). The oscillations and associated negative values are relatively small (their most pronounced effect is in the automatic scaling of the vertical axis (for $u(0, z, t)$) in the plots of Figures H5.6a and H5.6b). Negative values did not appear in the output from DERV2 because the two-point upwind approximations used to approximate the convective derivative u_z in eq. (H5.7) preclude oscillations, but they introduce numerical diffusion (Schiesser 1991, 133).

(2) The solution at $z_l = 30$ did not have the constant values of the solution from DERV2 because of the better resolution of the solution with respect to z obtained with the five-point, biased upwind approximations used in DSS034 (rather than the two-point, upwind approximations used in DERV2)

28.0	.7312	.7503	.8016	.8701	.9397	1.0000
29.0	.7482	.7661	.8141	.8783	.9435	1.0000
30.0	.7573	.7746	.8209	.8828	.9456	1.0000

(3) As a corollary to (2), the numerical solution from DERV6 is probably more accurate than the solution from DERV2 since fourth-order correct approximations of the spatial derivatives (in DSS034) were used in DERV6 whereas only second-order correct approximations were used in DERV2. This conclusion is supported by the fact that $u(0, 30, 1) = .7573$ from DERV6 is closer to the analytical solution, eq. (H5.6), to be discussed subsequently, than is $u(0, 30, 1) = .7313$ from DERV2. Also, the solution from DERV6 is free of unrealistic numerical effects, as indicated in Figures H5.6a and H5.6b.

In any case, we would expect that the solution from boundary condition (H5.12) is better than from boundary conditions (H5.23) or (H5.24), as discussed previously when considering DERV3. We now consider DERV7, which is the counterpart of DERV3

```
      SUBROUTINE DERV7
      IMPLICIT DOUBLE PRECISION (A-H,O-Z)
      PARAMETER (NR=11,NZ=31)
      COMMON/T/           T,       NSTOP,       NORUN
     +      /Y/    U(NR,NZ)
     +      /F/    UT(NR,NZ)
     +      /G/         R(NR),      Z(NZ),       V(NR)
     +      /C/            PE,         RO,          ZL,
     +                     DR,        DRS,          DZ,        DZS
     +      /I/            IP,      NCASE
C...
```

Heat transport

Graetz problem with constant wall temperature, zl = 30, DERV6

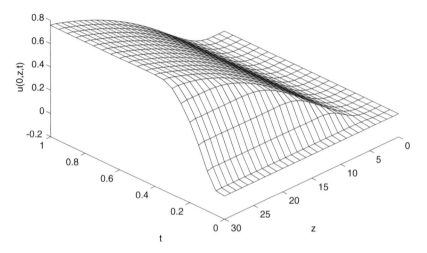

Figure H5.6a. Solution from Subroutine DERV6 for $z_l = 30$

Graetz problem with constant wall temperature, zl = 90, DERV6

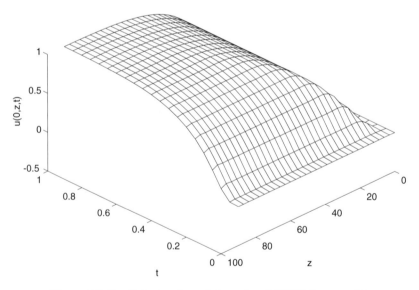

Figure H5.6b. Solution from Subroutine DERV6 for $z_l = 90$

```
C...     ARRAYS FOR UR, URR, UZ, UZZ
         DIMENSION UR(NR,NZ), URR(NR,NZ), UZ(NR,NZ), UZZ(NR,NZ)
C...
C...     Z = 0
         DO 1 IR=1,NR
            U(IR,1)=0.0D0
            UT(IR,1)=0.0D0
1        CONTINUE
C...
C...     R = R0, Z NE 0
         DO 2 JZ=2,NZ
            U(NR,JZ)=1.0D0
            UT(NR,JZ)=0.0D0
2        CONTINUE
C...
C...     UR
         CALL DSS034(0.0D0,R0,NR,NZ,1,U,UR,0.0D0)
C...
C...     R = 0
         DO 3 JZ=1,NZ
            UR(1,JZ)=0.0D0
3        CONTINUE
C...
C...     URR
         CALL DSS034(0.0D0,R0,NR,NZ,1,UR,URR,0.0D0)
C...
C...     UZ (FOR CONDUCTION)
         CALL DSS034(0.0D0,ZL,NR,NZ,2,U,UZ,0.0D0)
C...
C...     UZZ (FOR CONDUCTION)
         CALL DSS034(0.0D0,ZL,NR,NZ,2,UZ,UZZ,0.0D0)
C...
C...     UZ (FOR CONVECTION)
         CALL DSS034(0.0D0,ZL,NR,NZ,2,U,UZ,1.0D0)
C...
C...     R NE R0, Z NE 0, ZL
         DO 4 JZ=2,NZ-1
         DO 4 IR=1,NR-1
C...
C...        R = 0
            IF(IR.EQ.1)THEN
               UT(1,JZ)=-PE*V(1)*UZ(1,JZ)
     +                  +2.0D0*URR(1,JZ)
     +                  +UZZ(1,JZ)
C...
C...        R NE 0
```

```
              ELSE
     +        IF(IR.NE.1)THEN
                 UT(IR,JZ)=-PE*V(IR)*UZ(IR,JZ)
     +                     +URR(IR,JZ)+(1.0D0/R(IR))*UR(IR,JZ)
     +                     +UZZ(IR,JZ)
              END IF
4       CONTINUE
C...
C...    R NE R0, Z = ZL
        DO 5 IR=1,NR-1
C...
C...       R = 0
           IF(IR.EQ.1)THEN
              UT(1,NZ)=-PE*V(1)*UZ(1,NZ)
     +                  +2.0D0*URR(1,NZ)
C...
C...       R NE 0
           ELSE
     +        IF(IR.NE.1)THEN
                 UT(IR,NZ)=-PE*V(IR)*UZ(IR,NZ)
     +                     +URR(IR,NZ)+(1.0D0/R(IR))*UR(IR,NZ)
              END IF
5       CONTINUE
        RETURN
        END
```

We can note the following points about DERV7:

(1) The programming is similar to DERV6. The only essential difference is the implementation of boundary condition (H5.12) rather than (H5.24) in DO loop 5

```
C...
C...    R NE R0, Z = ZL
        DO 5 IR=1,NR-1
C...
C...       R = 0
           IF(IR.EQ.1)THEN
              UT(1,NZ)=-PE*V(1)*UZ(1,NZ)
     +                  +2.0D0*URR(1,NZ)
C...
C...       R NE 0
           ELSE
     +        IF(IR.NE.1)THEN
                 UT(IR,NZ)=-PE*V(IR)*UZ(IR,NZ)
     +                     +URR(IR,NZ)+(1.0D0/R(IR))*UR(IR,NZ)
              END IF
5       CONTINUE
```

(2) As expected, since the radial derivative is in boundary condition (H5.12) (but not (H5.24)), we must consider two cases:

(2.1) For $r = 0$, the programming of eq. (H5.12) is based on the use of the special case (from l'Hospital's rule, as developed previously in the discussion of subroutine DERV1)

$$u_{rr} + \frac{1}{r} u_u = 2u_{rr}$$

(2.2) Otherwise, for $r \neq 0$, eq. (H5.12) is programmed directly.

Again, the same INITAL (with NCASE = 7), PRINT, and data file were used. The output from DERV 7 is abbreviated below:

```
t =     1.00

  u(r,z,t)

    z    r = 0.0   r = 0.2   r = 0.4   r = 0.6   r = 0.8   r = 1.0
   .0     .0000     .0000     .0000     .0000     .0000     .0000
  1.0    -.0002    -.0010     .0013     .0752     .4036    1.0000
                      .
                      .
                      .
 25.0     .6776     .7004     .7619     .8442     .9276    1.0000
 26.0     .6966     .7181     .7760     .8534     .9319    1.0000
 27.0     .7145     .7348     .7892     .8620     .9360    1.0000
 28.0     .7314     .7504     .8017     .8702     .9397    1.0000
 29.0     .7472     .7652     .8134     .8779     .9433    1.0000
 30.0     .7621     .7791     .8244     .8851     .9467    1.0000

t =     1.00

  u(r,z,t)

    z    r = 0.0   r = 0.2   r = 0.4   r = 0.6   r = 0.8   r = 1.0
   .0     .0000     .0000     .0000     .0000     .0000     .0000
  3.0    -.0025     .0144     .0889     .2778     .6066    1.0000
                      .
                      .
                      .
 75.0     .9846     .9857     .9887     .9926     .9966    1.0000
 78.0     .9872     .9881     .9906     .9938     .9971    1.0000
 81.0     .9893     .9901     .9921     .9948     .9976    1.0000
 84.0     .9911     .9917     .9934     .9956     .9980    1.0000
 87.0     .9924     .9930     .9944     .9963     .9983    1.0000
 90.0     .9935     .9939     .9951     .9967     .9985    1.0000
```

266   Heat transport

We can note the following points about this output:

(1) The small negative values are again present initially due to the numerical oscillation of the five-point, biased upwind approximations used in DSS034 to compute the convective derivative u_z of eq. (H5.7) (see the discussion of DERV5 for elaboration of this point).

(2) The exiting temperature for $z = z_l$ is higher than from DERV6 ($u(0, 30, 1)$ = .7573 for DERV6, $u(0, 30, 1) = .7621$ for DERV7), which is due to additional heat conducted radially (i.e., from the radial heat conduction term of eq. (H5.12)).

The numerical solution is smooth for both $z_l = 30$ and 90, as indicated in Figures H5.7a and H5.7b.

Finally, we consider DERV8, which is the counterpart of DERV4

```
      SUBROUTINE DERV8
      IMPLICIT DOUBLE PRECISION (A-H,O-Z)
      PARAMETER (NR=11,NZ=31)
      COMMON/T/         T,        NSTOP,      NORUN
     +      /Y/    U(NR,NZ)
     +      /F/    UT(NR,NZ)
     +      /G/         R(NR),    Z(NZ),      V(NR)
     +      /C/         PE,       RO,         ZL,
     +                  DR,       DRS,        DZ,        DZS
     +      /I/         IP,       NCASE
C...
C...  ARRAYS FOR UR, URR, UZ, UZZ
      DIMENSION UR(NR,NZ), URR(NR,NZ), UZ(NR,NZ), UZZ(NR,NZ)
C...
C...  Z = 0
      DO 1 IR=1,NR
         U(IR,1)=0.0D0
         UT(IR,1)=0.0D0
1     CONTINUE
C...
C...  R = RO, Z NE 0
      DO 2 JZ=2,NZ
         U(NR,JZ)=1.0D0
         UT(NR,JZ)=0.0D0
2     CONTINUE
C...
C...  UR
      CALL DSS034(0.0D0,RO,NR,NZ,1,U,UR,0.0D0)
C...
C...  R = 0
      DO 3 JZ=1,NZ
```

The Graetz problem with constant wall temperature

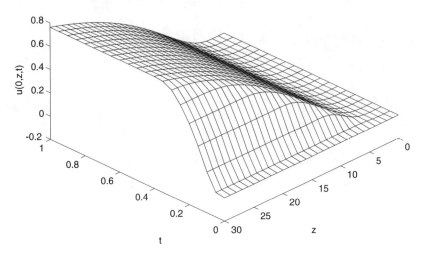

Figure H5.7a. Solution from Subroutine DERV7 for $z_l = 30$

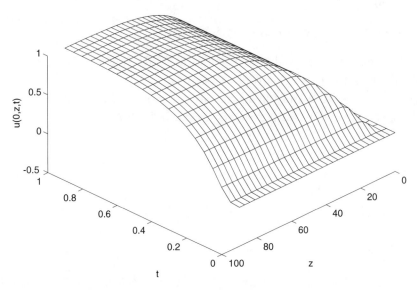

Figure H5.7b. Solution from Subroutine DERV7 for $z_l = 90$

```
              UR(1,JZ)=0.0D0
3       CONTINUE
C...
C...    URR
        CALL DSS034(0.0D0,R0,NR,NZ,1,UR,URR,0.0D0)
C...
C...    UZ (FOR CONVECTION)
        CALL DSS034(0.0D0,ZL,NR,NZ,2,U,UZ,1.0D0)
C...
C...    R NE R0
        DO 4 JZ=2,NZ
        DO 4 IR=1,NR-1
C...
C...       R = 0
           IF(IR.EQ.1)THEN
              UT(1,JZ)=-PE*V(1)*UZ(1,JZ)
     +                 +2.0D0*URR(1,JZ)
C...
C...       R NE 0
           ELSE
     +     IF(IR.NE.1)THEN
              UT(IR,JZ)=-PE*V(IR)*UZ(IR,JZ)
     +                  +URR(IR,JZ)+(1.0D0/R(IR))*UR(IR,JZ)
           END IF
4       CONTINUE
        RETURN
        END
```

We can note the following points about DERV8:

(1) The coding is essentially the same as in DERV7, except that the coding of eq. (H5.7) does not include u_{zz} (in DO loops 4).

(2) Since the PDE is only first order in z, boundary condition (H5.12) is not required, and the coding for it does not appear in DERV8 as it did in DERV7. In other words, the coding of the PDE is done entirely in DO loops 4, and special coding for $z = z_l$ is not required.

As before, the same INITAL (with NCASE = 8), PRINT, and data file were used. The output from DERV8 is abbreviated below:

```
t =    1.00

   u(r,z,t)

   z    r = 0.0   r = 0.2   r = 0.4   r = 0.6   r = 0.8   r = 1.0
   .0    .0000     .0000     .0000     .0000     .0000     .0000
  1.0   -.0002    -.0011     .0010     .0766     .4094    1.0000
```

The Graetz problem with constant wall temperature

.					.	
.					.	
.					.	
25.0	.6783	.7011	.7625	.8445	.9278	1.0000
26.0	.6973	.7188	.7765	.8537	.9321	1.0000
27.0	.7151	.7354	.7897	.8624	.9361	1.0000
28.0	.7320	.7510	.8022	.8705	.9399	1.0000
29.0	.7478	.7657	.8138	.8782	.9434	1.0000
30.0	.7627	.7796	.8249	.8854	.9468	1.0000

t =   1.00

  u(r,z,t)

z	r = 0.0	r = 0.2	r = 0.4	r = 0.6	r = 0.8	r = 1.0
.0	.0000	.0000	.0000	.0000	.0000	.0000
3.0	-.0028	.0143	.0892	.2790	.6082	1.0000
.					.	
.					.	
.					.	
75.0	.9847	.9858	.9887	.9926	.9966	1.0000
78.0	.9872	.9882	.9906	.9938	.9971	1.0000
81.0	.9894	.9901	.9921	.9949	.9976	1.0000
84.0	.9911	.9917	.9934	.9957	.9980	1.0000
87.0	.9925	.9930	.9944	.9963	.9983	1.0000
90.0	.9935	.9940	.9952	.9968	.9985	1.0000

We can note the following points about this output:

(1) As expected, the solutions from DERV7 and DERV8 are in close agreement since the problem is strongly convective (with $P_e = 60$), and therefore the axial conductive term u_{zz} in eq. (H5.7) has little effect. For example,

T(0,30,1)	T(0,90,1)	
.9891	.7467	DERV3
.9891	.7471	DERV4
.9935	.7621	DERV7
.9935	.7627	DERV8

(2) We have included in (1) representative values from DERV3 and DERV4, as well as from DERV7 and DERV8, to give an indication of

(2.1) The effect of using second-order approximations for the spatial derivatives of eq. (H5.7) (DERV3 and DERV4) compared with the fourth-order approximations of DSS034 (DERV7 and DERV8).

(2.2) The effect of axial conduction (in DERV3 and DERV7, and not in DERV4 and DERV8) for this strongly convective problem ($P_e = 60$); as we concluded previously, the effect is small.

(3) Of course, to more completely investigate the accuracy of any of the previous solutions, we could have

(3.1) Varied the number of axial and radial grid points from 31 and 11, respectively, and observed the effect on the numerical solution. Ideally, we would observe that the solution varied only in the third or fourth figure, for example, which generally is good enough for scientific and engineering applications; but if this were not the case, the number of grid points could, at least in principle, be increased until convergence to the required number of figures was achieved. This check on spatial convergence is generally an important part of assessing the accuracy of a numerical solution to a PDE problem. We did not do it here only to keep the discussion to reasonable length, and also, because the subsequent analysis of another numerical solution and the analytical solution of eq. (H5.6) indicates that the solutions computed by DERV7 and DERV8 are accurate to three figures.

(3.2) The error tolerance of the ODE integrator, in this case RKF45, could be varied to observe the effect of this tolerance on the temporal convergence (this tolerance is defined in the data file used with DERV1 to DERV8). Again, this should generally be done, but to conserve space in the preceding discussion, we assumed the tolerance of 0.00001 was sufficient for a solution with values that have a maximum value of one. Also, generally the error tolerance of the initial value ODE integration can be chosen to be substantially smaller than the spatial discretization error, for example, the error discussed in (3.1), so that the latter generally determines the overall solution accuracy.

Finally, the solutions from DERV8 are plotted in Figures H5.8a and H5.8b.

As expected, the plotted solutions of Figures 5.7a and 5.8a, and of Figures 5.7b and 5.8b, are in close agreement.

Next, we consider an alternative method of solution of the original Graetz problem, eq. (H5.1), without u_{zz}

$$P_e v(r) u_z = u_{rr} + \frac{1}{r} u_r \tag{H5.25}$$

with boundary conditions (H5.2), (H5.3), and (H5.4)

$$u_r(0, z) = 0, \qquad u(1, z) = 1$$
$$u(r, 0) = 0$$

We also compare the numerical solution of eqs. (H5.25) and (5.2) to (H5.4), to be discussed next, with (a) the analytical series solution, eq. (H5.6), and (b) the solution of the dynamic Graetz problem, eqs. (H5.7) to (H5.12), discussed previously.

The Graetz problem with constant wall temperature

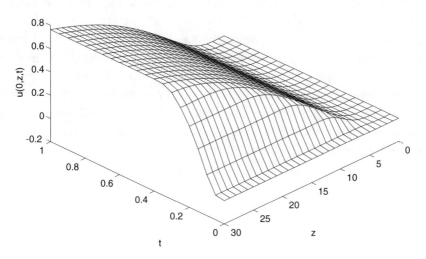

Figure H5.8a. Solution from Subroutine DERV8 for $z_l = 30$

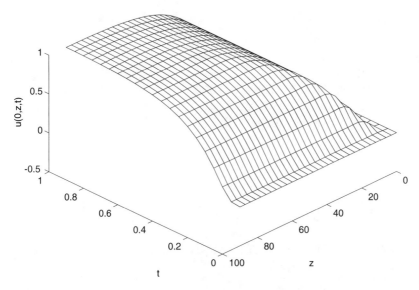

Figure H5.8b. Solution from Subroutine DERV8 for $z_l = 90$

To develop a numerical solution to eqs. (H5.25) and (H5.2) to (H5.4), we first note that eq. (H5.25) is first order in z, that is, it is an initial value problem in z. Thus, we can treat z as a "time-like" variable and apply an ODE integrator to it, subject to initial condition (H5.4). The problem then becomes one-dimensional in r. With this approach in mind, we can program subroutine INITAL to implement eq. (5.4) on a 21-point grid in r. (The increase in the number of radial grid points from 11 in the previous subroutines DERV1 to DERV8 to the present 21 was done to investigate the accuracy of the discretization in r, i.e., more points in r should give better accuracy in the numerical solution; also, increasing the number of radial grid points with this formulation increases the total number of ODEs to only 21, since we do not now have the multiplier effect of the axial grid, which previously would have required the use of $31 \times 21 = 651$ ODEs, e.g., with a 31-point axial grid.) Subroutine INITAL is listed below:

```
      SUBROUTINE INITAL
      IMPLICIT DOUBLE PRECISION (A-H,O-Z)
      PARAMETER(NR=21)
      COMMON/T/       Z,    NSTOP,    NORUN
     +      /Y/    U(NR)
     +      /F/    UZ(NR)
     +      /S/    UR(NR), URR(NR),   V(NR),         PE,
     +             R0,     DR,        DRS,           R(NR),
     +             ZL
     +      /I/    IP
C...
C...
C...  PROBLEM PARAMETERS
C...
C...    PECLET NUMBER
        IF(NORUN.EQ.1)PE= 60.0D0
        IF(NORUN.EQ.2)PE=120.0D0
        IF(NORUN.EQ.3)PE=180.0D0
C...
C...    RADIUS
        R0=1.0D0
C...
C...    LENGTH
        ZL=30.0D0
C...
C...    RADIAL GRID, VELOCITY PROFILE
        DR=R0/DFLOAT(NR-1)
        DRS=DR**2
        DO 1 I=1,NR
          R(I)=DR*DFLOAT(I-1)
```

```
              V(I)=(1.0D0-R(I)**2)
1        CONTINUE
C...
C...     INITIAL CONDITION
         DO 2 I=1,NR
              U(I)=0.0D0
2        CONTINUE
C...
C...     INITIAL DERIVATIVES
         CALL DERV
         IP=0
         RETURN
         END
```

We can note the following points about INITAL:

(1) The COMMON area defines 21 ODEs (in /Y/ and /F/); u of eq. (H5.25) is in array U(NR), and u_z is in UZ(NR). Note also that the initial value variable in /T/ is now designated Z.

```
         PARAMETER(NR=21)
         COMMON/T/       Z,    NSTOP,    NORUN
     +        /Y/   U(NR)
     +        /F/   UZ(NR)
     +        /S/   UR(NR), URR(NR),   V(NR),       PE,
     +              R0,        DR,     DRS,      R(NR),
     +              ZL
     +        /I/   IP
```

Also, u_r and u_{rr} of eq. (H5.25) are in arrays UR(NR) and URR(NR), respectively.

(2) The parameters of the problem are then defined

```
C...
C...
C...     PROBLEM PARAMETERS
C...
C...        PECLET NUMBER
            IF(NORUN.EQ.1)PE= 60.0D0
            IF(NORUN.EQ.2)PE=120.0D0
            IF(NORUN.EQ.3)PE=180.0D0
C...
C...        RADIUS
            R0=1.0D0
C...
C...        LENGTH
            ZL=30.0D0
```

In this case, solutions will be computed for three values of the Peclet number, $P_e = 60, 120,$ and 180; thus, the problem system becomes progressively more strongly convective for these three successive values of P_e. Only one length, $z_l = 30$, is used, and the objective then is to observe the solution at $z = 30$ with increasing P_e.

(3) The radial grid is then defined in DO loop 1

```
C...
C...   RADIAL GRID, VELOCITY PROFILE
       DR=RO/DFLOAT(NR-1)
       DRS=DR**2
       DO 1 I=1,NR
          R(I)=DR*DFLOAT(I-1)
          V(I)=(1.0D0-R(I)**2)
1      CONTINUE
```

(4) Initial condition (H5.4) is defined in DO loop 2

```
C...
C...   INITIAL CONDITION
       DO 2 I=1,NR
          U(I)=0.0D0
2      CONTINUE
```

(5) Finally, the initial derivatives are computed by a call to DERV, and a counter for printing and plotting in subroutine PRINT is initialized.

```
C...
C...   INITIAL DERIVATIVES
       CALL DERV
       IP=0
```

Now that the initial condition for eq. (H5.25) has been defined numerically, the numerical integration can begin. The ODEs that approximate the PDE, eq. (H5.25), on the 21-point radial grid are programmed in DERV

```
       SUBROUTINE DERV
       IMPLICIT DOUBLE PRECISION (A-H,O-Z)
       PARAMETER(NR=21)
       COMMON/T/       Z,      NSTOP,    NORUN
      +       /Y/    U(NR)
      +       /F/    UZ(NR)
      +       /S/    UR(NR), URR(NR),   V(NR),      PE,
      +              RO,     DR,        DRS,        R(NR),
      +              ZL
      +       /I/    IP
C...
C...   BOUNDARY CONDITION AT R = RO = 1
```

```
            U(NR)=1.0D0
            UZ(NR)=0.0D0
            NU=1
C...
C...    UR
            CALL DSS004(0.0D0,R0,NR,U,UR)
C...
C...    BOUNDARY CONDITION AT R = 0
            UR(1)=0.0D0
            NL=2
C...
C...    URR
            CALL DSS044(0.0D0,R0,NR,U,UR,URR,NL,NU)
C...
C...    PDE
            DO 1 I=1,NR-1
C...
C...        R = 0
            IF(I.EQ.1)THEN
            UZ(I)=1.0D0/(PE*2.0D0*V(I))*2.0D0*URR(I)
C...
C...        R NE 0
            ELSE IF(I.NE.1)THEN
            UZ(I)=1.0D0/(PE*2.0D0*V(I))*(URR(I)
         +(1.0D0/R(I))*UR(I))
            END IF
1           CONTINUE
            RETURN
            END
```

We can note the following points about DERV:

(1) The COMMON area is the same as in INITAL.

(2) Boundary condition (H5.3) is programmed first

```
C...
C...    BOUNDARY CONDITION AT R = R0 = 1
            U(NR)=1.0D0
            UZ(NR)=0.0D0
            NU=1
```

This is a Dirichlet boundary condition (since the dependent variable is defined at $r = 1$), which is designated with NU = 1 (to be used subsequently in the call to spatial differentiation routine DSS044).

(3) Subroutine DSS004, which computes first derivatives on a one-dimensional domain using five-point, centered finite differences, is called to compute u_r (in UR)

from u (in U)

```
C...
C...   UR
       CALL DSS004(0.0D0,R0,NR,U,UR)
```

DSS004 is discussed in detail elsewhere (Schiesser 1991, 105–11), so the arguments will be reviewed just briefly. The first and second arguments define the interval of the independent variable, in this case $0 \leq r \leq r_0$, with $r_0 = 1$. The third argument is the number of grid points, in this case NR $= 21$. The fourth argument is a one-dimensional array containing the dependent variable to be differentiated, in this case u ($=$ U, an input), and the fifth argument is the computed first derivative of the dependent variable, in this case u_r ($=$ UR, an output). Note that only centered finite differences are programmed in DSS004, and therefore this subroutine should be applied only to derivatives that represent conduction or diffusion (and not convection).

(4) Boundary condition (H5.2), which is a Neumann boundary condition designated with NL $= 1$, is programmed as

```
C...
C...   BOUNDARY CONDITION AT R = 0
       UR(1)=0.0D0
       NL=2
```

(5) Now that the two boundary conditions in r are defined numerically, the second-order radial derivative in eq. (H5.25), u_{rr} (in URR), can be computed by a call to DSS044

```
C...
C...   URR
       CALL DSS044(0.0D0,R0,NR,U,UR,URR,NL,NU)
```

Subroutine DSS044 computes second derivatives on a one-dimensional domain using five-point, centered finite differences. DSS044 is discussed in detail elsewhere (Schiesser 1991, 111–22), so the arguments will be reviewed just briefly. The first and second arguments define the interval of the independent variable, in this case $0 \leq r \leq r_0$, with $r_0 = 1$. The third argument is the number of grid points, in this case NR $= 21$. The fourth and fifth arguments are one-dimensional arrays containing the dependent variable to be differentiated and its first derivative, in this case u ($=$ U, input) and u_r ($=$ UR, input). The sixth argument is the computed second derivative of the dependent variable, in this case u_{rr} ($=$ URR, an output). Note that only centered finite differences are programmed in DSS004, and therefore this subroutine should be applied only to derivatives that represent conduction or diffusion (and not convection, which generally is modeled with first derivatives).

Also, although the array of first derivatives at NR grid points, UR, is an input to DSS044, only the end values of the first derivative are actually used in DSS044, that is, UR(1) and UR(NR), if Neumann boundary conditions are specified, NL = 2 and/or NU = 2. In the present case, NL = 2, so the first derivative at $r = 0$ (UR(1)) set from boundary condition (H5.2) is the only first derivative used in DSS044.

(6) Now that both radial derivatives in eq. (H5.25), u_r and u_{rr}, are computed, eq. (H5.25) can be programmed

```
C...
C...    PDE
        DO 1 I=1,NR-1
C...
C...      R = 0
          IF(I.EQ.1)THEN
          UZ(I)=1.0D0/(PE*2.0D0*V(I))*2.0D0*URR(I)
C...
C...      R NE 0
          ELSE IF(I.NE.1)THEN
          UZ(I)=1.0D0/(PE*2.0D0*V(I))*(URR(I)
     +    +(1.0D0/R(I))*UR(I))
          END IF
1       CONTINUE
```

Note that again we have to consider two cases:

(6.1) At $r = 0$, the group $\frac{1}{r}u_r$ is indeterminate, and as before, we use

$$u_{rr} + \frac{1}{r}u_r = 2u_{rr}$$

(6.2) For $r > 0$, eq. (H5.25) can be programmed directly (after division by $P_e v(r)$). Note also that the condition $v(1) = 0$ does not cause a problem since boundary condition (H5.3) has already been programmed, and therefore the grid point at $r = 1$ is not included in DO loop 1, that is, the ODE at $r = 1$ was programmed as UZ(NR) = 0.0D0.

The numerical integration of the 21 ODEs was performed by RKF45. Subroutine PRINT called by the main program in Appendix A (which also calls RKF45) to print and plot the solution is listed below:

```
        SUBROUTINE PRINT(NI,NO)
        IMPLICIT DOUBLE PRECISION (A-H,O-Z)
        PARAMETER(NR=21)
        COMMON/T/      Z,    NSTOP,    NORUN
     +        /Y/    U(NR)
     +        /F/    UZ(NR)
     +        /S/    UR(NR),  URR(NR),   V(NR),       PE,
     +               RO,      DR,       DRS,      R(NR),
```

```
     +                      ZL
     +            /I/       IP
C...
C...    MONITOR THE OUTPUT
        WRITE(*,*)NORUN,Z
C...
C...    OPEN A MATLAB FILE FOR PLOTTING
        IP=IP+1
        IF(IP.EQ.1)THEN
            OPEN(1,FILE='h5a.out')
            OPEN(2,FILE='h5b.out')
        END IF
C...
C...    WRITE THE NUMERICAL AND AND ANALYTICAL SOLUTIONS AT
C...    SELECTED Z
        IF(((IP-1)*(IP-51)*(IP-101)).EQ.0)THEN
C...
C...        NUMERICAL SOLUTION FOR 0 LE R LE R0
            NRI=4
            WRITE(NO,1)Z,(U(I),I=1,NR,NRI),
     +                  (UZ(I),I=1,NR,NRI)
1           FORMAT(//,' z = ',F4.1,/,
     +      12X,'r=0.0',4X,'r=0.2',4X,'r=0.4',
     +       4X,'r=0.6',4X,'r=0.8',4X,'r=1.0',
     +      /,' u(r,z)',6F8.5,
     +      /,' uz(r,z)',6F8.5,/)
C...
C...        SERIES SOLUTION AT R = 0
            CALL SERIES(PE,Z,UA)
            WRITE(NO,5)PE,Z,UA
5           FORMAT(/,' Series solution, Ua(0,z)',//,
     +          2X,'   Pe = ',F5.0  ,/,
     +          2X,'    z = ',F5.1  ,/,
     +          2X,'   Ua = ',F8.5 ,//)
        END IF
C...
C...    WRITE THE SOLUTION FOR MATLAB PLOTTING
        WRITE(1,2)Z,(U(I),I=1,NR)
2       FORMAT(22F8.4)
C...
C...    COMPUTE THE SERIES SOLUTION TO BE SUPERIMPOSED ON THE
C...    NUMERICAL SOLUTION (AT THE END OF THE THIRD SOLUTION)
        IF((IP.EQ.101).AND.(NORUN.EQ.3))THEN
            DO 3 NCASE=1,3
                IF(NCASE.EQ.1)PE= 60.0D0
                IF(NCASE.EQ.2)PE=120.0D0
```

```
            IF(NCASE.EQ.3)PE=180.0D0
         DO 4 I=1,101
            ZA=ZL*DFLOAT(I-1)/DFLOAT(101-1)
            CALL SERIES(PE,ZA,UA)
            WRITE(2,2)ZA,UA
4        CONTINUE
3     CONTINUE
      END IF
      RETURN
      END
```

We can note the following points about PRINT:

(1) The COMMON area is the same as in INITAL and PRINT.

(2) After monitoring the solution on unit ∗, files are opened at the beginning of the solution for plotting by Matlab

```
C...
C...  MONITOR THE OUTPUT
      WRITE(*,*)NORUN,Z
C...
C...  OPEN A MATLAB FILE FOR PLOTTING
      IP=IP+1
      IF(IP.EQ.1)THEN
         OPEN(1,FILE='h5a.dat')
         OPEN(2,FILE='h5b.dat')
      END IF
```

(3) The numerical solution is then written during the first, 51st, and 101st calls to PRINT (the data file to follow specifies that PRINT is called for $0 \leq z \leq 30$ at intervals of 0.3 or a total of 101 calls)

```
C...
C...  WRITE THE NUMERICAL AND AND ANALYTICAL SOLUTIONS AT
C...  SELECTED Z
      IF(((IP-1)*(IP-51)*(IP-101)).EQ.0)THEN
C...
C...     NUMERICAL SOLUTION FOR 0 LE R LE R0
         NRI=4
         WRITE(NO,1)Z,(U(I),I=1,NR,NRI),
     +               (UZ(I),I=1,NR,NRI)
1        FORMAT(//,' z = ',F4.1,/,
     +     12X,'r=0.0',4X,'r=0.2',4X,'r=0.4',
     +      4X,'r=0.6',4X,'r=0.8',4X,'r=1.0',
     +     /,'  u(r,z)',6F8.5,
     +     /,' uz(r,z)',6F8.5,/)
```

Note that the solution, $u(r, z)$, and its derivative, $u_z(r, z)$, are printed at six values of r, that is, $r = 0, 0.2, 0.4, \ldots, 1$ (since NRI $= 4$).

(4) The series solution of eq. (H5.6) is then evaluated for $r = 0$, and the current value of z by a call to subroutine SERIES

```
C...
C...      SERIES SOLUTION AT R = 0
          CALL SERIES(PE,Z,UA)
          WRITE(NO,5)PE,Z,UA
5         FORMAT(/,' Series solution, Ua(0,z)',//,
     +          2X,'      Pe = ',F5.0  ,/,
     +          2X,'      z  = ',F5.1  ,/,
     +          2X,'      Ua = ',F8.5  ,//)
      END IF
```

(5) The solution at the 21 values of r is then written to file 1 for subsequent Matlab plotting

```
C...
C...      WRITE THE SOLUTION FOR MATLAB PLOTTING
          WRITE(1,2)Z,(U(I),I=1,NR)
2         FORMAT(22F8.4)
```

(6) Finally, the series solution of eq. (H5.6) is computed for $0 \leq z \leq 30$ at intervals of 0.3 to coincide with the numerical solution

```
C...
C...      COMPUTE THE SERIES SOLUTION TO BE SUPERIMPOSED ON THE
C...      NUMERICAL SOLUTION (AT THE END OF THE THIRD SOLUTION)
          IF((IP.EQ.101).AND.(NORUN.EQ.3))THEN
              DO 3 NCASE=1,3
                  IF(NCASE.EQ.1)PE= 60.0D0
                  IF(NCASE.EQ.2)PE=120.0D0
                  IF(NCASE.EQ.3)PE=180.0D0
                  DO 4 I=1,101
                      ZA=ZL*DFLOAT(I-1)/DFLOAT(101-1)
                      CALL SERIES(PE,ZA,UA)
                      WRITE(2,2)ZA,UA
4                 CONTINUE
3             CONTINUE
```

Note that the series solution is evaluated at $P_e = 60, 120$, and 180, and is written to file 2 for comparison with the numerical solution.

Subroutine SERIES follows:

```
          SUBROUTINE SERIES(PE,Z,U)
C...
```

```
C...     The Graetz problem stated by Jakob (1949) is
C...
C...        2*vavg*v(r)*ua   = D*(ua     + (1/r)*ua )           (1)
C...                     z        rr            r
C...
C...        v(r) = 1 - r**(2)
C...
C...     where ua = (T - Ts)/(To - Ts)
C...
C...     For the numerical solution, u = (T - To)/(Ts - To).
C...     u can be expressed in terms of ua.  Thus,
C...
C...     T = Ts + ua*(To - Ts)
C...
C...     u = (Ts + ua*(To - Ts) - To)/(Ts - To) = 1 - ua
C...
C...     The analytical (series) solution given by Jakob,
C...     eq. (22-45), is
C...
C...     ua = 1.477*exp(-3.658*(1/Pe)*z)*R0(r)
C...
C...        - 0.810*exp(22.178*(1/Pe)*z)*R1(r)
C...
C...        + 0.385*exp(-53.05*(1/Pe)*z)*R2(r)
C...
C...     where z = z/ro, r = r/ro, (the LHS z and r are
C...     dimensionless),R0(0) = R1(0) = R2(0) = 1 (from Jakob,
C...     Table,22-1, page 455),
C...     Pe = ro*vavg/D  (D = k/(rho*Cp))
C...
C...     Jakob, M., Heat Transfer, John Wiley and  Sons, Inc.
C...     New York, Vol. 1, 1949
C...
         IMPLICIT DOUBLE PRECISION (A-H,O-Z)
C...
C...     RATIO OF Z/PE USED IN SERIES SOLUTION
         ZPE=Z/PE
C...
C...     UA(0,Z)
         TERM1= 1.477D0*DEXP( -3.658D0*ZPE)
         TERM2=-0.810D0*DEXP(-22.178D0*ZPE)
         TERM3= 0.385D0*DEXP( -53.05D0*ZPE)
         UA=TERM1+TERM2+TERM3
C...
C...     FOR COMPARISON WITH THE NUMERICAL SOLUTION, U = 1 - UA
         U=1.0D0-UA
```

```
C...
C...   END OF CALCULATION
       RETURN
       END
```

We can note the following points about SERIES:

(1) The arguments are for the Peclet number, P_e, axial position, z, and the solution u.

(2) The comments at the beginning define the PDE (eq. (H5.25)), except that the series solution reported by Jakob (1949) is for $u_a = 1 - u$. Also, the series solution is eq. (H5.6) with $r = 0$ (so that the radial eigenfunctions are $r_0(0) = r_1(0) = r_2(0) = 1$).

(3) The ratio z/P_e, which appears in the three exponentials in z, is evaluated first

```
C...
C...   RATIO OF Z/PE USED IN SERIES SOLUTION
       ZPE=Z/PE
```

(4) The series solution of eq. (H5.6) is then evaluated

```
C...
C...   UA(0,Z)
       TERM1= 1.477D0*DEXP( -3.658D0*ZPE)
       TERM2=-0.810D0*DEXP(-22.178D0*ZPE)
       TERM3= 0.385D0*DEXP( -53.05D0*ZPE)
       UA=TERM1+TERM2+TERM3
```

(5) Finally, the dimensionless temperature of the series solution, u_a, is converted to the dimensionless temperature of eq. (H5.20)

```
C...
C...   FOR COMPARISON WITH THE NUMERICAL SOLUTION, U = 1 - UA
       U=1.0D0-UA
```

This completes the programming. The data file for the three solutions ($P_e = 60, 120, 180$) is

```
Graetz problem, Pe = 60
0.           30.0        0.3
   21                              0.00001
Graetz problem, Pe = 120
0.           30.0        0.3
   21                              0.00001
Graetz problem, Pe = 180
0.           30.0        0.3
```

```
       21                    0.00001
END OF RUNS
```

Again, a tolerance of 0.00001 for the 21 ODEs is specified (but, of course, this is for the integration in z rather than the integration in t of DERV1 to DERV8). PRINT is called 101 times corresponding to $0 \leq z \leq 30$ at intervals of 0.3.

The output from PRINT is listed below:

```
RUN NO. -     1   Graetz problem, Pe = 60

INITIAL T -    .000D+00

  FINAL T -    .300D+02

  PRINT T -    .300D+00

NUMBER OF DIFFERENTIAL EQUATIONS -    21

MAXIMUM INTEGRATION ERROR -    .100D-04

z =    .0
             r=0.0     r=0.2     r=0.4     r=0.6     r=0.8     r=1.0
u(r,z)      .00000    .00000    .00000    .00000    .00000   1.00000
uz(r,z)     .00000    .00000    .00000    .00000    .00000    .00000

Series solution, Ua(0,z)

      Pe =    60.
       z =     .0
      Ua =   -.05200

z = 15.0
             r=0.0     r=0.2     r=0.4     r=0.6     r=0.8     r=1.0
u(r,z)      .41124    .45212    .56285    .71278    .86641   1.00000
uz(r,z)     .03493    .03282    .02675    .01792    .00832    .00000

Series solution, Ua(0,z)

      Pe =    60.
       z =   15.0
      Ua =    .41131

z = 30.0
             r=0.0     r=0.2     r=0.4     r=0.6     r=0.8     r=1.0
u(r,z)      .76279    .77965    .82491    .88539    .94680   1.00000
uz(r,z)     .01445    .01343    .01067    .00699    .00325    .00000
```

Series solution, Ua(0,z)

    Pe =   60.
     z =   30.0
    Ua =   .76284

RUN NO. -   2 Graetz problem, Pe = 120

INITIAL T -   .000D+00

  FINAL T -   .300D+02

  PRINT T -   .300D+00

NUMBER OF DIFFERENTIAL EQUATIONS - 21

MAXIMUM INTEGRATION ERROR -   .100D-04

z =   .0

	r=0.0	r=0.2	r=0.4	r=0.6	r=0.8	r=1.0
u(r,z)	.00000	.00000	.00000	.00000	.00000	.00000
uz(r,z)	.00000	.00000	.00000	.00000	.00000	.00000

Series solution, Ua(0,z)

    Pe = 120.
     z =   .0
    Ua = -.05200

z = 15.0

	r=0.0	r=0.2	r=0.4	r=0.6	r=0.8	r=1.0
u(r,z)	.11438	.16165	.30481	.52690	.77611	1.00000
uz(r,z)	.01949	.02092	.02195	.01778	.00909	.00000

Series solution, Ua(0,z)

    Pe = 120.
     z =   15.0
    Ua =   .11516

z = 30.0

	r=0.0	r=0.2	r=0.4	r=0.6	r=0.8	r=1.0
u(r,z)	.41124	.45212	.56285	.71278	.86641	1.00000
uz(r,z)	.01747	.01641	.01337	.00896	.00420	.00000

Series solution, Ua(0,z)

         Pe =   120.
          z =   30.0
         Ua =   .41131

RUN NO. -   3 Graetz problem, Pe = 180

INITIAL T -   .000D+00

  FINAL T -   .300D+02

  PRINT T -   .300D+00

NUMBER OF DIFFERENTIAL EQUATIONS -   21

MAXIMUM INTEGRATION ERROR -   .100D-04

z =   .0
           r=0.0   r=0.2   r=0.4   r=0.6   r=0.8   r=1.0
 u(r,z)   .00000   .00000   .00000   .00000   .00000   1.00000
 uz(r,z) .00000   .00000   .00000   .00000   .00000   .00000

Series solution, Ua(0,z)

         Pe =   180.
          z =   .0
         Ua = -.05200

z = 15.0
           r=0.0   r=0.2   r=0.4   r=0.6   r=0.8   r=1.0
 u(r,z)   .03200   .06400   .18475   .41893   .71855   1.00000
 uz(r,z) .00813   .01138   .01739   .01768   .00989   .00000

Series solution, Ua(0,z)

         Pe =   180.
          z =   15.0
         Ua =   .03405

z =   30.0
           r=0.0   r=0.2   r=0.4   r=0.6   r=0.8   r=1.0
 u(r,z)   .21690   .26626   .40544   .60377   .81441   1.00000
 uz(r,z) .01389   .01368   .01230   .00893   .00435   .00000

```
Series solution, Ua(0,z)

    Pe =   180.
     z =   30.0
    Ua =   .21725
```

We can note the following points about this output:

(1) The series solution of eq. (H5.6) is inaccurate at $z = 0$; the boundary (initial?) condition, eq. (H5.4), specifies $u(r, 0) = 0$, but the value from eq. (H5.6) is $-.05200$. This error is due to the slow convergence of the series at $z = 0$, so that more terms are required to closely satisfy boundary condition (H5.4).

(2) Fortunately, the series converges rapidly for $z > 0$, as demonstrated by the following comparison of the numerical and series solutions:

	$P_e = 60$	$P_e = 120$	$P_e = 180$
T(0, 30)(numerical)	.76279	.41124	.21690
T(0, 30)(analytical)	.76284	.41131	.21725

The rapid convergence of the series of eq. (H5.6) is due to the sharp decrease in the values of the three terms

$$e^{-3.658(1/P_e)z}, \; e^{-22.178(1/P_e)z}, \; e^{-53.05(1/P_e)z}$$

for $z/P_e = 30/60 = 0.5$. In fact, the first term of the series dominates, and accurate values of $u_a(r, z)$ are possible using just the first term; that is, the problem system of eqs. (H5.25) and (H5.2) to (H5.4) has the *dominant eigenvalue* -3.658. Also, the solution is invariant with respect to z/P_e, for example, the solution is the same for $z = 90$, $P_e = 180$ as for $z = 30$, $P_e = 60$ since in both cases $z/P_e = 0.5$.

We can also make a comparison between the steady state Graetz problem, eqs. (H5.25) and (H5.2) to (H5.4), the dynamic Graetz problem, eqs. (H5.7) to (H5.12), and the series solution, eq. (H5.6) (all at $P_e = 60$).

$u(0, 30, 1)$	.7621	(dynamic, DERV7)
$u(0, 30, 1)$	.7627	(dynamic, DERV8)
$u(0, 30)$	.76279	(steady state)
$u(0, 30)$	.76284	(series)

We mention again that Finlayson (1992, 302–7) reported a solution $T(0, 30) = 1$ for what appears to be the steady state problem (computed by finite elements).

Finally, the preceding solution to the steady state problem, eqs. (H5.25) and (H5.2) to (H5.4), is plotted in Figure H5.9a, and the centerline temperature, $T(0, z)$, from the numerical and analytical solutions is plotted vs z for $P_e = 60, 120, 180$ in Figure H5.9b. The numerical values used in these plots were read from files h5a.out and h5b.out written in the preceding subroutine PRINT for the steady state

The Graetz problem with constant wall temperature 287

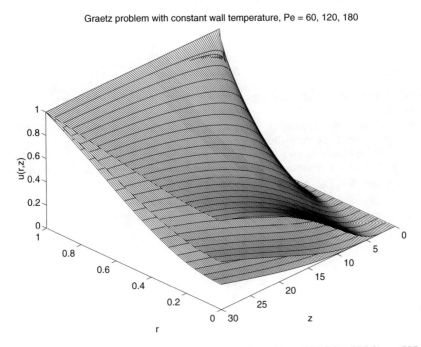

Figure H5.9a. Steady State Numerical Solutions from Eqs. (H5.26), (H5.2) to (H5.4)

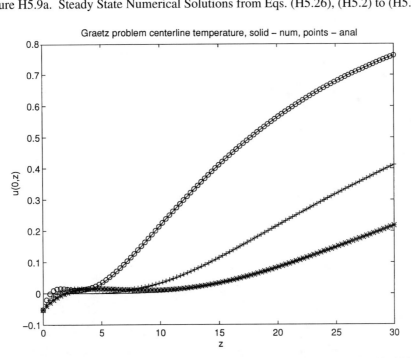

Figure H5.9b. Centerline Steady State Numerical and Series Solutions, $P_e = 60, 120, 180$

problem. The Matlab "m" files that produced Figures H5.1a to H5.9b are available as explained in the preface.

References

Douglas, J. 1984. Personal communication, Department of Mathematics, University of Chicago.
Finlayson, B. A. 1992. *Numerical Methods for Problems with Moving Fronts*. Seattle, WA: Ravenna Park Publishing.
Jakob, M. 1949. *Heat Transfer*. New York: John Wiley and Sons.
Schiesser, W. E. 1991. *The Numerical Method of Lines Integration of Partial Differential Equations*. San Diego: Academic Press.

H6

Heat exchanger dynamics

The tubular heat exchanger illustrated in Figure H6.1 is modeled by the following *first-order, linear hyperbolic partial differential equation (PDE)*:

$$\frac{\partial T}{\partial t} = -v\frac{\partial T}{\partial z} + \frac{4U}{\rho C_p D}(T_a - T) \tag{H6.1}$$

where

T	tube-side temperature
T_a	annulus temperature
z	position along the exchanger
t	time
v	linear fluid velocity
U	overall heat transfer coefficient
ρ	tube-side fluid density
C_p	tube-side fluid specific heat
D	tube diameter

$T(z, t)$ and $T_a(t)$ can be considered as perturbation variables, that is, variables that are departures from the normal steady state values and therefore are zero for $t < 0$. Thus, eq. (H6.1) can be integrated by the method of Laplace transforms. The initial and boundary conditions for eq. (H6.1) are

$$T(z, 0) = T_0(z), \tag{H6.2}$$

$$T(0, t) = T_e(t) \tag{H6.3}$$

where $T_0(z)$ and $T_e(t)$ are prescribed functions of z and t (and $T_e(t) = 0$ for $t < 0$). Note also that the annulus temperature can be a function of time, $T_a(t)$.

The problem consists of the following steps:

(1) Derive the analytical solution to eqs. (H6.1) to (H6.3).

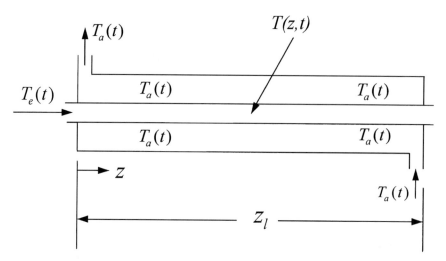

Figure H6.1. Single-Pass Heat Exchanger

(2) Specialize the analytical solution of (1) for $T_e(t) = 100$, $T_a(t) = 100$, $T_0(z) = 0$.

(3) Write a computer program that computes a numerical solution for (a) $T_e(t) = 100$, $T_a(t) = 0$; (b) $T_e(t) = 0$, $T_a(t) = 100$; and (c) $T_e(t) = 100$, $T_a(t) = 100$ (three cases) and compare the numerical solution with the analytical solution of (2). Use $v = 10$, $U = 0.1$, $\rho = C_p = 1$, $D = 2$, $z_l = 100$ (length of the exchanger).

Solution

We start by deriving eq. (H6.1). If an energy balance is written for an incremental section of the exchanger of Figure H6.1 of length Δz and cross-sectional area A_c,

$$A_c \Delta z \rho C_p \frac{\partial T}{\partial t} = A_c v \rho C_p \{T|_z - T|_{z+\Delta z}\} + \pi D \Delta z U (T_a - T)$$

Division by $A_c \Delta z C_p$ with $A_c = \pi D^2 / 4$ gives

$$\frac{\partial T}{\partial t} = -v \frac{\{T|_{z+\Delta z} - T|_z\}}{\Delta z} + \frac{4U}{\rho C_p D}(T_a - T)$$

Equation (H6.1) follows in the limit $z \to 0$.

Going on to the analytical solution, if the Laplace transform of $T(z, t)$ with respect to t is defined as

$$L_t\{T(z, t)\} = \int_0^\infty T(z, t) e^{-st} dt = \overline{T}(z, s)$$

Equations (H6.1) to (H6.3) transform to

$$s\overline{T} - T_0(z) = -v\frac{d\overline{T}}{dz} + (1/\tau)(\overline{T}_a - \overline{T}) \qquad (H6.4)$$

$$\overline{T}(0, s) = \overline{T}_e(s) \qquad (H6.5)$$

where $\overline{T}_a(s)$ is the Laplace transform of $T_a(t)$ and $\tau = \rho C_p D/(4U)$ (a characteristic time for heat transfer). Equation (H6.4) rearranges to

$$\frac{d\overline{T}}{dz} + \{s/v + 1/(\tau v)\}\overline{T} = 1/(\tau v)\overline{T}_a + T_0(z)/v \qquad (H6.6)$$

The solution to eq. (H6.6) subject to initial condition (H6.5) is easily derived using standard methods, for example, an integrating factor. However, we now derive a solution by taking a second Laplace transform with respect to z (since z is a semi-infinite independent variable)

$$L_z\{\overline{T}(z,s)\} = \int_0^\infty \overline{T}(z,s)e^{-pz}dz = \overline{\overline{T}}(p,s)$$

Equations (H6.5) and (H6.6) then transform to

$$p\overline{\overline{T}} - \overline{T}_e(s) + \{s/v + 1/(\tau v)\}\overline{\overline{T}} = 1/(\tau v)\overline{T}_a(s)/p + \overline{T}_0(p)/v \qquad (H6.7)$$

or

$$\overline{\overline{T}} = \frac{\overline{T}_e(s) + 1/(\tau v)\overline{T}_a(s)/p + \overline{T}_0(p)/v}{p + s/v + 1/(\tau v)} \qquad (H6.8)$$

Note that the original PDE problem, eqs. (H6.1), (H6.2), and (H6.3), has been reduced to algebraic eq. (H6.8). Also, the contributions of the three input temperatures, $T_a(t)$, $T_e(t)$, and $T_0(z)$, are clear from the right hand side (RHS) of eq. (H6.8).

$T(z, t)$ is then obtained from $\overline{\overline{T}}(p, s)$ of eq. (H6.8) by inverse Laplace transforms with respect to z and t. If we consider first the inverse transform with respect to z of each of the three RHS terms of eq. (H6.8),

$$\overline{f}_e(z,s) = L_z^{-1}\left\{\frac{\overline{T}_e(s)}{p + s/v + 1/(\tau v)}\right\}$$

$$= \{e^{-\{s/v+1/(\tau v)\}z}\}\overline{T}_e(s) \qquad (H6.9)$$

$$\overline{f}_a(z,s) = L_z^{-1}\left\{\frac{1/(\tau v)\overline{T}_a(s)/p}{p + s/v + 1/(\tau v)}\right\}$$

$$= \frac{1/(\tau v)}{s/v + 1/(\tau v)}\{1 - e^{-\{s/v+1/(\tau v)\}z}\}\overline{T}_a(s) \qquad (H6.10)$$

$$\overline{f}_0(z,s) = L_z^{-1}\left\{\frac{\overline{T}_0(p)/v}{p+s/v+1/(\tau v)}\right\}$$

$$= (1/v)\int_0^z e^{-\{s/v+1/(\tau v)\}(z-\lambda)}T_0(\lambda)d\lambda \tag{H6.11}$$

The RHS integral in eq. (H6.11) comes from the Laplace transform convolution (i.e., the inverse of the product of two functions in p).

$T(z,t)$ then comes from the inverse transform of $\overline{T}(z,s)$,

$$T(z,t) = L_t^{-1}\{\overline{f}_e(z,s) + \overline{f}_a(z,s) + \overline{f}_0(z,s)\}$$

$$= f_e(z,t) + f_a(z,t) + f_0(z,t) \tag{H6.12}$$

Whereas, in principle, the inversion of eq. (H6.12) can be carried out for general boundary and initial temperatures, $T_a(t)$, $T_e(t)$, and $T_0(z)$, in practice, the resulting solution, $T(z,t)$, would be complicated; also, general boundary and initial conditions are not generally required to represent realistic physical conditions. Typically, the initial temperature, $T_0(z)$, is a constant, T_{0c}, so that eq. (H6.11) becomes

$$\overline{f}_0(z,s) = (1/v)\int_0^z e^{-\{s/v+1/(\tau v)\}(z-\lambda)}T_0(\lambda)d\lambda$$

$$= (1/v)T_{0c}u(z)\frac{1}{\{s/v+1/(\tau v)\}}e^{-\{s/v+1/(\tau v)\}(z-\lambda)}\Big|_0^z$$

$$= (1/v)T_{0c}u(z)\frac{1}{\{s/v+1/(\tau v)\}}\{1 - e^{-\{s/v+1/(\tau v)\}z}\} \tag{H6.13}$$

where $u(z)$ is the unit step function

$$u(z) = 0, z < 0$$

$$= 1, z > 0 \tag{H6.14}$$

We now consider the t inversion of eqs. (H6.9), (H6.10), and (H6.13), starting with eq. (H6.9).

$$f_e(z,t) = L_t^{-1}\{\overline{f}_e(z,s)\} = L_t^{-1}\{e^{-\{s/v+1/(\tau v)\}z}\overline{T}_e(s)\}$$

$$= e^{-z/(\tau v)}L_t^{-1}\{e^{-s(z/v)}\overline{T}_e(s)\} = e^{-z/(\tau v)}u(t - z/v)T_e(t - z/v) \tag{H6.15}$$

where $u(t - z/v)$ is the unit step function defined by eq. (H6.14). Thus, the entering temperature, $T_e(t)$, leaves the system at z, displaced in time by $-z/v$, and attenuated by $\exp\{-z/(\tau v)\}$.

If $T_e(t)$ is a constant, T_{ec}, eq. (H6.15) reduces to

$$f_e(z,t) = T_{ec}u(t - z/v)e^{-z/(\tau v)} \tag{H6.16}$$

so that the constant entering temperature, T_{ec}, has no effect for $t < z/v$, and for $t > z/v$ is attenuated by $\exp\{-z/(\tau v)\}$. For the case of no heat transfer, $U = 0$, $\tau = \infty$ and eq. (H6.16) reduces to

$$f_e(z, t) = T_{ec} u(t - z/v)$$

that is, the constant temperature T_{ec} simply "flows" through the system as a step function, as expected.

Next, the inversion of eq. (H6.10) gives

$$f_a(z, t) = L_t^{-1}\{\overline{f}_a(z, s)\} = L_t^{-1}\left\{\frac{1/(\tau v)}{s/v + 1/(\tau v)}\{1 - e^{-\{s/v + 1/(\tau v)\}z}\}\overline{T}_a(s)\right\}$$

$$= (1/\tau)\int_0^t e^{-(1/\tau)(t-\lambda)} T_a(\lambda) d\lambda - (1/\tau) e^{-z/(\tau v)} \int_0^t u(t - z/v - \lambda)$$

$$\times e^{-(1/\tau)(t-z/v-\lambda)} T_a(\lambda) d\lambda$$

$$= (1/\tau)\int_0^t e^{-(1/\tau)(t-\lambda)} T_a(\lambda) d\lambda - (1/\tau) e^{-z/(\tau v)} u(t - z/v)$$

$$\times \int_0^{t-z/v} e^{-(1/\tau)(t-z/v-\lambda)} T_a(\lambda) d\lambda$$

$$= (1/\tau) e^{-t/\tau} \int_0^t e^{\lambda/\tau} T_a(\lambda) d\lambda - (1/\tau) u(t - z/v)$$

$$\times \int_0^{t-z/v} e^{-(1/\tau)(t-\lambda)} T_a(\lambda) d\lambda \tag{H6.17}$$

If $T_a(t)$ is a constant, T_{ac}, eq. (H6.17) gives

$$f_a(z, t) = T_{ac} e^{-(1/\tau)(t-\lambda)}\big|_0^t - T_{ac} u(t - z/v) e^{-(1/\tau)(t-\lambda)}\big|_0^{t-z/v}$$

$$= T_{ac} u(t)\{1 - e^{-t/\tau}\} - T_{ac} u(t - z/v)\{e^{-z/(\tau v)} - e^{-t/\tau}\} \tag{H6.18}$$

Note that $f_a(z, t)$ from eq. (H6.18) has the limiting values

$$f_a(z, 0) = 0, \qquad f_a(z, \infty) = T_{ac}(1 - e^{-z/(\tau v)})$$

which are reasonable physically since initially the annulus temperature will have no effect on the temperature $T(z, t)$, and after infinite time, $T(z, \infty)$ equals T_{ac} attenuated by the factor $(1 - e^{-z/(\tau v)})$ (with $T_e(z, t) = 0$). Also, for the case of no heat transfer, $U = 0$, $\tau = \infty$, and eq. (H6.18) reduces to

$$f_a(z, t) = T_{ac} u(t)\{1 - e^0\} - T_{ac} u(t - z/v)\{e^0 - e^0\} = 0$$

as expected, since the annulus temperature can have no effect on $T(z, t)$.

Finally, we consider the inversion of eq. (H6.13)

$$f_0(z,t) = L_t^{-1}\{\overline{f}_0(z,s)\} = L_t^{-1}\left\{\overline{T}_{0c}(s)\frac{1}{s+1/\tau}\{1-e^{-\{s/v+1/(\tau v)\}z}\}\right\}$$

$$= T_{0c}u(z)\{e^{-t/\tau} - e^{-z/(\tau v)}u(t-z/v)e^{-(t-z/v)/\tau}\} \quad \text{(H6.19)}$$

$f_0(z,t)$ from eq. (H6.19) has the properties

$$f_0(z,0) = T_{0c}u(z), \qquad f_0(z,\infty) = 0$$

which are reasonable physically, since initially the temperature is T_{0c}, and after infinite time, $T(z,\infty) = 0$ since the initial temperature, T_{0c}, has "flowed out" of the exchanger (with $T_e(t) = T_a(t) = 0$). For the case of no heat transfer, $U = 0, \tau = 0$, and eq. (H6.19) reduces to

$$f_0(z,0) = T_{0c}u(z)\{1 - u(t-z/v)\}$$

as expected, that is, the temperature remains at the initial temperature, T_{0c}, for $t < z/v$, then is zero for $t > z/v$.

We now take as the analytical solution, for comparison with the subsequent numerical solution, a combination of eqs. (H6.12), (H6.16), (H6.18), and (H6.19)

$$T(z,t) = f_e(z,t) + f_a(z,t) + f_0(z,t) = T_{ec}u(t-z/v)e^{-z/(\tau v)}$$
$$+ T_{ac}u(t)\{1 - e^{-t/\tau}\} - T_{ac}u(t-z/v)\{e^{-z/(\tau v)} - e^{-t/\tau}\}$$
$$+ T_{0c}u(z)\{e^{-t/\tau} - e^{-z/(\tau v)}u(t-z/v)e^{-(t-z/v)/\tau}\} \quad \text{(H6.20)}$$

that is, we have taken the constant temperature cases for $T_e(t)(= T_{ec})$, $T_a(t)(= T_{ac})$, and $T_0(z)(= T_{0c})$, which are physically realistic. Specifically, we consider the calculation of analytical and numerical solutions for $T_{ec} = 100$, $T_{ac} = 100$, $T_{0c} = 0$.

The coding of eqs. (H6.1) to (H6.3) starts with subroutine INITAL to define initial condition (H6.2)

```
      SUBROUTINE INITAL
      IMPLICIT DOUBLE PRECISION (A-H,O-Z)
      PARAMETER(NZ=101)
      COMMON/T/     TIME,      NSTOP,      NORUN
     1     /Y/      T(NZ)
     2     /F/      TT(NZ)
     3     /S/      TZ(NZ)
     4     /P/         V,         ZL,       TAU,
     5                TEC,        TAC,       TOC
     6     /I/        IP,       NCASE
C...
C... BOUNDARY CONDITION (AT Z = 0)
C...
```

```
C...          NCASE = 1 - CONSTANT
C...
C...          NCASE = 2 - VARIABLE
C...
      NCASE=1
C...
C...    PARAMETERS
C...
C...          LENGTH
      ZL=100.0D0
C...
C...          VELOCITY
      V=10.0D0
C...
C...          DENSITY
      RHO=1.0D0
C...
C...          SPECIFIC HEAT
      CP=1.0D0
C...
C...          TUBE DIAMETER
      D=2.0D0
C...
C...          HEAT TRANSFER COEFFICIENT
      U=0.1D0
C...
C...    TAU
      TAU=(RHO*CP*D)/(4.0D0*U)
C...
C...    ENTERING TEMPERATURE, ANNULUS TEMPERATURE
      IF(NORUN.EQ.1)THEN
         TEC=100.0D0
         TAC=0.0D0
      ELSE
     +IF(NORUN.EQ.2)THEN
         TEC=0.0D0
         TAC=100.0D0
      ELSE
     +IF(NORUN.EQ.3)THEN
         TEC=100.0D0
         TAC=100.0D0
      END IF
C...
C...    INITIAL CONDITION
      TOC=0.0D0
      DO 1 I=1,NZ
```

```
          T(I)=TOC
1     CONTINUE
C...
C...  INITIALIZE COUNTER FOR THE OUTPUT
      IP=0
      RETURN
      END
```

We can note the following points about INITAL:

(1) Double precision coding is used. Equation (H6.1) is approximated on a 101-point grid in z

```
      IMPLICIT DOUBLE PRECISION (A-H,O-Z)
      PARAMETER(NZ=101)
      COMMON/T/    TIME,      NSTOP,       NORUN
     1      /Y/    T(NZ)
     2      /F/    TT(NZ)
     3      /S/    TZ(NZ)
     4      /P/         V,       ZL,         TAU,
     5                TEC,      TAC,         TOC
     6      /I/        IP,    NCASE
```

The temperature, $T(z, t)$, of eq. (H6.1) is in array T(NZ), and the partial derivatives $\partial T/\partial t$ and $\partial T/\partial z$ in eq. (H6.1) are in arrays TT(NZ) and TZ(NZ), respectively. Note that T(NZ) is in COMMON/Y/ and TT(NZ) is in COMMON/F/ so that in the method of lines (MOL) approximation of eq. (H6.1), 101 ordinary differential equations (ODEs) are integrated with respect to t.

(2) Boundary condition (H6.3) has two forms:

```
C...
C...  BOUNDARY CONDITION (AT Z = 0)
C...
C...     NCASE = 1 - CONSTANT
C...
C...     NCASE = 2 - VARIABLE
C...
      NCASE=1
```

For NCASE $= 1$, $T_e(t)$ in eq. (H6.3) is a constant, and for NCASE $= 2$, $T_e(t)$ is an exponentially increasing function of t (discussed subsequently).

(3) The model parameters are then defined numerically

```
C...
C...  PARAMETERS
C...
C...     LENGTH
```

```
              ZL=100.0D0
C...
C...          VELOCITY
              V=10.0D0
C...
C...          DENSITY
              RHO=1.0D0
C...
C...          SPECIFIC HEAT
              CP=1.0D0
C...
C...          TUBE DIAMETER
              D=2.0D0
C...
C...          HEAT TRANSFER COEFFICIENT
              U=0.1D0
C...
C...          TAU
              TAU=(RHO*CP*D)/(4.0D0*U)
```

The characteristic time for heat transfer, $\tau = C_p D/(4U)$, is also computed.

(4) The three cases are then programmed using the run counter, NORUN

```
C...
C...          ENTERING TEMPERATURE, ANNULUS TEMPERATURE
              IF(NORUN.EQ.1)THEN
                  TEC=100.0D0
                  TAC=0.0D0
              ELSE
             +IF(NORUN.EQ.2)THEN
                  TEC=0.0D0
                  TAC=100.0D0
              ELSE
             +IF(NORUN.EQ.3)THEN
                  TEC=100.0D0
                  TAC=100.0D0
              END IF
```

These values of the boundary and annulus temperatures, TEC and TAC, were selected to illustrate some points about the solution to eq. (H6.1) discussed subsequently.

(5) Initial condition (H6.2) is then programmed

```
C...
C...          INITIAL CONDITION
              TOC=0.0D0
              DO 1 I=1,NZ
```

```
              T(I)=TOC
1       CONTINUE
```

Finally, an integer counter, which is used in subroutine PRINT is initialized.

Subroutine DERV is listed below:

```
        SUBROUTINE DERV
        IMPLICIT DOUBLE PRECISION (A-H,O-Z)
        PARAMETER(NZ=101)
        COMMON/T/      TIME,       NSTOP,      NORUN
     1       /Y/       T(NZ)
     2       /F/       TT(NZ)
     3       /S/       TZ(NZ)
     4       /P/        V,          ZL,         TAU,
     5                 TEC,         TAC,        TOC
     6       /I/        IP,        NCASE
C...
C...    BOUNDARY CONDITION
        IF(NCASE.EQ.1)T(1)=TEC
        IF(NCASE.EQ.2)T(1)=TEC*FBC(TIME)
        TT(1)=0.0D0
C...
C...    DERIVATIVE TZ
C...
C...        FIVE POINT BIASED UPWIND
            CALL DSS020(0.0D0,ZL,NZ,T,TZ,V)
C...
C...        VARIABLE ORDER UPWIND
C...        CALL DSS024(0.0D0,ZL,NZ,T,TZ,V)
C...
C...    PDE
        DO 1 I=2,NZ
            TT(I)=-V*TZ(I)-(1.0D0/TAU)*(T(I)-TAC)
1       CONTINUE
        RETURN
        END
```

We can note the following points about subroutine DERV:

(1) The COMMON area is the same as in subroutine INITAL.

(2) Boundary condition (H6.3) is programmed first

```
C...
C...    BOUNDARY CONDITION
        IF(NCASE.EQ.1)T(1)=TEC
        IF(NCASE.EQ.2)T(1)=TEC*FBC(TIME)
        TT(1)=0.0D0
```

For NCASE = 1, the entering temperature, T(1), is a constant, TEC (= 100 set in INITAL). For NCASE = 2, the entering temperature, T(1), increases exponentially with t according to function FBC, which is listed below:

```
      DOUBLE PRECISION FUNCTION FBC(TIME)
C...
C...  FUNCTION FBC DEFINES THE VARIABLE (TIME DEPENDENT)
C...  BOUNDARY CONDITION
C...
      IMPLICIT DOUBLE PRECISION (A-H,O-Z)
C...
C...  CHARACTERISTIC TIME FOR THE EXPONENTIAL FUNCTION
      TAUBC=1.0D0
C...
C...  EXPONENTIAL INCREASE OF THE BOUNDARY TEMPERATURE WITH
C...  TIME
      IF(TIME.LE.0.0D0)FBC=0.0D0
      IF(TIME.GT.0.0D0)FBC=1.0D0-DEXP(-TIME/TAUBC)
      RETURN
      END
```

FBC implements the function

$$T_e(0, t) = 0, \quad t < 0$$
$$T_e(0, t) = 1 - e^{-t/\tau_{bc}}, \quad t > 0 \quad \text{(H6.21)}$$

where τ_{bc} (= TAUBC) = 1. Thus, FBC provides a smooth (exponential) transition from the initial condition $T_0(z, 0) = 0$ at the boundary $z = 0$ to the boundary temperature $T_e(0, t) = T_{ec}$, which makes the calculation of a numerical solution to eq. (H6.1) substantially easier than for the case of an instantaneous change in the boundary temperature from $T_0(z, 0) = 0$ to $T_e(0, t) = T_{ec}$. The execution of the code for NCASE = 2 (which uses function FBC) is considered subsequently.

(3) The partial derivative, $\partial T/\partial z$ (= TZ), in eq. (H6.1) is then computed by a call to subroutine DSS020

```
C...
C...  DERIVATIVE TZ
C...
C...       FIVE POINT BIASED UPWIND
           CALL DSS020(0.0D0,ZL,NZ,T,TZ,V)
C...
C...       VARIABLE ORDER UPWIND
C...       CALL DSS024(0.0D0,ZL,NZ,T,TZ,V)
```

Note that the call to DSS024 is commented out for the initial execution of the code. We will return subsequently to the calculation of a solution using DSS024.

300   Heat transport

Subroutine DSS020 is based on the use of five-point, biased upwind approximations for the calculation of spatial derivatives such as $\partial T/\partial z$; the details of these approximations are given by Schiesser (1991).

(4) Equation (H6.1) is then programmed over the grid of NZ points

```
C...
C...    PDE
        DO 1 I=2,NZ
           TT(I)=-V*TZ(I)-(1.0D0/TAU)*(T(I)-TAC)
1       CONTINUE
```

Note the close resemblance of the coding to the PDE (eq. (H6.1)), which is one of the significant advantages of the numerical method of lines.

Subroutines INITAL and DERV contain the essential coding of the problem defined by eqs. (H6.1) to (H6.3). The only remaining requirement is to present the numerical solution in a useful format. This is done in subroutine PRINT

```
        SUBROUTINE PRINT(NI,NO)
        IMPLICIT DOUBLE PRECISION (A-H,O-Z)
        PARAMETER(NZ=101)
        COMMON/T/    TIME,       NSTOP,      NORUN
       1      /Y/    T(NZ)
       2      /F/    TT(NZ)
       3      /S/    TZ(NZ)
       4      /P/         V,       ZL,        TAU,
       5                TEC,      TAC,        TOC
       6      /I/        IP,    NCASE
C...
C...    MONITOR OUTPUT (ON UNIT *)
        IF(TIME.LT.0.01D0)WRITE(*,4)NCASE,TEC,TAC,TOC,TAU
4       FORMAT(//,' NCASE = ',I3,//,
       +            ' TEC = ',F7.2,'    TAC = ',F7.2,
       +            ' TOC = ',F7.2,'    TAU = ',F7.2,/)
        WRITE(*,3)NORUN,TIME,T(NZ),EXACT(NCASE)
3       FORMAT(' NORUN = ',I2,'    TIME = ',F6.2,
       +         ' T(NZ) = ',F7.2,'    TA = ',F7.2)
C...
C...    PRINT A HEADING FOR THE NUMERICAL AND EXACT SOLUTIONS
        IP=IP+1
        IF(IP.EQ.1)THEN
           WRITE(NO,1)
1          FORMAT(' TUBULAR HEAT EXCHANGER',//,
       1            9X,'t',5X,'T(zl,t)',4X,'Ta(zl,t)',8X,'Diff')
        END IF
C...
C...    PRINT THE NUMERICAL AND ANALYTICAL SOLUTIONS, AND THE
C...    DIFFERENCE
```

```
      TA=EXACT(NCASE)
      DIFF=T(NZ)-TA
      WRITE(NO,2)TIME,T(NZ),TA,DIFF
2     FORMAT(F10.2,3F12.2)
C...
C...  MATLAB PLOTTING OF THE SOLUTION
      CALL PLOTM
      RETURN
      END
```

We can note the following points about PRINT:

(1) The COMMON area is the same as in INITAL and DERV.

(2) The progress of the solution is monitored on the screen of the computer by writing to unit * (which designates the screen for most computers)

```
C...
C...  MONITOR OUTPUT (ON UNIT *)
      IF(TIME.LT.0.01D0)WRITE(*,4)NCASE,TEC,TAC,TOC,TAU
4     FORMAT(//,' NCASE = ',I3,//,
     +          ' TEC = ',F7.2,'    TAC = ',F7.2,
     +          ' TOC = ',F7.2,'    TAU = ',F7.2,/)
      WRITE(*,3)NORUN,TIME,T(NZ),EXACT(NCASE)
3     FORMAT(' NORUN = ',I2,'    TIME = ',F6.2,
     +       ' T(NZ) = ',F7.2,'    TA = ',F7.2)
```

This procedure of following the progress of the solution is quite useful, particularly during the initial development of the code. EXACT is a function that implements the analytical solution of eq. (H6.20).

(3) The integer counter IP is then incremented from the initial value of zero (set in INITAL) to one, and this value is used to write a heading for the solution

```
C...
C...  PRINT A HEADING FOR THE NUMERICAL AND EXACT SOLUTIONS
      IP=IP+1
      IF(IP.EQ.1)THEN
         WRITE(NO,1)
1        FORMAT(' TUBULAR HEAT EXCHANGER',//,
     1          9X,'t',5X,'T(zl,t)',4X,'Ta(zl,t)',8X,'Diff')
      END IF
```

(4) Again, the exact solution at $z = z_l$ ($= ZL = 100$ set in INITAL) is computed according to eq. (H6.20), by a call to function EXACT, and the numerical and analytical solutions at $z = z_l$, and their difference, are then displayed

```
C...
C...  PRINT THE NUMERICAL AND ANALYTICAL SOLUTIONS, AND THE
C...  DIFFERENCE
```

```
            TA=EXACT(NCASE)
            DIFF=T(NZ)-TA
            WRITE(NO,2)TIME,T(NZ),TA,DIFF
2           FORMAT(F10.2,3F12.2)
```

(5) Finally, a call to subroutine PLOTM writes the numerical solution to a file for subsequent plotting (by Matlab in this case, but any graphics system that can accept a file of numbers can be used)

```
C...
C...   MATLAB PLOTTING OF THE SOLUTION
       CALL PLOTM
```

Function EXACT, which is an implementation of eq. (H6.20), is listed below:

```
       DOUBLE PRECISION FUNCTION EXACT(NCASE)
C...
C...   FUNCTION EXACT COMPUTES THE EXACT SOLUTION TO THE PDE
C...   FOR A TUBULAR HEAT EXCHANGER
C...
C...      NCASE = 1 - CONSTANT BOUNDARY CONDITION TEMPERATURE
C...
C...      NCASE = 2 - VARIABLE BOUNDARY CONDITION TEMPERATURE
C...
       IMPLICIT DOUBLE PRECISION (A-H,O-Z)
       PARAMETER(NZ=101)
       COMMON/T/      TIME,       NSTOP,      NORUN
     1       /Y/      TD(NZ)
     2       /F/      TT(NZ)
     3       /S/      TZ(NZ)
     4       /P/         V,         ZL,        TAU,
     5                 TEC,        TAC,        TOC
     6       /I/         IP,       NDUM
C...
C...   PRECOMPUTE SOME EXPONENTIALS
       EXPZ=DEXP(-ZL/(TAU*V))
       EXPT=DEXP(-TIME/TAU)
C...
C...   SOLUTION WITH T LT Z/V
       IF((TIME-ZL/V).LE.0.0D0)THEN
           EXACT=TAC*(1.0D0-EXPT)+
     +           TOC*EXPT
C...
C...   SOLUTION WITH T GT Z/V
       ELSE
     +IF((TIME-ZL/V).GT.0.0D0)THEN
           EXPTD=DEXP(-(TIME-ZL/V)/TAU)
```

```
C...
C...        COMPLETE SOLUTION ACCORDING TO BOUNDARY CONDITION
C...
C...        CONSTANT BOUNDARY CONDITION
            IF(NCASE.EQ.1)THEN
                EXACT=TEC*EXPZ+
     +              TAC*((1.0D0-EXPT)-(EXPZ-EXPT))+
     +              TOC*(EXPT-EXPZ*EXPTD)
C...
C...        VARIABLE BOUNDARY CONDITION
            ELSE
     +      IF(NCASE.EQ.2)THEN
                EXACT=TEC*EXPZ*FBC(TIME-ZL/V)+
     +              TAC*((1.0D0-EXPT)-(EXPZ-EXPT))+
     +              TOC*(EXPT-EXPZ*EXPTD)
            END IF
          END IF
          RETURN
          END
```

We can note the following points about EXACT:

(1) Two cases are programmed, for NCASE = 1 (constant entering temperature) and NCASE = 2 (exponentially increasing entering temperature given by eq. (H6.21)). For either case, two exponentials are required in eq. (H6.20)

```
C...
C...    PRECOMPUTE SOME EXPONENTIALS
        EXPZ=DEXP(-ZL/(TAU*V))
        EXPT=DEXP(-TIME/TAU)
```

where $EXPZ = e^{-z_l/(\tau v)}$ and $EXPT = e^{-t/\tau}$.

(2) For $t > z/v$ (with $z = z_l$), eq. (H6.20) includes the exponential $EXPTD = e^{-(t-z/v)/\tau}$ (with $z = z_l$), which is programmed as

```
C...
C...    SOLUTION WITH T GT Z/V
        ELSE
     +  IF((TIME-ZL/V).GT.0.0D0)THEN
            EXPTD=DEXP(-(TIME-ZL/V)/TAU)
```

(3) For NCASE = 1, eq. (H6.20) (with $t < z/v$) is

$$T(z,t) = f_e(z,t) + f_a(z,t) + f_0(z,t)$$

$$= T_{ac}\{u(t)(1 - e^{-t/\tau})\} + T_{0c}u(z)\{e^{-t/\tau}\}$$

$$= T_{ac}(1 - e^{-t/\tau}) + T_{0c}e^{-t/\tau} \qquad (H6.22)$$

Equation (H6.22) is programmed as

```
C...
C...    SOLUTION WITH T LT Z/V
        IF((TIME-ZL/V).LE.0.0D0)THEN
            EXACT=TAC*(1.0D0-EXPT)+
     +          TOC*EXPT
```

(4) For NCASE = 1, eq. (H6.20) (with $t > z/v$) is

$$T(z,t) = f_e(z,t) + f_a(z,t) + f_0(z,t) = T_{ec}e^{-z/(\tau v)}$$
$$+ T_{ac}\{(1 - e^{-t/\tau}) - (e^{-z/(\tau v)} - e^{-(t-z/v)/\tau})\}$$
$$+ T_{0c}u(z)\{e^{-t/\tau} - e^{-z/(\tau v)}e^{-(t-z/v)/\tau}\} \quad (H6.23)$$

Equation (H6.23) is programmed as

```
C...
C...        CONSTANT BOUNDARY CONDITION
            IF(NCASE.EQ.1)THEN
                EXACT=TEC*EXPZ+
     +              TAC*((1.0D0-EXPT)-(EXPZ-EXPT))+
     +              TOC*(EXPT-EXPZ*EXPTD)
```

(5) For NCASE = 2, eq. (H6.20) must be modified to take into account the exponential boundary condition function of eq. (H6.21). This is done by using eq. (H6.15) in the $f_e(z,t)$ term of eq. (H6.20) (i.e., $T_e(t - z/v), t > z/v$ in place of the constant T_e)

$$T(z,t) = f_e(z,t) + f_a(z,t) + f_0(z,t) = T_{ec}(t - z/v)e^{-z/(\tau v)}$$
$$+ T_{ac}\{(1 - e^{-t/\tau}) - e^{-z/(\tau v)}(1 - e^{-(t-z/v)/\tau})\}$$
$$+ T_{0c}u(z)\{e^{-t/\tau} - e^{-z/(\tau v)}(1 - e^{-(t-z/v)/\tau})\} \quad (H6.24)$$

Equation (H6.24) is programmed as

```
C...
C...        VARIABLE BOUNDARY CONDITION
            ELSE
     +      IF(NCASE.EQ.2)THEN
                EXACT=TEC*EXPZ*FBC(TIME-ZL/V)+
     +              TAC*((1.0D0-EXPT)-(EXPZ-EXPT))+
     +              TOC*(EXPT-EXPZ*EXPTD)
```

Note that the term $T_{ec}(t - z/v)e^{-z/(\tau v)}$ in eq. (H6.23) is programmed as TEC*EXPZ*FBC(TIME-ZL/V) where the time variation of the entering temperature is programmed in function FBC as discussed previously.

Finally, to complete the discussion of the programming, subroutine PLOTM is listed below:

```
      SUBROUTINE PLOTM
      IMPLICIT DOUBLE PRECISION (A-H,O-Z)
      PARAMETER(NZ=101)
      COMMON/T/    TIME,      NSTOP,      NORUN
     1      /Y/    T(NZ)
     2      /F/    TT(NZ)
     3      /S/    TZ(NZ)
     4      /P/        V,        ZL,       TAU,
     5                TEC,       TAC,      TOC
     6      /I/       IP,      NCASE
C...
C...  OPEN A FILE FOR MATLAB PLOTTING
      IF((NORUN.EQ.1).AND.(IP.EQ.1))THEN
         OPEN(1,FILE='h6.out')
      END IF
C...
C...  WRITE THE SOLUTION TO THE FILE FOR MATLAB PLOTTING
      WRITE(1,1)TIME,T(NZ),EXACT(NCASE)
1     FORMAT(3F10.4)
      RETURN
      END
```

The essential feature of PLOTM is writing t and the numerical and analytical solutions at $z = z_l$ to file h6.out, which can then be plotted

```
      WRITE(1,1)TIME,T(NZ),EXACT(ZL,TIME)
```

The integration of the NZ (= 101) ODEs (i.e., the derivative vector TT(NZ) in COMMON/F/) is performed by subroutine RKF45 (Forsythe, Malcolm, and Moler 1977), which is called by main program RKF45M listed in Appendix A. The data file read by the main program is

```
HEAT EXCHANGER, INLET DISTURBANCE
0.          20.0         0.2
  101                             0.00001
HEAT EXCHANGER, ANNULUS TEMPERATURE DISTURBANCE
0.          20.0         0.2
  101                             0.00001
HEAT EXCHANGER, INLET AND ANNULUS TEMPERATURE DISTURBANCES
0.          20.0         0.2
  101                             0.00001
END OF RUNS
```

Note that three runs are programmed corresponding to the cases (programmed in INITAL): (a) $T_{ec} = 100$, $T_{ac} = 0$, $T_{0c} = 0$; (b) $T_{ec} = 0$, $T_{ac} = 100$; $T_{0c} = 0$; and

(c) $T_{ec} = 100$, $T_{ac} = 100$, $T_{0c} = 0$. In all three cases, $0 \leq t \leq 20$, with calls to PRINT at intervals of 0.2 (101 calls to PRINT), and 101 ODEs are integrated (i.e., the derivatives in TT(NZ) = TT(101) computed in DERV are integrated).

The output from these three runs is summarized below. For the first run

```
RUN NO. -   1  HEAT EXCHANGER, INLET DISTURBANCE

INITIAL T -    .000D+00

  FINAL T -    .200D+02

  PRINT T -    .200D+00

NUMBER OF DIFFERENTIAL EQUATIONS - 101

MAXIMUM INTEGRATION ERROR -     .100D-04

TUBULAR HEAT EXCHANGER
```

t	T(z1,t)	Ta(z1,t)	Diff
.00	.00	.00	.00
.20	.00	.00	.00
.40	.00	.00	.00
.	.		
.	.		
.	.		
6.60	.00	.00	.00
6.80	.00	.00	.00
7.00	.00	.00	.00
7.20	.01	.00	.01
7.40	-.01	.00	-.01
7.60	-.01	.00	-.01
7.80	.04	.00	.04
8.00	-.05	.00	-.05
8.20	-.05	.00	-.05
8.40	.20	.00	.20
8.60	-.08	.00	-.08
8.80	-.47	.00	-.47
9.00	.45	.00	.45
9.20	.98	.00	.98
9.40	-.94	.00	-.94
9.60	-2.03	.00	-2.03
9.80	1.63	.00	1.63
10.00	7.92	.00	7.92
10.20	12.33	13.53	-1.21
10.40	13.72	13.53	.19

```
10.60           13.71           13.53            .17
10.80           13.55           13.53            .02
11.00           13.52           13.53           -.01
11.20           13.53           13.53            .00
11.40           13.53           13.53            .00
12.00           13.53           13.53            .00
                   .               .
                   .               .
                   .               .
19.80           13.53           13.53            .00
20.00           13.53           13.53            .00
```

Note that the exact solution is a step at $t = 10$, that is, from eq. (H6.20)

$$T(z,t) = f_e(z,t) + f_a(z,t) + f_0(z,t)$$

$$= T_{ec}u(t - z/v)e^{-z/(\tau v)} = 100u(t - 100/10)e^{-100/\{(5)(10)\}}$$

$$= 100u(t - 10)0.1353 = 13.53u(t - 10) \tag{H6.25}$$

This is essentially an impossible solution to compute numerically, that is, it has a discontinuity at $t = 10$ so that $\partial T(z_l, t)/\partial t$ and $\partial T(z_l, t)/\partial z$ in eq. (H6.1) are infinite. The most serious error in the numerical solution occurs at $t = 9.60$ with a negative temperature of -2.03

```
 9.60           -2.03            .00           -2.03
```

The oscillation of the numerical solution in the neighborhood of $t = 10$ is evident in Figure H6.2.

Physically, the discontinuity of eq. (H6.25) occurs when the entering temperature $T_e(0, t) = 100$ reaches the far end of the exchanger, at $z = z_l$ (at $t = 10$ since $z_l = 100$ and $v = 10$). The entering temperature of 100 is attenuated to 13.53 because of heat transfer to the annulus which remains at $T_{ac} = 0$.

Abbreviated output is listed below for the second run

```
RUN NO. -    2   HEAT EXCHANGER, ANNULUS TEMPERATURE
                 DISTURBANCE

  INITIAL T -    .000D+00

    FINAL T -    .200D+02

    PRINT T -    .200D+00

NUMBER OF DIFFERENTIAL EQUATIONS -  101

MAXIMUM INTEGRATION ERROR -    .100D-04
```

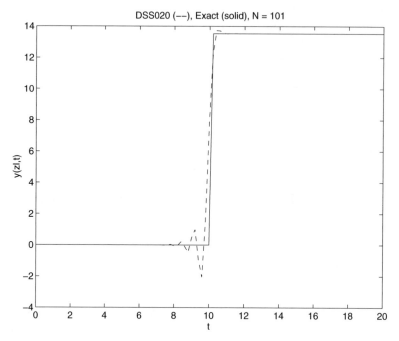

Figure H6.2. Numerical Solution to Eqs. (H6.1) to (H6.3) from DSS020 for the Case $T_{ec} = 100$, $T_{ac} = 0$, $T_0 = 0$

TUBULAR HEAT EXCHANGER

t	T(zl,t)	Ta(zl,t)	Diff
.00	.00	.00	.00
.20	3.92	3.92	.00
.40	7.69	7.69	.00
.60	11.31	11.31	.00
.80	14.79	14.79	.00
1.00	18.13	18.13	.00
.	.		
.	.		
.	.		
8.00	79.81	79.81	.00
8.20	80.60	80.60	.00
8.40	81.36	81.36	.00
8.60	82.09	82.09	-.01
8.80	82.80	82.80	.01
9.00	83.48	83.47	.01
9.20	84.09	84.12	-.02
9.40	84.71	84.74	-.03
9.60	85.38	85.34	.04

9.80	85.98	85.91	.06
10.00	86.34	86.47	-.13
.	.		
.	.		
.	.		
19.80	86.47	86.47	.00
20.00	86.47	86.47	.00

In this case, the difference between the numerical and analytical solutions does not exceed $-.13$. Thus, we might ask why the agreement in the second run is so much better than for the first run. This can be explained by considering the analytical solution, eq. (H6.20) (with $T_{ec} = 0$, $T_{ac} = 100$, $T_0 = 0$)

$$T(z,t) = f_e(z,t) + f_a(z,t) + f_0(z,t)$$
$$= T_{ac}\{u(t)(1 - e^{-t/\tau}) - u(t - z/v)(e^{-z/(\tau v)} - e^{-t/\tau})\} \quad \text{(H6.26)}$$

Again, the solution consists of two parts:

(1) For $t < z/v$, eq. (H6.26) is

$$T(z,t) = T_{ac}(1 - e^{-t/\tau}) \quad \text{(H6.27)}$$

which is a smooth exponential increase in temperature from $T(z,0) = 0$. Note, for example, (with $T_{ac} = 100$, $\tau = 5$)

$$T(z,5) = 100(1 - e^{-5/5}) = 63.21$$
$$T(z,10) = 100(1 - e^{-10/5}) = 86.47$$

(2) For $t > z/v$, eq. (H6.26) is

$$T(z,t) = T_{ac}\{(1 - e^{-t/\tau}) - (e^{-z/(\tau v)} - e^{-t/\tau})\}$$
$$= T_{ac}(1 - e^{-z/(\tau v)}) = 100(1 - e^{-100/\{(5)(10)\}}) = 86.47 \quad \text{(H6.28)}$$

so that the solution is independent of t.

In other words, the solution is initially a smooth exponential in t according to eq. (H6.27) and for large t becomes constant according to eq. (H6.28). Thus, the solution is smooth for all t and can therefore be calculated relatively easily compared to the first case, which has a discontinuity at $t = z/v$, according to eq. (H6.25); this smoothness explains why the numerical solution for the second case is more accurate. A plot of the numerical and analytical solutions is not presented since the differences between the two solutions are imperceptible when plotted. Physically, although the annulus temperatute $T_{ac} = 100$ constitutes a discontinuous change from the initial tube side temperature, $T_0 = 0$, the effect of this discontinuity is smoothed (into an exponential) by the heat transfer from the annulus to the tube side.

The third case is a combination of the first two cases, that is, $T_{ec} = T_{ac} = 100$, $T_0 = 0$. Since the original problem, eqs. (H6.1) to (H6.3), is linear, the solution for the third case is just the sum of the solutions for the first two cases (this superposition or addition of solutions is clear from eq. (H6.20)). The numerical solution is abbreviated below:

```
RUN NO. -    3   HEAT EXCHANGER, INLET AND ANNULUS TEMPERATURE
                 DISTURBANCES

INITIAL T -    .000D+00

  FINAL T -    .200D+02

  PRINT T -    .200D+00

NUMBER OF DIFFERENTIAL EQUATIONS -  101

MAXIMUM INTEGRATION ERROR -    .100D-04

TUBULAR HEAT EXCHANGER
```

t	T(zl,t)	Ta(zl,t)	Diff
.00	.00	.00	.00
.20	3.92	3.92	.00
.40	7.69	7.69	.00
.60	11.31	11.31	.00
.80	14.79	14.79	.00
1.00	18.13	18.13	.00
.	.	.	
.	.	.	
.	.	.	
5.00	63.21	63.21	.00
5.20	64.65	64.65	.00
5.40	66.04	66.04	.00
5.60	67.37	67.37	.00
5.80	68.65	68.65	.00
6.00	69.88	69.88	.00
.	.	.	
.	.	.	
.	.	.	
7.00	75.34	75.34	.00
7.20	76.31	76.31	.01
7.40	77.23	77.24	-.01
7.60	78.12	78.13	-.01
7.80	79.02	78.99	.04
8.00	79.76	79.81	-.05

8.20	80.56	80.60	−.05
8.40	81.57	81.36	.20
8.60	82.01	82.09	−.08
8.80	82.33	82.80	−.47
9.00	83.93	83.47	.46
9.20	85.08	84.12	.96
9.40	83.77	84.74	−.97
9.60	83.35	85.34	−1.99
9.80	87.60	85.91	1.69
10.00	94.26	86.47	7.80
10.20	98.80	100.00	−1.20
10.40	100.20	100.00	.20
10.60	100.18	100.00	.18
10.80	100.02	100.00	.02
11.00	99.99	100.00	−.01
11.20	100.00	100.00	.00
11.40	100.00	100.00	.00
.	.	.	
.	.	.	
.	.	.	
19.80	100.00	100.00	.00
20.00	100.00	100.00	.00

As expected, the temperature increases exponentially according to eq. (H6.27) for $t < z/v$ (i.e., $t < 10$). At $t = z/v = 10$, the discontinuity of eq. (H6.25) occurs and produces the maximum error of 7.80 in the numerical solution

10.00	94.26	86.47	7.80

However, this error, and the other errors in the neighborhood of $t = 10$, are on a rapidly rising portion of the solution, so that when the two solutions are plotted, as in Figure H6.3, the agreement appears to be acceptable.

In conclusion, we have observed that the performance of the method of lines applied to eqs. (H6.1) to (H6.3) is dependent upon the particular initial and boundary conditions. A discontinuity between the initial tube-side temperature and the entering temperature (the first case) propagates along the exchanger to the exit at $z = z_l$ and thereby causes difficulties in calculating the temperature $T(z_l, t)$ (as in Figure H6.2). For the second case of a change in the annulus temperature from the initial tube-side temperature, the discontinuity is smoothed into an exponential. For the third case, the properties of the first two cases appear simultaneously, and the net effect is to produce an acceptable solution.

Clearly the largest source of error is the discontinuity of the first case, and we might therefore consider how the adverse effects of this discontinuity might be reduced. To this end, another approximation of the convective derivative, $\partial T/\partial z$, in

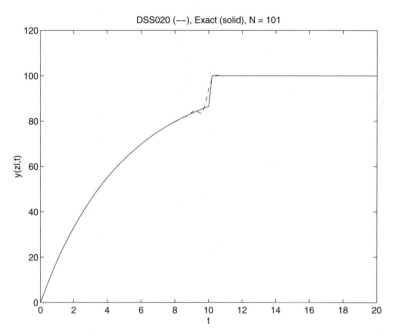

Figure H6.3. Numerical Solution to Eqs. (H6.1) to (H6.3) from DSS020 for the Case $T_{ec} = 100$, $T_{ac} = 100$, $T_0 = 0$

eq. (H6.1) is programmed in subroutine DSS024. We therefore return to subroutine DERV, comment out the call to DSS020, and activate the call to DSS024 (all other preceding code remains the same). The three solutions from DSS024 are discussed below. For the first case, we obtain

```
RUN NO. -    1  HEAT EXCHANGER, INLET DISTURBANCE

INITIAL T -    .000D+00

  FINAL T -    .200D+02

  PRINT T -    .200D+00

NUMBER OF DIFFERENTIAL EQUATIONS - 101

MAXIMUM INTEGRATION ERROR -    .100D-04

TUBULAR HEAT EXCHANGER
         t       T(z1,t)     Ta(z1,t)         Diff
        .00         .00          .00           .00
        .20         .00          .00           .00
        .40         .00          .00           .00
```

Heat exchanger dynamics

.60	.00	.00	.00
.80	.00	.00	.00
1.00	.00	.00	.00
.	.		
.	.		
.	.		
7.00	.00	.00	.00
7.20	.00	.00	.00
7.40	.00	.00	.00
7.60	.00	.00	.00
7.80	.00	.00	.00
8.00	.01	.00	.01
8.20	.01	.00	.01
8.40	-.03	.00	-.03
8.60	-.01	.00	-.01
8.80	.10	.00	.10
9.00	.06	.00	.06
9.20	-.18	.00	-.18
9.40	-.01	.00	-.01
9.60	1.45	.00	1.45
9.80	4.31	.00	4.31
10.00	7.66	.00	7.66
10.20	10.44	13.53	-3.09
10.40	12.19	13.53	-1.34
10.60	13.05	13.53	-.48
10.80	13.40	13.53	-.13
11.00	13.51	13.53	-.02
11.20	13.55	13.53	.02
11.40	13.56	13.53	.02
11.60	13.56	13.53	.03
11.80	13.56	13.53	.03
12.00	13.56	13.53	.03
.	.		
.	.		
.	.		
19.80	13.56	13.53	.03
20.00	13.56	13.53	.03

The performance of DSS024 is substantially better than that of DSS020 (compare the solutions in the neighborhood of the discontinuity at $t = 10$). The plot of the numerical and analytical solutions in Figure H6.4 substantiates this conclusion (compare Figures H6.2 and H6.4).

DSS024 does produces an offset of .03 in the solution, which could be reduced by further refinement of the finite difference approximations for computing spatial derivatives in DSS024.

314   Heat transport

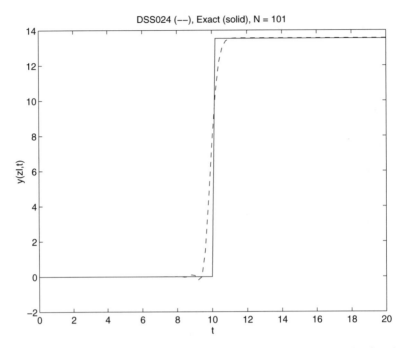

Figure H6.4. Numerical Solution to Eqs. (H6.1) to (H6.3) from DSS024 for the Case $T_{ec} = 100$, $T_{ac} = 0$, $T_0 = 0$

The output from DSS024 for the second case is abbreviated below:

```
RUN NO. -   2 HEAT EXCHANGER, ANNULUS TEMPERATURE DISTURBANCE

  INITIAL T -    .000D+00

   FINAL T -    .200D+02

   PRINT T -    .200D+00

NUMBER OF DIFFERENTIAL EQUATIONS - 101

MAXIMUM INTEGRATION ERROR -    .100D-04

TUBULAR HEAT EXCHANGER
```

t	T(z1,t)	Ta(z1,t)	Diff
.00	.00	.00	.00
.20	3.92	3.92	.00
.40	7.69	7.69	.00
.60	11.31	11.31	.00
.80	14.79	14.79	.00

1.00	18.13	18.13	.00
.	.	.	
.	.	.	
.	.	.	
5.80	68.65	68.65	.00
6.00	69.88	69.88	.00
.	.	.	
.	.	.	
.	.	.	
9.00	83.47	83.47	.00
9.20	84.12	84.12	.00
9.40	84.74	84.74	.00
9.60	85.32	85.34	-.02
9.80	85.79	85.91	-.13
10.00	86.11	86.47	-.36
10.20	86.29	86.47	-.17
10.40	86.38	86.47	-.08
.	.	.	
.	.	.	
.	.	.	
19.80	86.44	86.47	-.03
20.00	86.44	86.47	-.03

Again, the agreement between the numerical and analytical solutions is so close that any differences are imperceptible when the two solutions are plotted.

For the third case, the solution from DSS024 is

```
RUN NO. -   3   HEAT EXCHANGER, INLET AND ANNULUS TEMPERATURE
                DISTURBANCES

 INITIAL T -    .000D+00

   FINAL T -    .200D+02

   PRINT T -    .200D+00

NUMBER OF DIFFERENTIAL EQUATIONS - 101

MAXIMUM INTEGRATION ERROR -     .100D-04

TUBULAR HEAT EXCHANGER

           t      T(z1,t)      Ta(z1,t)         Diff
         .00          .00           .00          .00
         .20         3.92          3.92          .00
         .40         7.69          7.69          .00
```

.60	11.31	11.31	.00
.80	14.79	14.79	.00
1.00	18.13	18.13	.00
.	.		
.	.		
.	.		
5.80	68.65	68.65	.00
6.00	69.88	69.88	.00
.	.		
.	.		
.	.		
8.00	79.82	79.81	.01
8.20	80.61	80.60	.01
8.40	81.34	81.36	-.03
8.60	82.08	82.09	-.01
8.80	82.89	82.80	.10
9.00	83.53	83.47	.06
9.20	83.94	84.12	-.18
9.40	84.73	84.74	-.01
9.60	86.77	85.34	1.43
9.80	90.09	85.91	4.18
10.00	93.76	86.47	7.30
10.20	96.73	100.00	-3.27
10.40	98.57	100.00	-1.43
10.60	99.47	100.00	-.53
10.80	99.83	100.00	-.17
11.00	99.95	100.00	-.05
11.20	99.99	100.00	-.01
11.40	100.00	100.00	.00
11.60	100.00	100.00	.00
11.80	100.00	100.00	.00
12.00	100.00	100.00	.00
.	.		
.	.		
.	.		
19.80	100.00	100.00	.00
20.00	100.00	100.00	.00

Again, in the neighborhood of the discontinuity at $t = 10$, the differences between the numerical and analytical solutions appear large, but they occur in a portion of the solution that is changing rapidly, so that when the two solutions are plotted, as in Figure H6.5, these differences are relatively small. A comparison of Figures H6.3 and H6.5 again indicates that DSS024 gives better performance than DSS020.

As we concluded previously, the performance of the two differentiators, DSS020 and DSS024, is dependent upon the initial and boundary conditions. To demonstrate

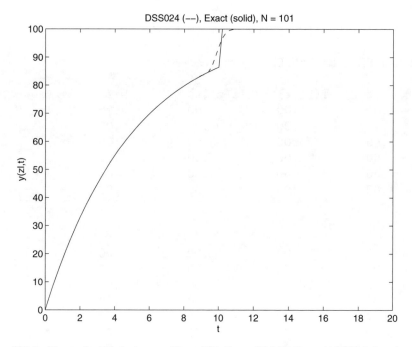

Figure H6.5. Numerical Solution to Eqs. (H6.1) to (H6.3) from DSS024 for the Case $T_{ec} = 100$, $T_{ac} = 100$, $T_0 = 0$

this point, we repeat the preceding runs, but with NCASE = 2 (set in INITAL), so that the exponential boundary condition of eq. (H6.21) (and implemented in function HBC) is used. This boundary condition provides a smooth transition of the boundary temperature, $T(0, t)$, from the initial value $T(0, 0) = T_{0c} = 0$ to the final value $T_{ec} = 100$ (in contrast with the preceding NCASE = 1 for which this transition was discontinuous). Thus, with the exponential boundary condition of eq. (H6.21), the numerical solution should be substantially easier to compute. Again, we consider two series of three runs, for DSS020 and DSS024 (by activating the call to each differentiator in subroutine DERV.

For DSS020 (and NCASE = 2), the numerical solution from the first run is

```
RUN NO. -   1   HEAT EXCHANGER, INLET DISTURBANCE

 INITIAL T -    .000D+00

   FINAL T -    .200D+02

   PRINT T -    .200D+00

NUMBER OF DIFFERENTIAL EQUATIONS - 101
```

MAXIMUM INTEGRATION ERROR - .100D-04

TUBULAR HEAT EXCHANGER

t	T(z1,t)	Ta(z1,t)	Diff
.00	.00	.00	.00
.20	.00	.00	.00
.40	.00	.00	.00
.60	.00	.00	.00
.80	.00	.00	.00
1.00	.00	.00	.00
.	.	.	
.	.	.	
.	.	.	
7.00	.00	.00	.00
7.20	.00	.00	.00
7.40	.00	.00	.00
7.60	.00	.00	.00
7.80	.00	.00	.00
8.00	.00	.00	.00
8.20	-.01	.00	-.01
8.40	.01	.00	.01
8.60	.03	.00	.03
8.80	-.04	.00	-.04
9.00	-.05	.00	-.05
9.20	.12	.00	.12
9.40	.13	.00	.13
9.60	-.22	.00	-.22
9.80	-.28	.00	-.28
10.00	.65	.00	.65
10.20	2.43	2.45	-.02
10.40	4.39	4.46	-.07
10.60	6.09	6.11	-.02
10.80	7.45	7.45	.00
11.00	8.56	8.55	.00
11.20	9.46	9.46	.00
11.40	10.20	10.20	.00
11.60	10.80	10.80	.00
11.80	11.30	11.30	.00
12.00	11.70	11.70	.00
.	.	.	
.	.	.	
.	.	.	
19.80	13.53	13.53	.00
20.00	13.53	13.53	.00

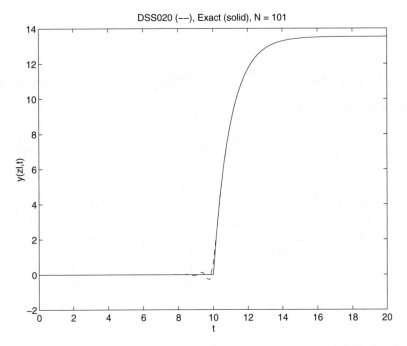

Figure H6.6. Numerical Solution to Eqs. (H6.1) to (H6.3) from DSS020 for the Case $T_{ec} = 100(1 - e^{-t/\tau_{bc}})$, $\tau_{bc} = 1$, $T_{ac} = 0$, $T_0 = 0$

We observe in this solution that the numerical oscillation near $t = 10$ is much less severe than for NCASE = 1; this conclusion can be confirmed by comparing Figures H6.2 and H6.6.

The smoothness of the solution and the improved agreement with the analytical solution is due to the use of boundary condition (H6.21) in place of a step change of $+100$ for $T(0, t)$ (from $T(0, 0) = 0$ to $T(0, t) = 100$). In other words, the performance of spatial differentiators such as DSS020 is dependent upon the nature of the imposed boundary conditions.

The solution for the second run is the same as for NCASE = 1 since the entering temperature, $T(0, t) = T_e(t) = 0$, has no effect in this case (only the annulus temperature, T_a, in eq. (H6.1) is nonzero). The solution for the third run is summarized below:

```
RUN NO. -    3   HEAT EXCHANGER, INLET AND ANNULUS TEMPERATURE
                 DISTURBANCES

   INITIAL T -   .000D+00

     FINAL T -   .200D+02
```

PRINT T -   .200D+00

NUMBER OF DIFFERENTIAL EQUATIONS - 101

MAXIMUM INTEGRATION ERROR -   .100D-04

TUBULAR HEAT EXCHANGER

t	$T(z1,t)$	$Ta(z1,t)$	Diff
.00	.00	.00	.00
.20	3.92	3.92	.00
.40	7.69	7.69	.00
.60	11.31	11.31	.00
.80	14.79	14.79	.00
1.00	18.13	18.13	.00
1.20	21.34	21.34	.00
1.40	24.42	24.42	.00
1.60	27.39	27.39	.00
1.80	30.23	30.23	.00
2.00	32.97	32.97	.00
.	.		
.	.		
.	.		
8.00	79.81	79.81	.00
8.20	80.59	80.60	-.01
8.40	81.37	81.36	.00
8.60	82.12	82.09	.02
8.80	82.76	82.80	-.03
9.00	83.43	83.47	-.04
9.20	84.22	84.12	.10
9.40	84.84	84.74	.10
9.60	85.16	85.34	-.18
9.80	85.69	85.91	-.22
10.00	86.99	86.47	.52
10.20	88.90	88.92	-.02
10.40	90.87	90.93	-.06
10.60	92.56	92.57	-.01
10.80	93.92	93.92	.00
11.00	95.02	95.02	.00
.	.	.	
.	.	.	
.	.	.	
19.80	100.00	100.00	.00
20.00	100.00	100.00	.00

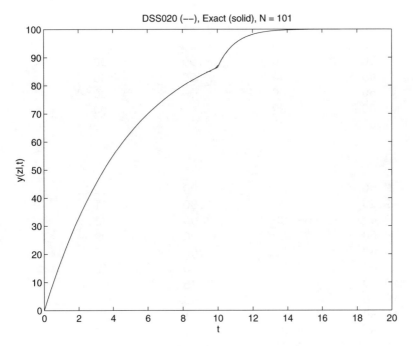

Figure H6.7. Numerical Solution to Eqs. (H6.1) to (H6.3) from DSS020 for the Case $T_{ec} = 100(1 - e^{-t/\tau_{bc}})$, $\tau_{bc} = 1$, $T_{ac} = 100$, $T_0 = 0$

As expected, the errors in this case are smaller than for NCASE = 1, since again the exponential function of eq. (H6.21) is smoother than the step function for NCASE = 1. The more accurate solution for NCASE = 3 is indicated by comparing Figures (H6.3) and (H6.7).

Finally, to complete the discussion, three solutions were computed using DSS024 and boundary condition (H6.21). The solution from the first run is summarized below:

```
RUN NO. -    1   HEAT EXCHANGER, INLET DISTURBANCE

 INITIAL T -    .000D+00

   FINAL T -    .200D+02

   PRINT T -    .200D+00

NUMBER OF DIFFERENTIAL EQUATIONS -  101

MAXIMUM INTEGRATION ERROR -    .100D-04

TUBULAR HEAT EXCHANGER
```

t	T(z1,t)	Ta(z1,t)	Diff
.00	.00	.00	.00
.20	.00	.00	.00
.40	.00	.00	.00
.60	.00	.00	.00
.80	.00	.00	.00
1.00	.00	.00	.00
.	.		
.	.		
.	.		
9.00	.02	.00	.02
9.20	.01	.00	.01
9.40	-.03	.00	-.03
9.60	.09	.00	.09
9.80	.59	.00	.59
10.00	1.58	.00	1.58
10.20	2.95	2.45	.50
10.40	4.49	4.46	.03
10.60	5.98	6.11	-.13
10.80	7.30	7.45	-.15
11.00	8.42	8.55	-.14
11.20	9.34	9.46	-.11
11.40	10.11	10.20	-.09
11.60	10.73	10.80	-.07
11.80	11.25	11.30	-.05
12.00	11.67	11.70	-.04
.	.	.	
.	.	.	
.	.	.	
19.80	13.56	13.53	.03
20.00	13.56	13.53	.03

As expected, the solution is better than for (a) NCASE = 1 and DSS024 (compare Figures (H6.4) and (H6.8)) since boundary condition (H6.21) is smoother than the step change in $T(0, t)$, and (b) NCASE = 2 and DSS020 (compare Figures (H6.6) and (H6.8)) since DSS024 is generally better than DSS020.

The solution for the second run is again unchanged (from NCASE = 1 and DSS024) since the entering temperature, $T(0, t) = T_e(t) = 0$, has no effect. The solution for the third run is summarized below:

RUN NO. -   3   HEAT EXCHANGER, INLET AND ANNULUS TEMPERATURE

                DISTURBANCES

INITIAL T -   .000D+00

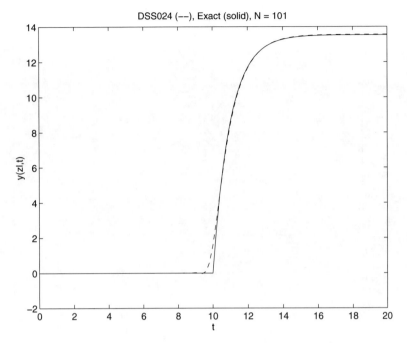

Figure H6.8. Numerical Solution to Eqs. (H6.1) to (H6.3) from DSS024 for the Case $T_{ec} = 100(1 - e^{-t/\tau_{bc}})$, $\tau_{bc} = 1$, $T_{ac} = 0$, $T_0 = 0$

```
    FINAL T -    .200D+02

    PRINT T -    .200D+00

NUMBER OF DIFFERENTIAL EQUATIONS -  101

MAXIMUM INTEGRATION ERROR -    .100D-04

TUBULAR HEAT EXCHANGER
```

t	T(zl,t)	Ta(zl,t)	Diff
.00	.00	.00	.00
.20	3.92	3.92	.00
.40	7.69	7.69	.00
.60	11.31	11.31	.00
.80	14.79	14.79	.00
1.00	18.13	18.13	.00
1.20	21.34	21.34	.00
1.40	24.42	24.42	.00
1.60	27.39	27.39	.00
1.80	30.23	30.23	.00

Heat transport

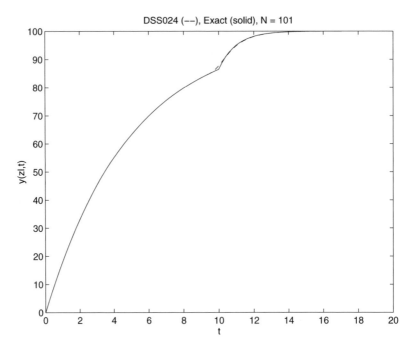

Figure H6.9. Numerical Solution to Eqs. (H6.1) to (H6.3) from DSS024 for the Case $T_{ec} = 100(1 - e^{-t/\tau_{bc}})$, $\tau_{bc} = 1$, $T_{ac} = 100$, $T_0 = 0$

2.00	32.97	32.97	.00
.	.	.	
.	.	.	
.	.	.	
9.00	83.49	83.47	.02
9.20	84.12	84.12	.00
9.40	84.72	84.74	-.02
9.60	85.41	85.34	.07
9.80	86.37	85.91	.46
10.00	87.69	86.47	1.22
10.20	89.25	88.92	.33
10.40	90.88	90.93	-.05
10.60	92.40	92.57	-.17
10.80	93.73	93.92	-.19
11.00	94.85	95.02	-.17
.	.	.	
.	.	.	
.	.	.	
19.80	100.00	100.00	.00
20.00	100.00	100.00	.00

As expected, the solution is better than for NCASE = 1 and DSS024 (compare Figures (H6.5) and (H6.9)) since boundary condition (H6.21) is smoother than the step change in $T(0, t)$. For NCASE = 2 and DSS020 the results are equivalent (compare Figures (H6.7) and (H6.9)).

In conclusion, we have observed that the performance of the spatial differentiators in the MOL solution of first-order hyperbolic PDEs, for example, eq. (H6.1) is dependent upon the initial and boundary conditions. If boundary conditions are imposed that introduce a discontinuous change from the initial conditions, the discontinuity can propagate (a characteristic of first-order hyperbolic PDEs) and numerical distortions in the solution can occur, for example, oscillations. The variable-order, finite difference approximations in DSS024 have generally been observed to minimize these distortions and produce numerical solutions of acceptable accuracy.

Finally, subroutine DSS024 is listed in Appendix C. Briefly, two-point, first-order, upwind finite difference approximations are used near the spatial boundaries, and five-point, fourth-order, biased upwind finite difference approximations are used at the interior points of the spatial grid. The two-point approximations minimize numerical oscillation, and the five-point approximations minimize numerical diffusion. Based on our experience, DSS024 is recommended for the MOL solution of first-order, hyperbolic PDEs.

References

Forsythe, G. E., M. A. Malcolm, and C. B. Moler. 1977. *Computer Methods for Mathematical Computations*. Englewood Cliffs, NJ: Prentice-Hall.

Schiesser, W. E. 1991. *The Numerical Method of Lines Integration of Partial Differential Equations*. San Diego: Academic Press.

Part three

Mass transport

MA1 A dynamic mass transfer model   329
MA2 Mass transfer with simultaneous convection and diffusion   370
MA3 Transient multicomponent diffusion   418

MA1

A dynamic mass transfer model

Consider the analysis of a mass transfer unit configured as a column through which two cocurrent or countercurrent streams flow, for example, a liquid and gas, as illustrated in Figure MA1.1.

Develop a mathematical model and computer code to determine the required height of the column for a given reduction in the gas phase composition.

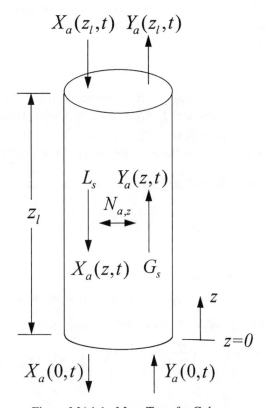

Figure MA1.1. Mass Transfer Column

Solution

A mass balance on one phase, subsequently designated the liquid phase, in an incremental volume of the column of length z and cross-sectional area S gives

$$S\Delta z H_L \frac{\partial X_a}{\partial t} = SL_s X_a|_{z+\Delta z} - SL_s X_a|_z + K_Y a(Y_a - Y_a^*)S\Delta z \quad \text{(MA1.1)}$$

where

X_a	liquid phase mol ratio of A (mols A/mol B)
	A = transferred component
	B = nontransferred component
Y_a	gas phase mol ratio of A (mols A/mols B)
Y_a^*	gas phase composition in equilibrium with liquid of composition X_a (mols A/mol B)
t	time (s)
z	axial position along the column (m)
S	column cross-sectional area (m^2)
H_L	liquid holdup (mols B/m^3)
L_s	liquid flow rate per unit column cross-sectional area (mols B/m^2-s)
K_Y	overall mass transfer coefficient (mols B/m^2-s)
a	volumetric mass transfer area (m^2/m^3)

Check on units:

$S\Delta z H_L \frac{\partial X_a}{\partial t}$	(m^2)(m)(mols B/m^3)(mols A/mol B)(1/s) = mols A/s
$SL_s X_a|_{z+\Delta z}$	(m^2)(mols B/m^2-s)(mols A/mol B) = mols A/s
$SL_s X_a|_z$	(m^2)(mols B/m^2-s)(mols A/mol B) = mols A/s
$K_Y a(Y_a - Y_a^*)S\Delta z$	(mols B/m^2-s)(m^2/m^3)(mols A/mol B)(m^3) = mols A/s

Note that we have made use of the basic mass transfer equation

$$N_{a,z} = K_Y(Y_a - Y_a^*) \quad \text{(MA1.2)}$$

Check on units:

$N_{a,z}$	mols A/m^2-s
$K_Y(Y_a - Y_a^*)$	(mols B/m^2-s)(mols A/mol B) = mols A/m^2-s

Division of eq. (MA1.1) by $S\Delta z$ gives (with L_s constant)

$$H_L \frac{\partial X_a}{\partial t} = L_s \left\{ \frac{X_a|_{z+\Delta z} - X_a|_z}{\Delta z} \right\} + K_Y a(Y_a - Y_a^*)$$

or with $\Delta z \to 0$,

$$H_L \frac{\partial X_a}{\partial t} = L_s \frac{\partial X_a}{\partial z} + K_Y a(Y_a - Y_a^*) \quad \text{(MA1.3)}$$

Equation (MA1.3) is the basic dynamic material balance for the liquid phase and can be classified mathematically as a *first-order, hyperbolic partial differential equation* (*PDE*).

Equation (MA1.3) could also be written in terms of a mass transfer rate expressed in liquid phase units

$$H_L \frac{\partial X_a}{\partial t} = L_s \frac{\partial X_a}{\partial z} + K_X a (X_a^* - X_a) \quad \text{(MA1.4)}$$

where K_X is defined in terms of a mass transfer flux analogous to eq. (MA1.2)

$$N_{a,z} = K_X (X_a^* - X_a) \quad \text{(MA1.5)}$$

Check on units:

$N_{a,z}$ mols A/m^2-s
$K_X (X_a^* - X_a)$ (mols B/m^2-s)(mols A/mol B) = mols A/m^2-s

We can now write material balances for the gas phase that are analogous to eqs. (MA1.3) and (MA1.4)

$$H_G \frac{\partial Y_a}{\partial t} = -G_s \frac{\partial Y_a}{\partial z} - K_Y a (Y_a - Y_a^*) \quad \text{(MA1.6)}$$

$$H_G \frac{\partial Y_a}{\partial t} = -G_s \frac{\partial Y_a}{\partial z} - K_X a (X_a^* - X_a) \quad \text{(MA1.7)}$$

Equations (MA1.3), (MA1.4), (MA1.6), and (MA1.7) are the basic dynamic mass transfer equations for a two-component system. For a multicomponent system, material balances would be written for each transferred component.

Additionally, we require an equilibrium relationship, for example, Henry's law:

$$Y_a^* = m X_a \quad \text{(MA1.8a)}$$

$$X_a^* = (1/m) Y_a \quad \text{(MA1.8b)}$$

or more generally

$$Y_a^* = f(X_a) \quad \text{(MA1.9a)}$$

$$X_a^* = f^{-1}(Y_a) \quad \text{(MA1.9b)}$$

We can now consider some important special cases of the preceding equations. At steady state, eqs. (MA1.3), (MA1.4), (MA1.6), and (MA1.7) become

$$0 = L_s \frac{dX_a}{dz} + K_Y a (Y_a - Y_a^*) \quad \text{(MA1.10)}$$

$$0 = L_s \frac{dX_a}{dz} + K_X a (X_a^* - X_a) \quad \text{(MA1.11)}$$

$$0 = -G_s \frac{dY_a}{dz} - K_Y a (Y_a - Y_a^*) \quad \text{(MA1.12)}$$

$$0 = -G_s \frac{dY_a}{dz} - K_X a (X_a^* - X_a) \quad \text{(MA1.13)}$$

Note that eqs. (MA1.10) to (MA1.13) are ordinary differential equations (ODEs), since they have only independent variable, z.

If eqs. (MA1.10) and (MA1.12) are added, we have

$$L_s \frac{dX_a}{dz} = G_s \frac{dY_a}{dz} \tag{MA1.14}$$

which is the basic ODE for the operating line. If L_s and G_s are constant (do not vary along the column and therefore do not vary with z), eq. (MA1.14) can be integrated as

$$L_s \int_{X_{a,1}}^{X_{a,2}} dX_a = G_s \int_{Y_{a,1}}^{Y_{a,2}} dY_a \tag{MA1.15}$$

or

$$L_s(X_{a,2} - X_{a,1}) = G_s(Y_{a,2} - Y_{a,1}) \tag{MA1.16}$$

where subscripts 1 and 2 denote the bottom and top of the column, respectively. Equation (MA1.16) can be used to calculate an exiting concentration if the two entering concentrations and the other exiting concentration are known.

Alternatively, eq. (MA1.14) can be integrated to some intermediate point in the column where the compositions are X_a and Y_a. Integrating from the bottom of the column (with compositions $X_{a,1}$, $Y_{a,1}$) to some intermediate point in the column (with compositions X_a, Y_a), we have from eq. (MA1.14)

$$L_s \int_{X_{a,1}}^{X_a} dX_a = G_s \int_{Y_{a,1}}^{Y_a} dY_a \tag{MA1.17}$$

or

$$L_s(X_a - X_{a,1}) = G_s(Y_a - Y_{a,1}) \tag{MA1.18}$$

Similarly, integrating from some intermediate point in the column (with compositions X_a, Y_a) to the top of the column (with compositions $X_{a,2}$, $Y_{a,2}$), we have from eq. (MA1.14)

$$L_s \int_{X_a}^{X_{a,2}} dX_a = G_s \int_{Y_a}^{Y_{a,2}} dY_a \tag{MA1.19}$$

or

$$L_s(X_{a,2} - X_a) = G_s(Y_{a,2} - Y_a) \tag{MA1.20}$$

Equations (MA1.18) and (MA1.20) define the operating line (i.e., the relationship between X_a and Y_a) in terms of the bottom and top compositions, respectively

$$Y_a = (L_s/G_s)(X_a - X_{a,1}) + Y_{a,1} \tag{MA1.21}$$

$$Y_a = (L_s/G_s)(X_a - X_{a,2}) + Y_{a,2} \tag{MA1.22}$$

We can now plot eq. (MA1.21) or eq. (MA1.22) on a $X_a - Y_a$ diagram along with the equilibrium relationship, for example, eqs. (MA1.8a) to (MA1.9b), to construct the conventional plot used to design mass transfer equipment (Welty, Wicks, and Wilson 1984, 688–98).

To carry out the steady state design of a mass transfer unit, we make use of eqs. (MA1.10) to (MA1.13). If we select eq. (MA1.12),

$$G_s \frac{dY_a}{dz} = K_Y a (Y_a^* - Y_a) \quad \text{(MA1.23)}$$

separation of variables followed by integration from the bottom to the top of the column gives

$$\int_0^z dz = z = (G_s/K_Y a) \int_{Y_{a,1}}^{Y_{a,2}} \frac{dY_a}{(Y_a^* - Y_a)} \quad \text{(MA1.24)}$$

Equation (MA1.24) is a conventional design equation that can be used to calculate the height of the column, z, the left hand side (LHS) integral, by evaluation of the right hand side (RHS) integral. Graphical integration can be used to evaluate the RHS (a poor procedure), or a numerical quadrature can be used, for example, Simpson's rule. The analogous design equation obtained from eq. (MA1.11) is

$$z = (L_s/K_X a) \int_{X_{a,1}}^{X_{a,2}} \frac{dX_a}{(X_a - X_a^*)} \quad \text{(MA1.25)}$$

Equations (MA1.24) and (MA1.25) can be written in the alternate forms

$$z = H_{OG} N_{OG} \quad \text{(MA1.26)}$$

$$z = H_{OL} N_{OL} \quad \text{(MA1.27)}$$

where

H_{OG} height of a gas phase transfer unit ($= (G_s/K_Y a)$) (m)
N_{OG} number of gas phase transfer units ($= \int_{Y_{a,1}}^{Y_{a,2}} \frac{dY_a}{(Y_a^* - Y_a)}$)
H_{OL} height of a liquid phase transfer unit ($= (L_s/K_X a)$)(m)
N_{OL} number of liquid phase transfer units ($= \int_{X_{a,1}}^{X_{a,2}} \frac{dX_a}{(X_a - X_a^*)}$)

Note that the integral defining a transfer unit can be interpreted as the change in the composition of the gas or liquid corresponding to the average driving force.

If the operating and equilibrium lines are straight, and if

$$\Delta = Y_a - Y_a^* \quad \text{(MA1.28)}$$

$$\Delta_1 = Y_{a,1} - Y_{a,1}^* \quad \text{(MA1.29)}$$

$$\Delta_2 = Y_{a,2} - Y_{a,2}^* \quad \text{(MA1.30)}$$

then we have

$$\frac{d\Delta}{dY_a} = \frac{\Delta_1 - \Delta_2}{Y_{a,1} - Y_{a,2}}$$

or

$$dY_a = \frac{Y_{a,1} - Y_{a,2}}{\Delta_1 - \Delta_2} d\Delta \tag{MA1.31}$$

Substituting eqs. (MA1.28) and (MA1.31) in eq. (MA1.24),

$$z = (G_s/K_Y a) \int_{Y_{a,1}}^{Y_{a,2}} \frac{dY_a}{(Y_a^* - Y_a)} = (G_s/K_Y a) \frac{Y_{a,1} - Y_{a,2}}{\Delta_1 - \Delta_2} \int_{Y_{a,2}}^{Y_{a,1}} \frac{d\Delta}{\Delta}$$

$$= (G_s/K_Y a) \frac{Y_{a,1} - Y_{a,2}}{\Delta_1 - \Delta_2} \ln \Delta \Big|_{Y_{a,2}}^{Y_{a,1}} = (G_s/K_Y a) \frac{Y_{a,1} - Y_{a,2}}{\Delta_1 - \Delta_2} \ln(\Delta_1/\Delta_2)$$

$$= (G_s/K_Y a) \frac{Y_{a,1} - Y_{a,2}}{\frac{\Delta_1 - \Delta_2}{\ln(\Delta_1/\Delta_2)}} = (G_s/K_Y a) \frac{Y_{a,1} - Y_{a,2}}{(Y_a - Y_a^*)_{lm}} \tag{MA1.32}$$

where

$$(Y_a - Y_a^*)_{lm} = \frac{\Delta_1 - \Delta_2}{\ln(\Delta_1/\Delta_2)} = \frac{(Y_a - Y_a^*)_1 - (Y_a - Y_a^*)_2}{\ln\{(Y_a - Y_a^*)_1/(Y_a - Y_a^*)_2\}} \tag{MA1.33}$$

is the log-mean concentration difference expressed in gas phase mol ratios.

Similarly, for the case of linear equilibrium and operating lines, eq. (MA1.25) can be written as

$$z = (L_s/K_X a) \int_{X_{a,1}}^{X_{a,2}} \frac{dX_a}{(X_a - X_a^*)} = (L_s/K_X a) \frac{(X_{a,1} - X_{a,2})}{(X_a - X_a^*)_{lm}} \tag{MA1.34}$$

where

$$(X_a - X_a^*)_{lm} = \frac{(X_a - X_a^*)_1 - (X_a - X_a^*)_2}{\ln\{(X_a - X_a^*)_1/(X_a - X_a^*)_2\}} \tag{MA1.35}$$

We now consider the use of some of the preceding equations for the design of a NH_3 absorber as presented by Welty, Wicks, and Wilson (1984, 693–5). NH_3 is to be absorbed from air at 20°C and atmospheric pressure in a countercurrent packed tower, 0.154 m in diameter, using NH_3-free water as the absorbent. The inlet gas rate is 11.04 m³/min, and the inlet water rate is 11.18 kg/min. Under these conditions, the overall capacity coefficent, $K_Y a$, can be assumed to be 73.8 kg mol/hr-m³.

The gas phase NH_3 concentration is to be reduced from 0.0825 mole fraction to 0.003 mole fraction. The tower will be cooled so that it operates isothermally at 20°C; the equilibrium data are given by Welty, Wicks, and Wilson (1984, 680). The length of the column to accomplish this change in the gas phase composition is to be determined.

The sizing of the column is done by three methods:

(1) Equations (MA1.3) and (MA1.6) are integrated to determine $X_a(z, t)$ and $Y_a(z, t)$ (the subscript "a" refers to the transferred component, which in this case is NH_3). The solution will be carried out numerically to steady state for a column of given height to observe if the correct exiting liquid and gas compositions are achieved. If not, the column height is adjusted and the solution is repeated until convergence to the specified exiting compositions is achieved.

(2) The RHS integral of eq. (MA1.24) is evaluated by numerical quadrature (i.e., Simpson's rule) to determine the column height, z. Note that the steady state material balance based on the gas phase composition, Y_a, is used (rather than eq. (MA1.25) for the liquid phase) since for this system most of the resistance to mass transfer is in the gas phase (which means the gas phase driving force in eq. (MA1.24), $(Y_a^* - Y_a)$, is relatively large and therefeore the accuracy of the calculated height from eq. (MA1.24) is relatively accurate).

(3) The log-mean concentration difference of eqs. (MA1.32) and (MA1.33) is used to compute the column height. Since the equilibrium relationship is not linear, eqs. (MA1.32) and (MA1.33) are not entirely appropriate for this calculation, and the point of this third approach is to observe how much error is introduced in using the log-mean concentration difference in this particular application.

All three methods are implemented in a single program, (discussed subsequently). Since in the first approach, we are integrating two simultaneous, first-order, hyperbolic partial differential equations (PDEs), (eqs. (MA1.3) and (MA1.6)), we work within a method of lines (MOL) framework (Schiesser 1991). Subroutine INITAL sets the initial conditions of eqs. (MA1.3) and (MA1.6). Since we are interested in only the steady state solution, we choose homogeneous (zero) initial conditions for eqs. (MA1.3) and (MA1.6), that is, no NH_3 in the liquid or gas phase initially

```
      SUBROUTINE INITAL
C...
C...  NH3 ABSORBER
C...
C...  DOUBLE PRECISION CODING IS USED
      IMPLICIT DOUBLE PRECISION (A-H,O-Z)
      DOUBLE PRECISION LS, KYA
C...
C...  COMMON AREA TO LINK SUBROUTINES
```

```
      PARAMETER(NZ=21)
      COMMON/T/       T,    NSTOP,    NORUN
     +     /Y/   XA(NZ),  YA(NZ)
     +     /F/   XAT(NZ), YAT(NZ)
     +     /S/   XAZ(NZ), YAZ(NZ), YAS(NZ)
     +     /C/      XA1,     YA1,     XA2,     YA2,
     +       ZL,    GS,      LS,     KYA,       S
C...
C...  COEFFICIENTS IN POLYNOMIAL FOR EQUILIBRIUM LINE
      COMMON/E/A(2)
C...
C...  INPUT/OUTPUT (I/O) UNIT NUMBERS
      COMMON/IO/ NI, NO
C...
C...  GIVEN END COMPOSITIONS (MOLE FRACTIONS)
C...
C...     ENTERING GAS
         YA1=0.0825D0
C...
C...     ENTERING LIQUID
         XA2=0.0D0
C...
C...     EXITING GAS
         YA2=0.003D0
C...
C...  CONVERT PRECEDING MOL FRACTIONS TO MOL RATIOS (BASED
C...  ON AIR AND WATER AS THE NONTRANSFERRED COMPONENTS)
C...
C...     ENTERING GAS
         YA1=YA1/(1.0D0-YA1)
C...
C...     ENTERING LIQUID
         XA2=XA2/(1.0D0-XA2)
C...
C...     EXITING GAS
         YA2=YA2/(1.0D0-YA2)
C...
C...  FLOW RATES (MOLS/M**2-MIN, SO THE COLUMN CROSS
C...  SECTIONAL AREA IS REQUIRED)
C...
C...     PI
         PI=4.0D0*DATAN(1.0D0)
C...
C...     COLUMN DIAMETER (M)
         D=0.154D0
```

```
C...
C...        COLUMN CROSS SECTIONAL AREA (M**2)
            S=PI*(D**2)/4.0D0
C...
C...        GAS VOLUMETRIC FLOW RATE (M**3/MIN)
            V=11.04D0
C...
C...        GAS CONSTANT (M**3-ATM/KG MOL-K) (NOTE - ONE KG MOL
C...        OF GAS OCCUPIES 22.4 M**3 AT 1 ATM AND 273.16 K)
            R=22.4D0*1.0D0/273.16D0
C...
C...        TEMPERATURE (K)
            TK=273.16D0+20.0D0
C...
C...        PRESSURE (ATM)
            P=1.0D0
C...
C...        GAS MOLAR FLOW RATE (KG MOL/MIN-M**2)
            G=V*P/(R*TK)*(1.0D0/S)
C...
C...        GAS MOLAR FLOW RATE ON SOLUTE FREE BASIS (KG MOLS
C...        AIR/MIN-M**2)
            GS=G*(1.0D0/(1.0D0+YA1))
C...
C...        LIQUID MOLAR FLOW RATE ON SOLUTE FREE BASIS(KG MOLS
C...        H2O/MIN-M**2)
            LS=11.18D0/18.0D0*(1.0D0/S)
C...
C...     EXITING LIQUID COMPOSITION
         XA1=OPLN(YA1,YA2,XA2,GS,LS)
C...
C...     COLUMN LENGTH (M)
         ZL=3.28D0
C...
C...     WRITE SUMMARY OF OPERATING CONDITIONS
         WRITE(NO,1)YA1,XA2,YA2,XA1,LS,GS
1        FORMAT(' Summary of Operating Conditions',//,
     +        '     Entering gas    = ',F7.4,/,
     +        '     Entering liquid = ',F7.4,/,
     +        '     Exiting gas     = ',F7.4,/,
     +        '     Exiting liquid  = ',F7.4,/,
     +        '     Liquid flow rate = ',F7.4,' kg mols
                                    H20/min-m**2',/,
     +        '     Gas flow rate   = ',F7.4,' kg mols
                                    air/min-m**2',/)
```

```
C...
C...   COEFFICIENTS IN THE LEAST SQUARES SECOND ORDER
C...   POLYNOMIAL APPROXIMATION OF THE EQUILIBRIUM DATA
C...   (THE COEFFICIENTS ARE WRITTEN INTO COMMON/E/ BY THIS
C...   CALL TO COEFF)
       CALL COEFF
C...
C...   MASS TRANSFER COEFFICIENT (KG MOL AIR/M**2-MIN)
       KYA=73.8D0
C...
C...   COLUMN HEIGHT USING LM MOL RATIO CONCENTRATION
C...   DIFFERENCE
       WRITE(NO,2)
2      FORMAT(///,'Column height from log mean concentration',
      +         'differenece')
       Z=HTLM(GS,KYA,XA1,YA1,XA2,YA2)
       WRITE(NO,3)Z
3      FORMAT(/,' Column height = ',F6.2,' m',//)
C...
C...   COLUMN HEIGHT USING INTEGRATION OF MOL RATIO
C...   CONCENTRATION DIFFERENCE
       WRITE(NO,4)
4      FORMAT(///,' Column height from numerical integration
      +         ' of', concentration difference')
       Z=HTIN(GS,KYA,YA1,YA2)
       WRITE(NO,5)Z
5      FORMAT(/,' Column height = ',F6.2,' m',//)
C...
C...   INITIAL CONDITIONS FOR PDE SOLUTION
       DO 10 I=1,NZ
          XA(I)=0.0D0
          YA(I)=0.0D0
10     CONTINUE
       RETURN
       END
```

We can note the following points about subroutine INITAL:

(1) A 21-point grid in z is defined for the MOL integration of eqs. (MA1.3) and (MA1.6)

```
C...
C...   DOUBLE PRECISION CODING IS USED
       IMPLICIT DOUBLE PRECISION (A-H,O-Z)
       DOUBLE PRECISION LS, KYA
C...
C...   COMMON AREA TO LINK SUBROUTINES
```

```
      PARAMETER(NZ=21)
      COMMON/T/        T,     NSTOP,      NORUN
     +      /Y/  XA(NZ),   YA(NZ)
     +      /F/  XAT(NZ),  YAT(NZ)
     +      /S/  XAZ(NZ),  YAZ(NZ),  YAS(NZ)
     +      /C/      XA1,      YA1,      XA2,     YA2,
     +      ZL,      GS,       LS,       KYA,      S
```

X_a and Y_a in eqs. (MA1.3) and (MA1.6) are in arrays XA(NZ) and YA(NZ), respectively, in COMMON/Y/. The temporal derivatives $\partial X_a/\partial t$ and $\partial Y_a/\partial t$ in eqs. (MA1.3) and (MA1.6) are in arrays XAT(NZ) and YAT(NZ) in COMMON/F/, and the spatial derivatives $\partial X_a/\partial z$ and $\partial Y_a/\partial z$ are in arrays XAZ(NZ) and YAZ(NZ) in COMMON/S/. The equilibrium gas composition, Y_a^*, in eqs. (MA1.3) and (MA1.6) is in array YAS(NZ) in COMMON/S/.

(2) The given end compositions are then defined numerically

```
C...
C...     GIVEN END COMPOSITIONS (MOLE FRACTIONS)
C...
C...          ENTERING GAS
              YA1=0.0825D0
C...
C...          ENTERING LIQUID
              XA2=0.0D0
C...
C...          EXITING GAS
              YA2=0.003D0
```

In particular, the column is to be sized so that the exiting gas composition does not exceed $Y_a(z_l, t = \infty) = 0.003/(1 - 0.003) = 0.00309 \approx 0.003 \, (= \text{YA2})$.

(3) A series of problem parameters is then defined numerically, or computed. In particular, the column cross-sectional area is required to compute the liquid and gas flow rates, L_s and G_s, which are based on a unit column cross-sectional area

```
C...
C...     FLOW RATES (MOLS/M**2-MIN, SO THE COLUMN CROSS
C...     SECTIONAL AREA IS REQUIRED)
C...
C...          PI
              PI=4.0D0*DATAN(1.0D0)
C...
C...          COLUMN DIAMETER (M)
              D=0.154D0
C...
C...          COLUMN CROSS SECTIONAL AREA (M**2)
```

```
              S=PI*(D**2)/4.0D0
C...
C...          GAS VOLUMETRIC FLOW RATE (M**3/MIN)
              V=11.04D0
C...
C...          GAS CONSTANT (M**3-ATM/KG MOL-K) (NOTE - ONE KG MOL
C...          OF GAS OCCUPIES 22.4 M**3 AT 1 ATM AND 273.16 K)
              R=22.4D0*1.0D0/273.16D0
C...
C...          TEMPERATURE (K)
              TK=273.16D0+20.0D0
C...
C...          PRESSURE (ATM)
              P=1.0D0
C...
C...          GAS MOLAR FLOW RATE (KG MOL/MIN-M**2)
              G=V*P/(R*TK)*(1.0D0/S)
C...
C...          GAS MOLAR FLOW RATE ON SOLUTE FREE BASIS (KG MOLS
C...          AIR/MIN-M**2)
              GS=G*(1.0D0/(1.0D0+YA1))
C...
C...          LIQUID MOLAR FLOW RATE ON SOLUTE FREE BASIS
C...          (KG MOLS H20/MIN-M**2)
              LS=11.18D0/18.0D0*(1.0D0/S)
```

(4) The exiting liquid composition is then computed by a call to function OPLN, which implements eq. (MA1.16)

```
C...
C...    EXITING LIQUID COMPOSITION
        XA1=OPLN(YA1,YA2,XA2,GS,LS)
```

Note that the three previous end concentrations, $Y_{a,1}$, $Y_{a,2}$, and $X_{a,2}$, and the liquid and gas flow rates, L_s and G_s, are the inputs to OPLN. The only unknown in eq. (MA1.16) is $X_{a,1}(=0.0590)$, the exiting liquid composition at the bottom of the column. Function OPLN is discussed subsequently.

(5) The column height (which will be checked by observing if the exiting concentrations, $Y_{a,2} = 0.003$, and the computed $X_{a,1}(=0.0590)$ from (4) above, are returned by eqs. (MA1.3) and (MA1.6) at steady state) is defined

```
C...
C...    COLUMN LENGTH (M)
        ZL=3.28D0
```

(6) The steady state operating conditions are then summarized in an output statement

```
C...
C...   WRITE SUMMARY OF OPERATING CONDITIONS
       WRITE(NO,1)YA1,XA2,YA2,XA1,LS,GS
1      FORMAT(' Summary of Operating Conditions',//,
      +       '       Entering gas      = ',F7.4,/,
      +       '       Entering liquid   = ',F7.4,/,
      +       '       Exiting gas       = ',F7.4,/,
      +       '       Exiting liquid    = ',F7.4,/,
      +       '       Liquid flow rate  = ',F7.4,' kg mols
                                              H20/min-m**2',/,
      +       '       Gas flow rate     = ',F7.4,' kg mols
                                              air/min-m**2',/)
```

(7) The equilibrium data for the NH_3-air-water system are approximated by a second-order, least-squares polynomial by a call to subroutine COEFF (discussed subsequently). The coefficients of the least-squares polynomial are written into array A(2), which is in COMMON/E/

```
C...
C...   COEFFICIENTS IN THE LEAST SQUARES SECOND ORDER
C...   POLYNOMIAL APPROXIMATION OF THE EQUILIBRIUM DATA
C...   (THE COEFFICIENTS ARE WRITTEN INTO COMMON/E/ BY THIS
C...   CALL TO COEFF)
       CALL COEFF
```

(8) The column height is then computed from the log-mean concentration difference according to eqs. (MA1.32) and (MA1.33) by a call to function HTLM (discussed subsequently)

```
C...
C... MASS TRANSFER COEFFICIENT (KG MOL AIR/M**2-MIN)
     KYA=73.8D0
C...
C... COLUMN HEIGHT USING LM MOL RATIO CONCENTRATION
C... DIFFERENCE
     WRITE(NO,2)
2    FORMAT(///,'Column height from log mean concentration',
    +             'differenece')
     Z=HTLM(GS,KYA,XA1,YA1,XA2,YA2)
     WRITE(NO,3)Z
3    FORMAT(/,' Column height = ',F6.2,' m',//)
```

The calculation of the column height from the log-mean concentration difference is independent of the MOL sizing of the column, and the two approaches, ideally, should give the same result. They do not, however, because the log-mean concentration applies only to the case when the equilibrium and operating lines are linear, and in this case the equilibrium line is not linear (as we shall observe when discussing the second-order polynomial fit of the equilibrium data).

(9) The column height is then computed by the evaluation of the RHS integral of eq. (MA1.24) by a call to function HTIN (discussed subsequently)

```
C...
C...   COLUMN HEIGHT USING INTEGRATION  OF MOL RATIO
C...   CONCENTRATION DIFFERENCE
       WRITE(NO,4)
4      FORMAT(///,' Column height from numerical integration
      +       of',' concentration difference')
       Z=HTIN(GS,KYA,YA1,YA2)
       WRITE(NO,5)Z
5      FORMAT(/,' Column height = ',F6.2,' m',//)
```

Function HTIN calls a routine for the Simpson's rule evaluation of integrals.

(10) Finally, homogeneous initial conditions for eqs. (MA1.3) and (MA1.6) are implemented in DO loop 10

```
C...
C...   INITIAL CONDITIONS FOR PDE SOLUTION
       DO 10 I=1,NZ
          XA(I)=0.0D0
          YA(I)=0.0D0
10     CONTINUE
```

At this point, subroutine INITAL, which is called by a main program similar to RKF45M in Appendix A, has set the initial conditions for the 42 ODEs that are the MOL approximation of eqs. (MA1.3) and (MA1.6). The temporal integration of these ODEs is then programmed in subroutine DERV listed below:

```
       SUBROUTINE DERV
C...
C...   DOUBLE PRECISION CODING IS USED
       IMPLICIT DOUBLE PRECISION (A-H,O-Z)
       DOUBLE PRECISION LS, KYA
C...
C...   COMMON AREA TO LINK SUBROUTINES
       PARAMETER(NZ=21)
       COMMON/T/       T,     NSTOP,     NORUN
      +      /Y/   XA(NZ),    YA(NZ)
      +      /F/   XAT(NZ),   YAT(NZ)
      +      /S/   XAZ(NZ),   YAZ(NZ),   YAS(NZ)
      +      /C/       XA1,       YA1,       XA2,       YA2,
      +            ZL,        GS,        LS,       KYA,         S
```

```
C...
C...   BOUNDARY CONDITION AT Z = 0
       YA(1)=YA1
C...
C...   BOUNDARY CONDITION AT Z = ZL
       XA(NZ)=XA2
C...
C...   XAZ, YAZ
       CALL DSS020(0.0D0,ZL,NZ,XA,XAZ,-LS)
       CALL DSS020(0.0D0,ZL,NZ,YA,YAZ, GS)
C...
C...   YA* GAS COMPOSITION (FROM EQUILIBRIUM LINE)
       DO 1 I=1,NZ
          YAS(I)=EQUIL(XA(I))
1      CONTINUE
C...
C...   PDES
       DO 2 I=1,NZ
          YAT(I)=-GS*YAZ(I)+KYA*(YAS(I)-YA(I))
          XAT(I)= LS*XAZ(I)-KYA*(YAS(I)-YA(I))
2      CONTINUE
        YAT(1)=0.0D0
       XAT(NZ)=0.0D0
       RETURN
       END
```

We can note the following points about DERV:

(1) The COMMON area is the same as in INITAL.

(2) The boundary condition for eq. (MA1.6) specifies the entering gas composition at the bottom of the column (set in INITAL and passed through COMMON/C/ to DERV

```
C...
C...   BOUNDARY CONDITION AT Z = 0
       YA(1)=YA1
```

Note that this boundary condition is set at grid point 1.

(3) Similarly, the boundary condition for eq. (MA1.3) specifies the entering liquid composition at the top of the column

```
C...
C...   BOUNDARY CONDITION AT Z = ZL
       XA(NZ)=XA2
```

Note that this boundary condition is set at grid point NZ.

(4) The spatial derivatives, $\partial X_a/\partial z$ and $\partial Y_a/\partial z$, in eqs. (MA1.3) and (MA1.6) respectively, are computed by calls to the spatial differentiator DSS020

```
C...
C...   XAZ, YAZ
       CALL DSS020(0.0D0,ZL,NZ,XA,XAZ,-LS)
       CALL DSS020(0.0D0,ZL,NZ,YA,YAZ, GS)
```

Note the signs of the flow rates, $-L_S$ and G_S, reflecting the countercurrent flow in the column of Figure MA1.1.

(5) The equilibrium gas composition, Y_a^*, along the spatial grid is then computed from the liquid composition, X_a, by calls to function EQUIL (discussed subsequently)

```
C...
C...   YA* GAS COMPOSITION (FROM EQUILIBRIUM LINE)
       DO 1 I=1,NZ
          YAS(I)=EQUIL(XA(I))
1      CONTINUE
```

(6) Then the 42 ODE temporal derivatives are computed in DO loop 2

```
C...
C...   PDES
       DO 2 I=1,NZ
          YAT(I)=-GS*YAZ(I)+KYA*(YAS(I)-YA(I))
          XAT(I)= LS*XAZ(I)-KYA*(YAS(I)-YA(I))
2      CONTINUE
```

Note the close resemblance of the coding in DO loop 2 to the PDEs, eqs. (MA1.3) and (MA1.6).

(7) Finally, since the entering liquid and gas compositions are specified as boundary conditions, their temporal derivatives are zeroed

```
       YAT(1)=0.0D0
       XAT(NZ)=0.0D0
```

This completes the programming of the 42 ODEs. The numerical solution to eqs. (MA1.3) and (MA1.6) is then written and plotted by subroutine PRINT listed below:

```
       SUBROUTINE PRINT(NI,NO)
C...
C...   DOUBLE PRECISION CODING IS USED
       IMPLICIT DOUBLE PRECISION (A-H,O-Z)
       DOUBLE PRECISION LS, KYA
C...
C...   COMMON AREA TO LINK SUBROUTINES
       PARAMETER(NZ=21)
```

```
      COMMON/T/         T,     NSTOP,    NORUN
     +     /Y/  XA(NZ),    YA(NZ)
     +     /F/  XAT(NZ), YAT(NZ)
     +     /S/  XAZ(NZ), YAZ(NZ), YAS(NZ)
     +     /C/       XA1,      YA1,      XA2,      YA2,
     +           ZL,       GS,       LS,      KYA,        S
C...
C...  ARRAYS FOR PLOTTING
      PARAMETER(NP=21)
      DIMENSION TP(NP), XAP(NP), YAP(NP)
C...
C...  MONITOR THE OUTPUT
      WRITE(*,*)T,YA(NZ)
C...
C...  WRITE A HEADING FOR THE SOLUTION
      IF(T.LT.0.001D0)THEN
         WRITE(NO,1)
1        FORMAT(/,9X,'t',3X,'Xa(L,t)',3X,'Xa(0,t)',
     +                 3X,'Ya(0,t)',3X,'Ya(L,t)',/)
         IP=0
      END IF
C...
C...  WRITE THE SOLUTION
      WRITE(NO,2)T,XA(NZ),XA(1),YA(1),YA(NZ)
2     FORMAT(F10.2,4F10.5)
C...
C...  STORE THE SOLUTION FOR PLOTTING
      IP=IP+1
      TP(IP)=T
      XAP(IP)=XA(1)
      YAP(IP)=YA(NZ)
C...
C...  PLOT THE SOLUTION
      IF(IP.LT.NP)RETURN
C...
C...     BOTTOM LIQUID COMPOSITION VS T
         CALL SPLOTS(1,NP,TP,XAP)
         WRITE(NO,3)
3        FORMAT(/,' Xa(0,t) vs t')
C...
C...     TOP GAS COMPOSITION VS T
         CALL SPLOTS(1,NP,TP,YAP)
         WRITE(NO,4)
4        FORMAT(/,' Ya(L,t) vs t')
      RETURN
      END
```

We can note the following points about PRINT:

(1) The COMMON area is the same as for INITAL and DERV.

(2) Arrays are defined for 21-point plotting of the numerical solution

```
C...
C...    ARRAYS FOR PLOTTING
        PARAMETER(NP=21)
        DIMENSION TP(NP), XAP(NP), YAP(NP)
```

(3) After displaying the progress of the solution by writing to unit *, a heading is written for the solution

```
C...
C...    MONITOR THE OUTPUT
        WRITE(*,*)T,YA(NZ)
C...
C...    WRITE A HEADING FOR THE SOLUTION
        IF(T.LT.0.001D0)THEN
            WRITE(NO,1)
1           FORMAT(/,9X,'t',3X,'Xa(L,t)',3X,'Xa(0,t)',
       +                 3X,'Ya(0,t)',3X,'Ya(L,t)',/)
            IP=0
        END IF
```

(4) The end concentrations are then written as a function of time

```
C...
C...    WRITE THE SOLUTION
        WRITE(NO,2)T,XA(NZ),XA(1),YA(1),YA(NZ)
2       FORMAT(F10.2,4F10.5)
```

(5) The exiting concentrations, $X_a(0, t)$ and $Y_a(z_l, t)$, are stored for subsequent plotting

```
C...
C...    STORE THE SOLUTION FOR PLOTTING
        IP=IP+1
        TP(IP)=T
        XAP(IP)=XA(1)
        YAP(IP)=YA(NZ)
```

(6) Finally, the exiting concentrations are plotted as a function of time by calls to the point plotting routine SPLOTS

```
C...
C...    PLOT THE SOLUTION
        IF(IP.LT.NP)RETURN
C...
C...        BOTTOM LIQUID COMPOSITION VS T
```

```
              CALL SPLOTS(1,NP,TP,XAP)
              WRITE(NO,3)
3             FORMAT(/,' Xa(0,t) vs t')
C...
C...          TOP GAS COMPOSITION VS T
              CALL SPLOTS(1,NP,TP,YAP)
              WRITE(NO,4)
4             FORMAT(/,' Ya(L,t) vs t')
```

The data read by the main program for the integration of the 42 ODEs are

```
NH3 absorber
0.            2.0         0.1
     42                        0.00001
END OF RUNS
```

Note that time runs over the interval $0 \leq t \leq 2$ (in min since the problem parameters are all in min) with calls to PRINT at an interval of 0.1 (21 calls to PRINT). The 42 ODEs are integrated with an error of 0.00001.

These data raise the question of the time scale for eqs. (MA1.3) and (MA1.6). In particular, the time scale is determined by the liquid and gas holdups, H_L and H_G, which were not defined in the preceding coding. In effect, when eqs. (MA1.3) and (MA1.6) were programmed in DO loop 2 of subroutine DERV, we used the values $H_L = H_G = 1$, which then set the time scale of the solution to $0 \leq t \leq 2$ (determined by trial and error execution of the program), hence the time interval used in the preceding data file. In this way (using $H_L = H_G = 1$) the time scale was set arbitrarily, which is acceptable since we are only interested in the final steady state solution.

The output from the execution of the preceding program (subroutines INITAL, DERV, and PRINT, plus data) is summarized below:

```
RUN NO. - 1 NH3 absorber

INITIAL T - .000D+00

FINAL T - .200D+01

PRINT T - .100D+00

NUMBER OF DIFFERENTIAL EQUATIONS - 42

MAXIMUM INTEGRATION ERROR - .100D-04
Summary of Operating Conditions

    Entering gas    = .0899
    Entering liquid = .0000
    Exiting gas     = .0030
```

```
Exiting liquid = .0590
Liquid flow rate = 33.3456 kg mols H20/min-m**2
Gas flow rate  = 22.6208 kg mols air/min-m**2

      a1 = 1.3098   a2 = -2.8390

  i   XNH3    YNH3    YNH3    o/o
      tab     tab     calc    diff
  1   .0164   .0210   .0207   -1.3
  2   .0252   .0320   .0312   -2.5
  3   .0349   .0420   .0423    .6
  4   .0455   .0530   .0537   1.4
  5   .0722   .0800   .0798   -.3
```

Column height from log mean concentration differenece

Log-mean concentration difference = .009706

Number of transfer units = 8.95

Height of transfer unit = .307 m

Column height = 2.74 m

Column height from numerical integration of concentration difference
Number of transfer units = 10.72

Height of transfer unit = .307 m

Column height = 3.28 m

```
t     Xa(L,t)   Xa(0,t)   Ya(0,t)   Ya(L,t)

 .00   .00000   .00000    .08992    .00000
 .10   .00000   .04812    .08992    .00000
 .20   .00000   .05328    .08992    .00020
 .30   .00000   .05549    .08992    .00096
 .40   .00000   .05673    .08992    .00165
 .50   .00000   .05751    .08992    .00213
 .60   .00000   .05802    .08992    .00245
 .70   .00000   .05835    .08992    .00265
 .80   .00000   .05857    .08992    .00278
 .90   .00000   .05871    .08992    .00287
1.00   .00000   .05880    .08992    .00292
         .        .          .
         .        .          .
         .        .          .
1.90   .00000   .05895    .08992    .00302
2.00   .00000   .05896    .08992    .00302
```

```
           ..1....1....1....1....1....1....1....1....1....1....1..
 .590D-01+          1 1 1   1 1   1 1   1 1   1 1   1   1 1   1 1   1 1  +I
        -       1                                                       -I
        -     1                                                         -I
        -                                                               -I
        -                                                               -I
 .393D-01+                                                              +I
        -                                                               -I
        -                                                               -I
        -                                                               -I
        -                                                               -I
 .197D-01+                                                              +I
        -                                                               -I
        -                                                               -I
        -                                                               -I
        -  -1                                                           -I
-.260D-17+                                                              +I
           ..1....1....1....1....1....1....1....1....1....1....1..
            .000D+00  .40D+00   .80D+00   .12D+01   .16D+01   .20D+01
```

Xa(0,t) vs t

```
           ..1....1....1....1....1....1....1....1....1....1....1..
 .302D-02+                         1  1 1   1   1 1   1 1   1 1   1 1   +I
        -                       1 1                                     -I
        -                      1                                        -I
        -                                                               -I
        -                   1                                           -I
 .201D-02+                                                              +I
        -              1                                                -I
        -                                                               -I
        -                                                               -I
        -                                                               -I
 .100D-02+        1                                                     +I
        -                                                               -I
        -                                                               -I
        -      1                                                        -I
        -    1 1                                                        -I
-.278D-05+                                                              +I
           ..1....1....1....1....1....1....1....1....1....1....1..
            .000D+00  .40D+00   .80D+00   .12D+01   .16D+01   .20D+01
```

Ya(L,t) vs t

We can note the following points about the output:

(1) The steady state operating conditions are printed first

```
Summary of Operating Conditions

    Entering gas     =   .0899
    Entering liquid  =   .0000
```

```
Exiting gas       =   .0030
Exiting liquid    =   .0590
Liquid flow rate  = 33.3456 kg mols H20/min-m**2
Gas flow rate     = 22.6208 kg mols air/min-m**2
```

(2) The coefficients of the second-order, least-squares polynomial fitted to the equilbrium data are then printed

```
a1 =   1.3098    a2 =   -2.8390
```

The second-order polynomial for the equilibrium data is therefore

$$Y_a^* = 1.3098 X_a - 2.8390 X_a^2 \tag{MA1.36}$$

Note that:

(2.1) The polynomial of eq. (MA1.36) passes through the point $X_a = 0$, $Y_a = 0$, as required by physical considerations (i.e., zero NH_3 concentration in the liquid phase corresponds to zero concentration in the gas phase). The mathematical details for ensuring the polynomial goes through the origin are considered subsequently.

(2.2) The polynomial is nonlinear (the equilibrium "line" or curve is not linear).

(2.3) The equilibrium curve is concave downward since $d^2 Y_a / d X_a^2 < 0$, which explains why the log-mean concentration difference underestimates the column height (discussed subsequently).

(3) The maximum percent difference between the tabulated equilibrium data and the calculated values from eq. (MA1.36) is -2.5, which is considered adequate for the column design, that is, the calculation of Y_a^* in eqs. (MA1.3) and (MA1.6). The table comparing the tabulated and computed equilibrium data was printed by function EQUIL (discussed subsequently).

(4) The column height calculated with the log-mean concentration difference, that is, from eqs. (MA1.32) and (MA1.33) programmed in function HTLM, is 2.74 m, which is less than the value of 3.35 m ($= 11.0$ ft) reported by Welty, Wicks, and Wilson 1984, 695). This underestimate is due to

(4.1) The operating line of eq. (MA1.18) positioned above the equilibrium line in a plot of Y_a versus X_a, which is the case for absorption (for desorption, the operating line is below the equilibrium line (Welty, Wicks, and Wilson 1984, 678–9).

(4.2) The concave equilibrium curve of eq. (MA1.36), which means that the driving force for mass transfer, $Y_a^* - Y_a$, in eqs. (MA1.3) and (MA1.6) (and also eq. (MA1.24)) is less than if the equilibrium curve is linear. Thus, the mass transfer rate for the nonlinear equilibrium curve of eq. (MA1.36) is less than for a linear equilibrium curve (as required by the log-mean concentration difference), and therefore the column height is greater (due to the nonlinear equilibrium curve). The reader should make a sketch of the operating line of eq. (MA1.18) and the equilibrium curve of eq. (MA1.36) to confirm this conclusion.

(5) The height of the column calculated from eq. (MA1.24) (programmed in subroutine HTIN) is 3.28 m, which is probably more accurate than the value of 3.35 m reported by Welty, Wicks, and Wilson (1984, 695) since the Simpson's rule integration in function HTIN is generally more accurate than the graphical integration used by Welty, Wicks, and Wilson. In any case, the agreement (3.28 m versus 3.35 m) is close, and probably within the error introduced by the uncertainties in the problem parameters, particularly the mass transfer coefficient, $K_Y a$, and to some extent, the equilibrium data represented by eq. (MA1.36).

(6) The PDE solution approaches the correct steady state exiting concentrations based on a column length of 3.28 m. Thus, at the end of the solution, when steady state has been reached

```
    1.90      .00000      .05895      .08992      .00302
    2.00      .00000      .05896      .08992      .00302
```

Note in particular that the exiting gas composition, $Y_a(z_l, t)$, is .00302, which is close to the prescribed value of 0.003. Also, the exiting liquid concentration computed by the overall material balance of eq. (MA1.16) is $X_a(0, t) = 0.059$, compared with the PDE value of .05896.

(7) The approach of the solutions of eqs. (MA1.3) and (MA1.6) to steady state is clear from the point plots of $X_a(0, t)$ and $Y_a(z_l, t)$ versus t produced by the calls to SPLOTS. Also, the time scale of $0 \leq t \leq 2$ is adequate to define the complete transient.

Finally, we might consider why the PDE formulation is used (which requires a trial and error specification of the column height to produce the correct exiting concentrations) when eq. (MA1.24) can be used to compute z directly. There are several significant advantages of the PDE approach:

(1) Equation (MA1.24) and, more generally, the conventional steady state equations used to design mass transfer equipment are limited to essentially binary systems, for example, NH_3 in air and H_2O, with one transferred component, for example, NH_3. In practice, mass transfer operations are usually multicomponent, that is, more than one transferred component, so that the conventional design procedures cannot be used. However, the PDE formulation can be readily extended to multicomponent systems by writing PDEs, such as eqs. (MA1.3) and (MA1.6), for each component. Thus, we would integrate a system of simultaneous PDEs, with two PDEs for each of the transferred components. This approach presupposes that multicomponent equilibrium data and mass transfer coefficients are available.

(2) Equation (MA1.24) and, more generally, the conventional steady state equations used to design mass transfer equipment are limited to essentially isothermal systems, so that any heat effects due to the absorption are ignored. Again, in practice, absorption systems are often nonisothermal. The PDE formulation can be

applied to this situation by writing PDE energy balances for the liquid and gas, which are integrated along with the component balance PDEs. This approach presupposes that multicomponent equilibrium data for a range of temperatures are available.

(3) Equation (MA1.24) and, more generally, the conventional steady state equations used to design mass transfer equipment are of little use in designing control systems, that is, unsteady state (dynamic) models are required. In the PDE formulation, the dynamic behavior of the mass transfer system is computed. If controller equations are added to the model PDEs, the effectiveness of a proposed control system can be evaluated (but correct liquid and gas holdups, H_L and H_G, would be required to define the correct time scale).

We conclude this analysis with a discussion of the several routines called previously. Function OPLN is a straightforward implementation of the operating line equation (MA1.18)

```
      DOUBLE PRECISION FUNCTION OPLN(YA,YA2,XA2,GS,LS)
C...
C...  Function OPLN computes the bulk liquid composition
C...  from the corresponding gas composition using the
C...  equation for the operating line based on the top
C...  compositions (at end 2)
C...
C...     Xa = (Gs/Ls)*(Ya - Ya2) + Xa2
C...
      IMPLICIT DOUBLE PRECISION (A-H,O-Z)
      DOUBLE PRECISION LS
C...
C...  EQUATION FOR THE OPERATING LINE BASED ON THE
C...  COMPOSITIONS AT THE TOP OF THE COLUMN (END 2)
      XA=(GS/LS)*(YA-YA2)+XA2
C...
C...  EVALUATE FUNCTION OPLN
      OPLN=XA
      RETURN
      END
```

Note that X_a is a linear function of Y_a, since G_s and L_s are constant.

Function EQUIL for the calculation of Y_a^* in eqs. (MA1.3), (MA1.6), and (MA1.24) is listed below:

```
      DOUBLE PRECISION FUNCTION EQUIL(XA)
C...
C...  Function EQUIL computes the gas phase composition in
C...  equilibroum with the liquid in the column, using the
```

```
C...      least squares second order polynomial evaluated by
C...      subroutine COEFF.
C...
          IMPLICIT DOUBLE PRECISION (A-H,O-Z)
C...
C...      COEFFICIENTS IN POLYNOMIAL FOR EQUILIBRIUM LINE
          COMMON/E/A(2)
C...
C...      SECOND ORDER LEAST SQUARES POLYNOMIAL
          EQUIL=A(1)*XA+A(2)*(XA**2)
          RETURN
          END
```

Note that eq. (MA1.36) computes Y_a^* ($=$ EQUIL) from X_a. The least-squares coefficients a_1 and a_2 are computed by a call to subroutine COEFF and the subordinate routine POLY, listed below:

```
          SUBROUTINE COEFF
C...
C...      Subroutine COEFF tests the least squares second order
C...      polynomial fit of the NH3-H20-air equilibrium data in
C...      Welty, et al, (1984), Example 2, pp 680-681.
C...
C...      DOUBLE PRECISION CODING IS USED
          IMPLICIT DOUBLE PRECISION (A-H,O-Z)
C...
C...      COEFFICIENTS IN POLYNOMIAL FOR EQUILIBRIUM LINE
          COMMON/E/A(2)
C...
C...      INPUT/OUTPUT (I/O) UNIT NUMBERS
          COMMON/IO/ NI, NO
C...
C...      ARRAYS FOR FIVE DATA POINTS
          PARAMETER(N=5)
          DIMENSION X(N), Y(N), YC(N)
C...
C...      EQUILIBRIUM DATA (FROM WELTY, ET AL)
          DATA X/0.0164D0, 0.0252D0, 0.0349D0, 0.0455D0,
                 0.0722D0/
          DATA Y/0.021D0 , 0.032D0 , 0.042D0 , 0.053D0 ,
C...             0.08D0  /
C...      EVALUATE COEFFICIENTS IN THE LEAST SQUARES, SECOND
C...      ORDER POLYNOMIAL
          CALL POLY(X,Y,N)
C...
C...      WRITE THE COEFFICIENTS
```

```
          WRITE(NO,1)A(1),A(2)
1         FORMAT(10X,' a1 = ',F8.4,'    a2 = ',F8.4,/)
C...
C...  HEADING FOR THE COMPUTED RESULTS
          WRITE(NO,2)
2         FORMAT(' i',6X,'XNH3',6X,'YNH3',6X,'YNH3',7X,'o/o',/,
     +            5X,7X,'tab', 7X,'tab', 6X,'calc',6X,'diff',/)
C...
C...  COMPUTE VALUES OF THE GAS PHASE COMPOSITION FOR
C...  COMPARISON WITH THE TABULATED VALUES
          DO 5 I=1,5
C...
C...      GAS PHASE COMPOSITION FROM THE SECOND ORDER
C...      POLYNOMIAL
          YC(I)=A(1)*X(I)+A(2)*X(I)**2
C...
C...      COMPUTE AND PRINT THE GAS PHASE COMPOSITIONS
          DIFF=(YC(I)-Y(I))/Y(I)*100.0D0
          WRITE(NO,4)I,X(I),Y(I),YC(I),DIFF
4         FORMAT(I5,3F10.4,F10.1)
C...
C...  NEXT POINT
5         CONTINUE
C...
C...  END OF CALCULATION
          RETURN
          END
```

We can note the following points about EQUIL:

(1) The COMMON areas, /E/ and /IO/, contain the computed least-squares polynomial coefficients (in A(2)) and the input/output unit numbers (NI, NO), respectively

```
C...
C...  COEFFICIENTS IN POLYNOMIAL FOR EQUILIBRIUM LINE
      COMMON/E/A(2)
C...
C...  INPUT/OUTPUT (I/O) UNIT NUMBERS
      COMMON/IO/ NI, NO
```

(2) The arrays for storing the equilibrium data are defined, then filled with five numerical values

```
C...
C...  ARRAYS FOR FIVE DATA POINTS
      PARAMETER(N=5)
```

```
      DIMENSION X(N), Y(N), YC(N)
C...
C...  EQUILIBRIUM DATA (FROM WELTY, ET AL)
      DATA X/0.0164D0, 0.0252D0, 0.0349D0, 0.0455D0,
           0.0722D0/
      DATA Y/0.021D0 , 0.032D0 , 0.042D0 , 0.053D0 ,
           0.08D0  /
```

(3) The coefficients a_1 and a_2 in the second-order, least-squares polynomial, that is, eq. (MA1.36), are computed by a call to subroutine POLY (discussed subsequently) and returned to COEFF through COMMON/E/

```
C...
C...  EVALUATE COEFFICIENTS IN THE LEAST SQUARES, SECOND
C...  ORDER POLYNOMIAL
      CALL POLY(X,Y,N)
```

(4) The coefficients are printed, then used to compare the values of Y_a^* computed from eq. (MA1.36) with the tabulated values

```
C...
C...  WRITE THE COEFFICIENTS
      WRITE(NO,1)A(1),A(2)
1     FORMAT(10X,' a1 = ',F8.4,'   a2 = ',F8.4,/)
C...
C...  HEADING FOR THE COMPUTED RESULTS
      WRITE(NO,2)
2     FORMAT(' i',6X,'XNH3',6X,'YNH3',6X,'YNH3',7X,'o/o',/,
     +         5X,7X,'tab', 7X,'tab', 6X,'calc',6X,'diff',/)
C...
C...  COMPUTE VALUES OF THE GAS PHASE COMPOSITION FOR
C...  COMPARISON WITH THE TABULATED VALUES
      DO 5 I=1,5
C...
C...     GAS PHASE COMPOSITION FROM THE SECOND ORDER
C...     POLYNOMIAL
         YC(I)=A(1)*X(I)+A(2)*X(I)**2
C...
C...     COMPUTE AND PRINT THE GAS PHASE COMPOSITIONS
         DIFF=(YC(I)-Y(I))/Y(I)*100.0D0
         WRITE(NO,4)I,X(I),Y(I),YC(I),DIFF
4        FORMAT(I5,3F10.4,F10.1)
C...
C...  NEXT POINT
5     CONTINUE
```

As noted previously in the discussion of the output, the maximum percent difference between the computed and tabulated values of Y_a^* is -2.5.

Subroutine POLY is listed below:

```
      SUBROUTINE POLY(X,Y,N)
C...
C...  Subroutine POLY computes the coefficients in the least
C...  squares second order polynomial approximating the
C...  equilibrium data.
C...
C...  Consider the least squares analysis of data using a
C...  polynomial model
C...
C...      y = a0 + a1*x + a2*(x**2) + ...                  (1)
C...
C...  For purposes of discussion, we consider a second order
C...  polynomial.  The extension to higher order polynomials
C...  is straightforward.
C...
C...  Application of the least squares principle to N data
C...  gives
C...
C...                      N
C...      min E(a0,a1,a2) = sum  (y(i) - ym(i))**2
C...                      i=1
C...
C...                      N
C...                   = sum  (y(i) - (a0 + a1*x(i)
C...                     i=1
C...                   + a2*(x(i)**2)))**2                 (2)
C...
C...  We can find the minimum of eq. (2), i.e., the values
C...  of a0,a1 and a2 which minimize E, by the usual methods
C...  of differential calculus.
C...
C...  Before proceeding, however, we consider the special
C...  case for which the model of eq. (1) is required to
C...  pass through the point y(0) = 0, x(0) = 0.  Thus we
C...  see a0 = 0 in eq. (1), and we therefore have to
C...  evaluate only a1 and a2 in eq. (2).
C...
C...  The required partial derivatives (denoted with a "d"
C...  because of the limitations of the character set) are
C...  set to zero
C...
C...  dE/da1 =
C...
C...         N
C...      2 sum(y(i) - (a1*x(i) + a2*(x(i)**2)))*(-x(i)) = 0 (3)
C...        i=1
```

```
C...
C...   dE/da2 =
C...
C...         N
C...     2 sum(y(i)-(a1*x(i)+a2*(x(i)**2)))*(-x(i)**2) = 0   (4)
C...        i=1
C...
C...   Eqs. (3) and (4) can be rearranged to
C...
C...         N              N              N
C...     a1 sum x(i)**2 + a2 sum x(i)**3 = sum x(i)*y(i)     (5)
C...        i=1            i=1            i=1
C...
C...         N            N            N
C...     a1 sum x(i)**3+a2 sum x(i)**4=sum (x(i)**2)*y(i)    (6)
C...        i=1          i=1          i=1
C...
C...   Eqs. (5) and (6) are two linear algebraic equations
C...   for a1 and a2.  The following code computes a1 and a2
C...   from a vector of data of length N.
C...
       IMPLICIT DOUBLE PRECISION (A-H,O-Z)
C...
C...   COEFFICIENTS IN POLYNOMIAL FOR EQUILIBRIUM LINE
       COMMON/E/A(2)
C...
C...   INPUT/OUTPUT (I/O) UNIT NUMBERS
       COMMON/IO/ NI, NO
C...
C...   ARRAYS FOR X, Y
       DIMENSION X(N), Y(N)
C...
C...   COMPUTE THE ELEMENTS OF THE COEFFICIENT MATRIX AND RHS
C...   VECTOR OF EQS. (5) AND (6)
       A11=0.0D0
       A12=0.0D0
       A21=0.0D0
       A22=0.0D0
        B1=0.0D0
        B2=0.0D0
       DO 1 I=1,N
          A11=A11+X(I)**2
          A12=A12+X(I)**3
          A21=A12
          A22=A22+X(I)**4
          B1=B1+X(I)*Y(I)
          B2=B2+(X(I)**2)*Y(I)
```

```
1       CONTINUE
C...
C...    COMPUTE A1 AND A2 IN EQ. (1) (BY DETERMINANTS)
        DET=A11*A22-A12*A21
C...
C...    TEST IF THE ALGEBRAIC EQUATIONS ARE NUMERICALLY
C...    SINGULAR
        IF(DABS(DET).LT.1.0D-20)THEN
C...
C...        DETERMINANT IS SMALL SO TERMINATE THE CALCULATION.
C...        THE PRECEDING THRESHOLD OF 1.0D-20 WAS SELECTED
C...        ARBITRARILY AND IS DETERMINED PRIMARILY BY THE WORD
C...        LENGTH OF THE COMPUTER
            WRITE(NO,2)DET
2           FORMAT('Determinant of coefficient matrix =',D11.3)
            STOP
        END IF
C...
C...    SYSTEM IS NOT NUMERICALLY SINGULAR (ILL-CONDITIONED),
C...    SO CONTINUE THE SOLUTION OF THE ALGEBRAIC SYSTEM
        DET1=B1*A22-B2*A12
        DET2=B2*A11-B1*A21
C...
C...    COMPUTE COEFFICIENTS A1 AND A2 FROM THE DETERMINANTS
C...    (BY KRAMER'S RULE, WHICH IS GENERALLY NOT RECOMMENDED
C...    FOR THE SOLUTION OF SYSTEMS OF LINEAR ALGEBRAIC
C...    EQUATIONS, BUT IS ADEQUATE FOR THIS SMALL 2 X 2
C...    PROBLEM)
        A(1)=DET1/DET
        A(2)=DET2/DET
        RETURN
        END
```

Subroutine POLY is based on the following least-squares analysis. The model for the equilibrium data is taken to be

$$y_m = a_1 x + a_2 x^2 \qquad (MA1.37)$$

where x is the independent variable, in this case the liquid mol ratio of NH_3 in water, X_a, and y_m is the dependent variable, in this case the gas mol ratio of NH_3 in air, Y_a. Note that eq. (MA1.37) has the important characteristic that it passes through the origin, that is, $x = 0$, $y_m = 0$.

In the least-squares approach to the use of the polynomial of eq. (MA1.37), we first formulate the sum of squares of the errors, $E(a_1, a_2)$, for the n data pairs, $(x_i, y_i), i = 1, 2, \ldots, n$

$$E(a_1, a_2) = \sum_{i=1}^{n}(y_i - y_{m,i})^2 = \sum_{i=1}^{n}\{y_i - (a_1 x_i + a_2 x_i^2)\}^2 \qquad (MA1.38)$$

We then compute the values of a_1 and a_2 that minimize $E(a_1, a_2)$. To determine the minimum, we differentiate $E(a_1, a_2)$, with respect to a_1 and a_2, and set the two derivatives equal to zero

$$\frac{\partial E(a_1, a_2)}{\partial a_1} = \sum_{i=1}^{n} 2\{y_i - (a_1 x_i + a_2 x_i^2)\}(-x_i) = 0$$

$$\frac{\partial E(a_1, a_2)}{\partial a_2} = \sum_{i=1}^{n} 2\{y_i - (a_1 x_i + a_2 x_i^2)\}(-x_i^2) = 0$$

or

$$a_1 \sum_{i=1}^{n} x_i^2 + a_2 \sum_{i=1}^{n} x_i^3 = \sum_{i=1}^{n} x_i y_i \tag{MA1.39}$$

$$a_1 \sum_{i=1}^{n} x_i^3 + a_2 \sum_{i=1}^{n} x_i^4 = \sum_{i=1}^{n} x_i^2 y_i \tag{MA1.40}$$

Equations (MA1.39) and (MA1.40) are two linear algebraic equations in a_1 and a_2, which can be written in matrix form as

$$\begin{bmatrix} \sum_{i=1}^{n} x_i^2 & \sum_{i=1}^{n} x_i^3 \\ \sum_{i=1}^{n} x_i^3 & \sum_{i=1}^{n} x_i^4 \end{bmatrix} \cdot \begin{bmatrix} a_1 \\ a_2 \end{bmatrix} = \begin{bmatrix} \sum_{i=1}^{n} x_i y_i \\ \sum_{i=1}^{n} x_i^2 y_i \end{bmatrix} \tag{MA1.41}$$

or more compactly

$$\begin{bmatrix} a_{11} & a_{12} \\ a_{21} & a_{22} \end{bmatrix} \cdot \begin{bmatrix} a_1 \\ a_2 \end{bmatrix} = \begin{bmatrix} b_1 \\ b_2 \end{bmatrix} \tag{MA1.42}$$

where $a_{11} = \sum_{i=1}^{n} x_i^2$, etc.

The 2×2 linear system of eq. (MA1.42) is programmed in POLY. We can note the following points about the programming:

(1) Again, as in subroutine EQUIL, the COMMON areas /E/ and /IO/ contain the computed least-squares polynomial coefficients, a_1 and a_2 (in A(2)), and the input/output unit numbers, (NI, NO), respectively.

(2) The coefficients a_{11} to b_2 in eq. (MA1.42) are zeroed

```
A11=0.0D0
A12=0.0D0
A21=0.0D0
A22=0.0D0
 B1=0.0D0
 B2=0.0D0
```

(3) These coefficients are then computed from the experimental data according to eqs. (MA1.41) and (MA1.42) in DO loop 1

```
      DO 1 I=1,N
         A11=A11+X(I)**2
         A12=A12+X(I)**3
         A21=A12
         A22=A22+X(I)**4
         B1=B1+X(I)*Y(I)
         B2=B2+(X(I)**2)*Y(I)
1     CONTINUE
```

(4) Equation (MA1.42) is then solved for a_1 and a_2 using determinants (Kramer's rule).

```
C...
C...   COMPUTE A1 AND A2 IN EQ. (1) (BY DETERMINANTS)
       DET=A11*A22-A12*A21
C...
C...   TEST IF THE ALGEBRAIC EQUATIONS ARE NUMERICALLY
C...   SINGULAR
       IF(DABS(DET).LT.1.0D-20)THEN
C...
C...      DETERMINANT IS SMALL SO TERMINATE THE CALCULATION.
C...      THE PRECEDING THRESHOLD OF 1.0D-20 WAS SELECTED
C...      ARBITRARILY AND IS DETERMINED PRIMARILY BY THE WORD
C...      LENGTH OF THE COMPUTER
          WRITE(NO,2)DET
2         FORMAT('Determinant of coefficient matrix =',D11.3)
          STOP
       END IF
```

First a check is made to determine if the 2×2 system of eq. (MA1.42) is ill-conditioned (near singular). If the system is well-conditioned, a_1 and a_2 are computed

```
C...
C...   SYSTEM IS NOT NUMERICALLY SINGULAR (ILL-CONDITIONED),
C...   SO CONTINUE THE SOLUTION OF THE ALGEBRAIC SYSTEM
       DET1=B1*A22-B2*A12
       DET2=B2*A11-B1*A21
C...
C...   COMPUTE COEFFICIENTS A1 AND A2 FROM THE DETERMINANTS
C...   (BY KRAMER'S RULE, WHICH IS GENERALLY NOT RECOMMENDED
C...   FOR THE SOLUTION OF SYSTEMS OF LINEAR ALGEBRAIC
C...   EQUATIONS, BUT IS ADEQUATE FOR THIS SMALL 2 X 2
C...   PROBLEM)
       A(1)=DET1/DET
       A(2)=DET2/DET
```

Note the division by the determinant ($=$ DET) of the two linear algebraic equations (eq. (MA1.41) or eq. (MA1.42)), which is the reason for checking the magnitude of the determinant.

Function HTLM for the calculation of the column height by the log-mean concentration difference of eqs. (MA1.32) and MA1.33) is listed below:

```
      DOUBLE PRECISION FUNCTION HTLM(GS,KYA,XA1,YA1,XA2,YA2)
C...
C...  Function HTLM compuytes the height of a column using
C...  the log-mean concentration difference, according to
C...  the equation
C...
C...                                        *
C...     z = (Gs/Kya)*(Ya1 - Ya2)/(Ya - Ya )
C...                                      lm
C...  where
C...                        *              *
C...            *      (Ya1 - Ya1 ) - (Ya2 - Ya2 )
C...     (Ya - Ya )  = ----------------------------
C...          lm                        *
C...                             (Ya1 - Ya1 )
C...                         ln ------------
C...                                    *
C...                             (Ya2 - Ya2 )
C...
      IMPLICIT DOUBLE PRECISION (A-H,O-Z)
      DOUBLE PRECISION KYA
C...
C...  COEFFICIENTS IN POLYNOMIAL FOR EQUILIBRIUM LINE
      COMMON/E/A(2)
C...
C...  INPUT/OUTPUT (I/O) UNIT NUMBERS
      COMMON/IO/ NI, NO
C...
C...  EQUILIBRIUM GAS PHASE COMPOSITION AT THE BOTTOM OF THE
C...  COLUMN
      YA1S=EQUIL(XA1)
C...
C...  EQUILIBRIUM GAS PHASE COMPOSITION AT THE TOP OF THE
C...  COLUMN
      YA2S=EQUIL(XA2)
C...
C...  LOG MEAN CONCENTRATION DIFFERENCE
      DELTA1=YA1-YA1S
      DELTA2=YA2-YA2S
      DYLM=(DELTA1-DELTA2)/DLOG(DELTA1/DELTA2)
C...
```

```
C...      WRITE LOG MEAN CONCENTRATION DIFFERENCE
          WRITE(NO,1)DYLM
1         FORMAT(//,'Log-mean concentration difference=',F9.6,/)
C...
C...      WRITE NUMBER OF TRANSFER UNITS
          WRITE(NO,2)(YA1-YA2)/DYLM
2         FORMAT(/,' Number of transfer units = ',F6.2,/)
C...
C...      WRITE HEIGHT OF TRANSFER UNIT (FT)
          WRITE(NO,3)GS/KYA
3         FORMAT(/,' Height of transfer unit = ',F6.3,' m',/)
C...
C...      COLUMN HEIGHT
          HTLM=(GS/KYA)*(YA1-YA2)/DYLM
          RETURN
          END
```

We can note the following points about HTLM:

(1) Again, as in subroutine EQUIL and function POLY, the COMMON areas /E/ and /IO/ contain the computed least-squares polynomial coefficients, a_1 and a_2 (in A(2)), and the input/output unit numbers, (NI, NO), respectively.

(2) y_a^* at the bottom and top of the column are computed from the corresponding values of x_a by calls to EQUIL.

```
C...
C...      EQUILIBRIUM GAS PHASE COMPOSITION AT THE BOTTOM OF THE
C...      COLUMN
          YA1S=EQUIL(XA1)
C...
C...      EQUILIBRIUM GAS PHASE COMPOSITION AT THE TOP OF THE
C...      COLUMN
          YA2S=EQUIL(XA2)
```

(3) The concentration differences in eq. (MA1.33) at the bottom and top of the column are computed next, $\Delta_1 = (Y_a - Y_a^*)_1$, $\Delta_2 = (Y_a - Y_a^*)_2$ (see also eqs. (MA1.29) and (MA1.30))

```
C...
C...      LOG MEAN CONCENTRATION DIFFERENCE
          DELTA1=YA1-YA1S
          DELTA2=YA2-YA2S
```

Then the log-mean concentration difference is computed according to eq. (MA1.33) and written as output.

```
      DYLM=(DELTA1-DELTA2)/DLOG(DELTA1/DELTA2)
C...
C...  WRITE LOG MEAN CONCENTRATION DIFFERENCE
      WRITE(NO,1)DYLM
1     FORMAT(//,'Log-mean concentration difference=',F9.6,/)
```

(4) The number of transfer units is computed and written as output

```
C...
C...  WRITE NUMBER OF TRANSFER UNITS
      WRITE(NO,2)(YA1-YA2)/DYLM
2     FORMAT(/,' Number of transfer units = ',F6.2,/)
```

(5) The height of a transfer unit is computed next and written as output

```
C...
C...  WRITE HEIGHT OF TRANSFER UNIT (FT)
      WRITE(NO,3)GS/KYA
3     FORMAT(/,' Height of transfer unit = ',F6.3,' m',/)
```

(6) Finally, the height of the column is computed according to eq. (MA1.32)

```
C...
C...  COLUMN HEIGHT
      HTLM=(GS/KYA)*(YA1-YA2)/DYLM
```

Function HTIN for computing the height of the column from eq. (MA1.24) is listed below:

```
      DOUBLE PRECISION FUNCTION HTIN(GS,KYA,YA1,YA2)
C...
C...  Function HTIN computes the height of a column using
C...  the design integral
C...
C...                Ya2    dYa
C...     z = (Gs/Kya) INT --------
C...                Ya1  Ya - Ya*
C...
C...
C...  The integration is done by Simpson's rule, implemented
C...  in function SIMP
C...
      IMPLICIT DOUBLE PRECISION (A-H,O-Z)
      DOUBLE PRECISION LS, KYA

C...  COEFFICIENTS IN POLYNOMIAL FOR EQUILIBRIUM LINE
      COMMON/E/A(2)
```

```
C...
C...   INPUT/OUTPUT (I/O) UNIT NUMBERS
       COMMON/IO/ NI, NO
C...
C...   EXTERNAL FUNCTION FOR INTEGRAND
       EXTERNAL YAINT

C...
C...   EVALUATE DESIGN INTEGRAL.  NOTE THAT A SIGN REVERSAL
C...   IS REQUIRED SINCE DYA IN THE INTEGRAL IS NEGATIVE
C...   WHILE 1/(YA - YA*) IS POSITIVE
       NP=100
       TU=-SIMP(YAINT,YA1,YA2,NP)
C...
C...   WRITE NUMBER OF TRANSFER UNITS
       WRITE(NO,2)TU
2      FORMAT(//,' Number of transfer units = ',F6.2,/)
C...
C...   WRITE HEIGHT OF TRANSFER UNIT (M)
       WRITE(NO,3)GS/KYA
3      FORMAT(/,' Height of transfer unit = ',F6.3,' m',/)
C...
C...   COLUMN HEIGHT
       HTIN=(GS/KYA)*TU
       RETURN
       END
```

We can note the following points about HTIN:

(1) Again, as in subroutine EQUIL and functions POLY and HTLM, the COMMON areas /E/ and /IO/ contain the computed least squares polynomial coefficients, a_1 and a_2 (in A(2)), and the input/output unit numbers, (NI, NO), respectively.

(2) A function, YAINT, to define the integrand $1/(Y_a - Y_a^*)$ in eq. (MA1.24) is declared an EXTERNAL since it is an argument of function SINT, which performs the Simpson's rule integration

```
C...
C...   EXTERNAL FUNCTION FOR INTEGRAND
       EXTERNAL YAINT
```

(3) The Simpson's rule integration of the RHS of eq. (MA1.24) is then performed with 100 panels (101 points in Y_a) by a call to function SIMP

```
C...
C...   EVALUATE DESIGN INTEGRAL.  NOTE THAT A SIGN REVERSAL
C...   IS REQUIRED SINCE DYA IN THE INTEGRAL IS NEGATIVE
```

```
C...   WHILE 1/(YA - YA*) IS POSITIVE
       NP=100
       TU=-SIMP(YAINT,YA1,YA2,NP)
```

Note that the integration is between the limits $Y_{a,1}$ (= YA1) and $Y_{a,2}$ (= YA2), which were computed in subroutine INITAL. Also, since $Y_{a,2} < Y_{a,1}$, the differential dY_a in the integral is negative, and thus the sign of the integral is changed so that the number of transfer units is positive.

(4) The integral is then the number of transfer units as indicated in eq. (MA1.26)

```
C...
C...   WRITE NUMBER OF TRANSFER UNITS
       WRITE(NO,2)TU
2      FORMAT(//,' Number of transfer units = ',F6.2,/)
```

(5) The height of a transfer unit is defined by eq. (MA1.26)

```
C...
C...   WRITE HEIGHT OF TRANSFER UNIT (M)
       WRITE(NO,3)GS/KYA
3      FORMAT(/,' Height of transfer unit = ',F6.3,' m',/)
```

(6) The height of the column is then the product (height of a transfer unit) (number of transfer units) according to eq. (MA1.26)

```
C...
C...   COLUMN HEIGHT
       HTIN=(GS/KYA)*TU
```

To complete this discussion of the column height calculation by numerical integration (according to eq. (MA1.24)), we consider functions YAINT and SIMP. The integrand of the RHS integral of eq. (MA1.24) is computed by function YAINT

```
       DOUBLE PRECISION FUNCTION YAINT(YA)
C...
C...   Function YAINT computes the integrand of the integral
C...
C...
C...       Ya2    dYa
C...       INT  --------
C...       Ya1  Ya - Ya*
C...
C...
       IMPLICIT DOUBLE PRECISION (A-H,O-Z)
       DOUBLE PRECISION LS, KYA
       COMMON/C/    XA1,      YA1,     XA2,       YA2,
      +     ZL,     GS,       LS,      KYA,       S
```

```
C...
C...     LIQUID COMPOSITION FROM THE OPERATING LINE
         XA=OPLN(YA,YA2,XA2,GS,LS)
C...
C...     EQUILIBRIUM GAS COMPOSITION
         YAS=EQUIL(XA)
C...
C...     INTEGRAND
         YAINT=1.0D0/(YA-YAS)
         RETURN
         END
```

We can note the following points about YAINT:

(1) The liquid composition at a point along the column is computed by a call to OPLN (which again implements the operating line eq. (MA1.20)

```
C...
C...     LIQUID COMPOSITION FROM THE OPERATING LINE
         XA=OPLN(YA,YA2,XA2,GS,LS)
```

(2) This liquid composition is then converted to an equilibrium gas composition, Y_a^*, by a call to EQUIL

```
C...
C...     EQUILIBRIUM GAS COMPOSITION
         YAS=EQUIL(XA)
```

(3) Then the integrand of eq. (MA1.24), $1/(Y_a - Y_a^*)$, is computed

```
C...
C...     INTEGRAND
         YAINT=1.0D0/(YA-YAS)
```

The Simpson's rule integration is programmed in function SIMP

```
         DOUBLE PRECISION FUNCTION SIMP(F,A,B,NP)
         IMPLICIT DOUBLE PRECISION (A-H,O-Z)
C...
C...     EXTERNAL FUNCTION FOR INTEGRAND
         EXTERNAL F
C...
C...     CHECK FOR AN EVEN NUMBER OF PANELS
         IF((NP/2*2).NE.NP)THEN
            WRITE(*,2)NP
 2          FORMAT(//,' NP = ',I3,' (NOT EVEN)')
            SIMP=0.0D0
            RETURN
         END IF
```

```
C...
C...   INTEGRATION INTERVAL
       DX=(B-A)/DFLOAT(NP)
C...
C...   INITIALIZE SUM
       SIMP=(F(A)+4.0D0*F(A+DX)+F(B))
C...
C...   CONTINUE SUM
       IF(NP.GT.2)THEN
           DO 1 I=2,NP-1,2
               SIMP=SIMP
     +              +2.0D0*F(A+DFLOAT(I  )*DX)
     +              +4.0D0*F(A+DFLOAT(I+1)*DX)

1          CONTINUE
       END IF
C...
C...   INTEGRAL
       SIMP=(DX/3.0D0)*SIMP
       RETURN
       END
```

We can note the following points about SIMP:

(1) The first argument, F, is an external, which, for the evaluation of the RHS integral of eq. (MA1.24), is function YAINT

```
C...
C...   EXTERNAL FUNCTION FOR INTEGRAND
       EXTERNAL F
```

that is, again, the first argument in the call to SIMP in HTIN

```
       TU=-SIMP(YAINT,YA1,YA2,NP)
```

(2) A check is then made that SIMP is called with an even number of panels (an odd number of values of the integrand)

```
C...
C...   CHECK FOR AN EVEN NUMBER OF PANELS
       IF((NP/2*2).NE.NP)THEN
           WRITE(*,2)NP
2          FORMAT(//,' NP = ',I3,' (NOT EVEN)')
           SIMP=0.0D0
           RETURN
       END IF
```

(3) The panel width (integration interval) is then computed using the upper and lower limits of the integral and the number of panels

```
C...
C...    INTEGRATION INTERVAL
        DX=(B-A)/DFLOAT(NP)
```

Note that this value can be negative of $B < A$, as in the present case ($Y_{a,2} < Y_{a,1}$).

(4) The Simpson's rule integration is then started by using the integrand at the lower and upper limits (F(A) and F(B), respectively), plus the first value of the integrand to the right of the lower limit (F(A+DX))

```
C...
C...    INITIALIZE SUM
        SIMP=(F(A)+4.0D0*F(A+DX)+F(B))
```

Note the calls to the integrand function, F, for computing the integrand (i.e., YAINT in this case).

(5) The weighted sum of the Simpson's rule integration is then completed by using the remaining interior values of the integrand

```
C...
C...    CONTINUE SUM
        IF(NP.GT.2)THEN
            DO 1 I=2,NP-1,2
                SIMP=SIMP
     +                  +2.0D0*F(A+DFLOAT(I  )*DX)
     +                  +4.0D0*F(A+DFLOAT(I+1)*DX)

1           CONTINUE
        END IF
```

(6) Finally, the integral is computed according to the formula

$$\int_a^b f(x)dx \approx (\Delta x/3)\{f(a) + 4f(a + \Delta x) + 2f(a + 2\Delta x) + \cdots + 4f(b - \Delta x) + f(b)\} \quad \text{(MA1.43)}$$

```
C...
C...    INTEGRAL
        SIMP=(DX/3.0D0)*SIMP
```

This completes the programming of eq. (MA1.24). We can now summarize the preceding results:

Method	Column Height
PDEs (eqs. (MA1.3), (MA1.6))	3.28 m
Log-mean concentration difference (eqs. (MA1.32, MA1.33))	2.74 m
Numerical integration (eq. MA1.24))	3.28 m
Welty, Wicks, and Wilson (1984, 695)	3.35 m

Although the steady state methods, particularly numerical integration, produced a result consistent with the PDE formulation, they are quite limited in the class of mass transfer systems to which they can be applied, as noted previously; that is, they require the assumptions of steady state, isothermal, binary operation, and none of these conditions is generally encounterd in practice. The PDE formulation, however, can be applied to unsteady state, nonisothermal, multicomponent systems and is therefore the preferred method of analysis.

References

Forsythe, G. E., M. A. Malcolm, and C. B. Moler. 1977. *Computer Methods for Mathematical Computations*. Englewood Cliffs, NJ: Prentice-Hall.

Schiesser, W. E. 1991. *The Numerical Method of Lines Integration of Partial Differential Equations*. San Diego: Academic Press.

Silebi, C. A., and W. E. Schiesser. 1992. *Dynamic Modeling of Transport Process Systems*. San Diego: Academic Press.

Welty, J. R., C. E. Wicks, and R. E. Wilson. 1984. *Fundamentals of Momentum, Heat, and Mass Transfer*. 3d ed. New York: John Wiley and Sons.

MA2

Mass transfer with simultaneous convection and diffusion

Develop a dynamic model for the system of Figure MA2.1, taking into account simultaneous convection and axial dispersion. Derive an analytical solution to the model and compute a numerical solution that can be compared with the analytical solution.

Solution

A mass balance on an incremental volume in Figure MA2.1 of length z and cross-sectional area S gives

$$S\Delta z \frac{\partial C}{\partial t} = -SD\frac{\partial C}{\partial z}\bigg|_z - \left\{ -SD\frac{\partial C}{\partial z}\bigg|_{z+\Delta z} \right\} + SvC|_z - SvC|_{z+\Delta z} \quad \text{(MA2.1)}$$

where

- C concentration (kg-mol/m^3)
- z axial position (m)
- t time (s)
- S cross-sectional area (m^2)
- D axial dispersion coefficient (m^2/s)
- v fluid velocity (m/s)

Division of eq. (MA2.1) by $S\Delta z$ gives

$$\frac{\partial C}{\partial t} = \frac{D\frac{\partial C}{\partial z}\big|_{z+\Delta z} - D\frac{\partial C}{\partial z}\big|_z}{\Delta z} - \frac{vC|_{z+\Delta z} - vC|_z}{\Delta z} \quad \text{(MA2.2)}$$

Division of eq. (MA2.2) with $z \to 0$ gives

$$\frac{\partial C}{\partial t} = D\frac{\partial^2 C}{\partial z^2} - v\frac{\partial C}{\partial z} \quad \text{(MA2.3)}$$

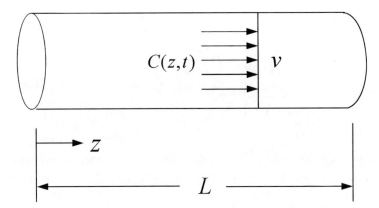

Figure MA2.1. One-Dimensional Mass Transfer System

We take as the initial and boundary conditions for eq. (MA2.3)

$$C(z, 0) = C_0 \qquad \text{(MA2.4)}$$

$$D\frac{\partial C(0, t)}{\partial z} = vC(0, t) \qquad \text{(MA2.5)}$$

$$\frac{\partial C(L, t)}{\partial z} = 0 \qquad \text{(MA2.6)}$$

where L is the total length of the system. Equations (MA2.3) to (MA2.6) constitute the model for the mass transfer system. Equation (MA2.4) is termed a *convective-diffusion* or a *hyperbolic-parabolic* partial differential equation (PDE) since it is partly hyperbolic (due to the convective term $-v\frac{\partial C}{\partial z}$) and partly parabolic (due to the diffusive term $D\frac{\partial^2 C}{\partial z^2}$). Additional terms might appear in a mass balance (for a mass transfer system) due to reaction, for example, or in an energy balance (for a heat transfer system) due to heat generation produced by a reaction. These mixed (i.e., mixed convective-diffusion or hyperbolic-parabolic) PDEs are therefore quite common in applications. We consider eq. (MA2.3) because it has only the essential terms for the dynamic analysis of a convective-diffusion system.

We now develop analytical and numerical solutions to eq. (MA2.3) with the one required initial condition, eq. (MA2.4) (required by the first-order derivative $\frac{\partial C}{\partial t}$), and the two required boundary conditions, eqs. (MA2.5) and (MA2.6) (required by the second-order derivative $\frac{\partial^2 C}{\partial z^2}$).

If dimensionless variables are defined as

$$\phi = C/C_0 \qquad \text{(MA2.7)}$$

$$z' = z/L \qquad \text{(MA2.8)}$$

$$t' = tD/L^2 \qquad \text{(MA2.9)}$$

substitution of eqs. (MA2.7) to (MA2.9) in eqs. (MA2.3) to (MA2.6) gives

$$\frac{C_0}{L^2/D}\frac{\partial \phi}{\partial t'} = D\frac{C_0}{L^2}\frac{\partial^2 \phi}{\partial z'^2} - v\frac{C_0}{L}\frac{\partial \phi}{\partial z'}$$

$$C_0\phi(z', 0) = C_0$$

$$D\frac{C_0}{L}\frac{\partial \phi(0, t)}{\partial z'} = vC_0\phi(0, t), \quad \frac{C_0}{L}\frac{\partial \phi(1, t)}{\partial z'} = 0$$

If we now drop the primes (′) with the understanding that we are working with the dimensionless variables of eqs. (MA2.7) to (MA2.9), we have as the model equations

$$\frac{\partial \phi}{\partial t} = \frac{\partial^2 \phi}{\partial z^2} - \frac{vL}{D}\frac{\partial \phi}{\partial z} \quad (MA2.10)$$

$$\phi(z, 0) = 1 \quad (MA2.11)$$

$$\frac{\partial \phi(0, t)}{\partial z} = \frac{vL}{D}\phi(0, t) \quad (MA2.12)$$

$$\frac{\partial \phi(1, t)}{\partial z} = 0 \quad (MA2.13)$$

Equations (MA2.10) to (MA2.13) are the required mathematical model. The dimensionless group $\frac{vL}{D}$ will be designated as P_e (the Peclet number), which is a measure of the relative strength of the convective (hyperbolic) term in eq. (MA2.10), $-\frac{vL}{D}\frac{\partial \phi}{\partial z} = -P_e\frac{\partial \phi}{\partial z}$, to the diffusive term, $\frac{\partial^2 \phi}{\partial z^2}$ (note that for large P_e, v is large relative to D, indicating a *strongly convective* or *strongly hyperbolic* system).

In order to derive an analytical solution to eqs. (MA2.10) to (MA2.13), we first use a change of variables

$$\phi(z, t) = u(z, t)e^{(P_e/2)z} \quad (MA2.14)$$

The purpose of the change from $\phi(z, t)$ to $u(z, t)$ defined by eq. (MA2.14) is essentially to eliminate the first derivative $-P_e\frac{\partial \phi}{\partial z}$ in eq. (MA2.10). The various derivatives in eqs. (MA2.10) to (MA2.13) become, through eq. (MA2.14),

$$\frac{\partial \phi}{\partial t} = \frac{\partial u}{\partial t}e^{(P_e/2)z}$$

$$\frac{\partial \phi}{\partial z} = \frac{\partial u}{\partial z}e^{(P_e/2)z} + u(P_e/2)e^{(P_e/2)z}$$

$$\frac{\partial^2 \phi}{\partial z^2} = \frac{\partial^2 u}{\partial z^2}e^{(P_e/2)z} + \frac{\partial u}{\partial z}(P_e/2)e^{(P_e/2)z}$$

$$+ \frac{\partial u}{\partial z}(P_e/2)e^{(P_e/2)z} + u(P_e/2)^2 e^{(P_e/2)z}$$

Substitution of these derivatives in eqs. (MA2.10) to (MA2.13) gives

$$\frac{\partial u}{\partial t}e^{(P_e/2)z} = \left\{\frac{\partial^2 u}{\partial z^2}e^{(P_e/2)z} + P_e\frac{\partial u}{\partial z}e^{(P_e/2)z} + u(P_e/2)^2 e^{(P_e/2)z}\right\}$$

$$- P_e\left\{\frac{\partial u}{\partial z}e^{(P_e/2)z} + u(P_e/2)e^{(P_e/2)z}\right\}$$

or

$$\frac{\partial u}{\partial t} = \frac{\partial^2 u}{\partial z^2} - (P_e^2/4)u \quad \text{(MA2.15)}$$

$$u(z, 0) = e^{-(P_e/2)z} \quad \text{(MA2.16)}$$

$$\frac{\partial \phi(0, t)}{\partial z} = \frac{\partial u(0, t)}{\partial z} + (P_e/2)u(0, t) = P_e\phi(0, t) = P_e u(0, t)$$

or

$$\frac{\partial u(0, t)}{\partial z} - (P_e/2)u(0, t) = 0 \quad \text{(MA2.17)}$$

Also, at $z = 1$,

$$\frac{\partial u(1, t)}{\partial z} + (P_e/2)u(1, t) = 0 \quad \text{(MA2.18)}$$

We now use $\beta = -P_e^2/4$ to simplify eq. (MA2.15).

To derive a series solution to eqs. (MA2.15) to (MA2.18), a separated solution is first assumed,

$$u(z, t) = \theta(t)\psi(z) \quad \text{(MA2.19)}$$

Substitution of eq. (MA2.19) in eq. (MA2.15) gives

$$\theta'(t)\psi(z) = \theta(t)\psi''(z) + \beta\theta(t)\psi(z)$$

or

$$\theta'(t)/\theta(t) = \frac{\psi''(z) + \beta\psi(z)}{\psi(z)} = -\lambda \quad \text{(MA2.20)}$$

where λ is a separation constant.

Equation (MA2.20) consists of two ordinary differential equations (ODEs)

$$\theta'(t) + \lambda\theta(t) = 0 \quad \text{(MA2.21)}$$

$$\psi''(z) + \beta\psi(z) + \lambda\psi(z) = 0 \quad \text{(MA2.22)}$$

with the general solutions

$$\theta(t) = ce^{-\lambda t} \quad \text{(MA2.23)}$$

$$\psi(z) = a\sin(\mu z) + b\cos(\mu z) \quad \text{(MA2.24)}$$

where $\mu = \sqrt{\beta + \lambda}$, and a, b, and c are constants to be determined. Thus, the solution to this point is

$$u(z,t) = ce^{-\lambda t}\{a\sin(\mu z) + b\cos(\mu z)\} \qquad \text{(MA2.25)}$$

If eq. (MA2.25) is substituted in boundary condition (MA2.17),

$$ce^{-\lambda t}\{a\mu\cos(\mu 0) - b\mu\sin(\mu 0)\} - (P_e/2)ce^{-\lambda t}\{a\sin(\mu 0) + b\cos(\mu 0)\} = 0$$

or, after cancelling common factors,

$$a\mu - (P_e/2)b = 0$$

and therefore

$$a = P_e/(2\mu)b \qquad \text{(MA2.26)}$$

If eq. (MA2.25) is substituted in boundary condition (MA2.18),

$$ce^{-\lambda t}\{a\mu\cos(\mu 1) - b\mu\sin(\mu 1)\} + (P_e/2)ce^{-\lambda t}\{a\sin(\mu 1) + b\cos(\mu 1)\} = 0$$

or, with the substitution of eq. (MA2.26),

$$\{[P_e/(2\mu)b]\mu\cos(\mu 1) - b\mu\sin(\mu 1)\} + (P_e/2)$$
$$\times\{[P_e/(2\mu)b]\sin(\mu 1) + b\cos(\mu 1)\} = 0$$
$$(P_e/2)\cos(\mu)/\sin(\mu) - \mu + (P_e/2)\{P_e/(2\mu) + \cos(\mu 1)/\sin(\mu 1)\} = 0$$
$$1/\tan(\mu) - (2\mu/P_e) + P_e/(2\mu) + 1/\tan(\mu) = 0$$
$$1/\tan(\mu) - (\mu/P_e) + P_e/(4\mu) = 0 \qquad \text{(MA2.27)}$$

Equation (MA2.27) is the characteristic equation for eqs. (MA2.15), (MA2.17), and (MA2.18), which defines the eigenvalues μ_i, $i = 1, 2, 3, \ldots$ for these equations.

Since equations (MA2.15), (MA2.17), and (MA2.18) are linear, a superposition of solutions is also a solution,

$$u(z,t) = \sum_{i=1}^{\infty} c_i e^{-\lambda_i t}\left\{\frac{P_e}{2\mu_i}\sin(\mu_i z) + \cos(\mu_i z)\right\} \qquad \text{(MA2.28)}$$

where c_i is a Fourier coefficient to be determined in accordance with initial condition (MA2.16) and $\lambda_i = \mu_i^2 - \beta$; that is, if we apply eq. (MA2.16) to eq. (MA2.28), we have

$$u(z,0) = e^{-(P_e/2)z} = \sum_{i=1}^{\infty} c_i\left\{\frac{P_e}{2\mu_i}\sin(\mu_i z) + \cos(\mu_i z)\right\} \qquad \text{(MA2.29)}$$

As the first step in determining c_i in eq. (MA2.29), we observe that eqs. (MA2.17), (MA2.18), and (MA2.22) are a special case of the Sturm–Liouville problem (Morse and Feshbach 1953, 719–29)

$$\frac{d}{dz}\left\{p(z)\frac{d\psi}{dz}\right\} + \{q(z) + \gamma r(z)\}\psi = 0 \quad \text{(MA2.30)}$$

$$a_1 \frac{d\psi(a)}{dz} + a_2 \psi(a) = 0 \quad \text{(MA2.31)}$$

$$b_1 \frac{d\psi(b)}{dz} + b_2 \psi(b) = 0 \quad \text{(MA2.32)}$$

Since the $\psi(z)$ of eqs. (MA2.30) to (MA2.32) have the orthogonality property

$$\int_a^b r(z)\psi_i(z)\psi_j(z)dz \quad \begin{matrix} = 0, i \neq j \\ \neq 0, i = j \end{matrix} \quad \text{(MA2.33)}$$

where $\psi_i(z)$ and $\psi_j(z)$ are the solutions of eqs. (MA2.30) to (MA2.32) for eigenvalues γ_i and γ_j, it follows that for eqs. (MA2.17), (MA2.18), and (MA2.22), with $p(z) = r(z) = 1, q(z) = 0, \gamma = \mu^2$,

$$\int_0^1 \psi_i(z)\psi_j(z)dz = \int_0^1 \left\{\frac{P_e}{2\mu_i}\sin(\mu_i z) + \cos(\mu_i z)\right\}$$

$$\times \left\{\frac{P_e}{2\mu_j}\sin(\mu_j z) + \cos(\mu_j z)\right\}dz \quad \begin{matrix} = 0, i \neq j \\ \neq 0, i = j \end{matrix} \quad \text{(MA2.34)}$$

We confirm eq. (MA2.34) for this special case of eqs. (MA2.30) to (MA2.33). Thus substitution of

$$\psi(z) = \frac{P_e}{2\mu}\sin(\mu z) + \cos(\mu z) \quad \text{(MA2.35)}$$

in eq. (MA2.22) gives (with $\mu^2 = \beta + \lambda$)

$$(-\mu^2)\left\{\frac{P_e}{2\mu}\sin(\mu z) + \cos(\mu z)\right\} + (\beta + \lambda)\left\{\frac{P_e}{2\mu}\sin(\mu z) + \cos(\mu z)\right\} = 0 \quad \text{(MA2.36)}$$

so that the eigenfunctions of eq. (MA2.35) satisfy eq. (MA2.22).

Then, following the usual Sturm–Liouville analysis, eq. (MA2.22) is written for two different eigenvalues, γ_i and γ_j

$$\psi_i'' + \gamma_i \psi_i = 0 \quad \text{(MA2.37)}$$

$$\psi_j'' + \gamma_j \psi_j = 0 \quad \text{(MA2.38)}$$

Multiplication of eqs. (MA2.37) and (MA2.38) by ψ_j and ψ_i, respectively, followed by subtraction of the two equations gives

$$\psi_j \psi_i'' - \psi_i \psi_j'' + (\gamma_i - \gamma_j) \psi_i \psi_j = 0 \tag{MA2.39}$$

If eq. (MA2.39) is integrated with respect to z,

$$\int_0^1 \{\psi_j \psi_i'' - \psi_i \psi_j''\} dz + (\gamma_i - \gamma_j) \int_0^1 \psi_i \psi_j dz = 0 \tag{MA2.40}$$

Integration by parts applied to the first integral gives

$$\int_0^1 \{\psi_j \psi_i'' - \psi_i \psi_j''\} dz = \{\psi_j \psi_i' - \psi_i \psi_j'\}\big|_0^1 - \int_0^1 \{\psi_j' \psi_i' - \psi_i' \psi_j'\} dz = 0 \tag{MA2.41}$$

If boundary conditions (MA2.17) and (MA2.18) are substituted in the first right hand side (RHS) term of eq. (MA2.41),

$$\{\psi_j(1) \psi_i'(1) - \psi_i(1) \psi_j'(1)\} - \{\psi_j(0) \psi_i'(0) - \psi_i(0) \psi_j'(0)\}$$
$$= \{\psi_j(1)(-P_e/2)\psi_i(1) - \psi_i(1)(-P_e/2)\psi_j(1)\}$$
$$- \{\psi_j(0)(P_e/2)\psi_i(0) - \psi_i(0)(P_e/2)\psi_j(0)\} = 0 \tag{MA2.42}$$

Since the second RHS integral of eq. (MA2.41) equals zero, from eq. (MA2.40) we have

$$(\gamma_i - \gamma_j) \int_0^1 \psi_i \psi_j dz = 0$$

and since for the case $i \neq j$, $\gamma_i \neq \gamma_j$ (the roots of eq. (MA2.27) are distinct)

$$\int_0^1 \psi_i \psi_j dz = 0 \tag{MA2.43}$$

which establishes the orthogonality of $\psi(z)$ given by eq. (MA2.35). We could have also used $\psi(z)$ from eq. (MA2.35) in eq. (MA2.42) (rather than boundary conditions (MA2.17) and (MA2.18)), but the algebra is rather complicated to arrive at the same final result.

Finally, we return to initial condition (MA2.16) expressed in eq. (MA2.29) (which is now written in terms of $\psi(z)$ of eq. (MA2.35) to facilitate the remaining analysis)

$$u(z, 0) = e^{-(P_e/2)z} = \sum_{i=1}^{\infty} c_i \psi_i(z) \tag{MA2.44}$$

If eq. (MA2.44) is multiplied by $\psi_j(z)$ and integrated with respect to z,

$$\int_0^1 \psi_j(z) e^{-(P_e/2)z} dz = \sum_{i=1}^{\infty} c_i \int_0^1 \psi_j(z) \psi_i(z) dz \tag{MA2.45}$$

Because of the orthogonality of eq. (MA2.43), eq. (MA2.45) gives the Fourier coefficients explicitly as

$$c_i = \frac{\int_0^1 \psi_i(z) e^{-(P_e/2)z} dz}{\int_0^1 \psi_i^2(z) dz} \tag{MA2.46}$$

The final solution to eqs. (MA2.10) to (MA2.13) consists of eqs. (MA2.14), (MA2.27), (MA2.28), (MA2.35), and (MA2.46). In principle, eq. (MA2.35) can be substituted in eq. (MA2.46), and the Fourier coefficients, c_i, can be evaluated by analytical integration (of the integrals in the RHS of eq. (MA2.46)). However, the integration is rather complicated, and therefore we will use numerical integration to compute the c_i.

Actually, in the subsequent programming, we consider the following minor variation on eqs. (MA2.10) to (MA2.13)

$$\frac{\partial \phi_1}{\partial t} = \frac{\partial^2 \phi_1}{\partial z^2} - P_e \frac{\partial \phi_1}{\partial z} \qquad (MA2.47)$$

$$\phi_1(z, 0) = 0 \qquad (MA2.48)$$

$$\frac{\partial \phi_1(0, t)}{\partial z} = P_e \{\phi_1(0, t) - 1\} \qquad (MA2.49)$$

$$\frac{\partial \phi_1(1, t)}{\partial z} = 0 \qquad (MA2.50)$$

Equations (MA2.47) to (MA2.50), which define $\phi_1(z, t)$, are a somewhat more realistic representation of a mass transfer system, for example, a reactor, than eqs. (MA2.10) to (MA2.13), which define the previous $\phi(z, t)$, since initially the concentration is zero (eq. (MA2.48)), and fluid of unit concentration then flows into the left end of the system (eq. (MA2.49)). However, eqs. (MA2.10) to (MA2.13) have the advantage in deriving an analytical solution that the boundary conditions, eqs. (MA2.12) and (MA2.13), are homogeneous, and the nonhomogeneous initial condition, eq. (MA2.10), can be expanded in a Fourier series, that is, eq. (MA2.44). These mathematical properties are helpful in deriving the analytical solution for $\phi(z, t)$.

The solution to eqs. (MA2.47) to (MA2.50) follows from the preceding solution, (eqs. (MA2.14), (MA2.27), (MA2.28), (MA2.35), and (MA2.46))

$$\phi_1(z, t) = 1 - \phi(z, t)$$

$$\phi_1(z, t) = 1 - e^{(P_e/2)z} u(z, t)$$

$$= 1 - e^{(P_e/2)z} \sum_{i=1}^{\infty} c_i e^{-\lambda_i t} \left\{ \frac{P_e}{2\mu_i} \sin(\mu_i z) + \cos(\mu_i z) \right\}$$

$$= e^{(P_e/2)z} \sum_{i=1}^{\infty} c_i \left\{ \frac{P_e}{2\mu_i} \sin(\mu_i z) + \cos(\mu_i z) \right\} - e^{(P_e/2)z}$$

$$\times \sum_{i=1}^{\infty} c_i e^{-\lambda_i t} \left\{ \frac{P_e}{2\mu_i} \sin(\mu_i z) + \cos(\mu_i z) \right\}$$

and finally

$$\phi_1(z,t) = e^{(P_e/2)z} \sum_{i=1}^{\infty} c_i (1 - e^{-\lambda_i t}) \left\{ \frac{P_e}{2\mu_i} \sin(\mu_i z) + \cos(\mu_i z) \right\} \quad \text{(MA2.51)}$$

where the eigenvalues μ_i, $i = 1, 2, \ldots$, are the roots of the characteristic equation

$$1/\tan(\mu) - \mu/Pe + Pe/(4\mu) = 0 \quad \text{(MA2.52)}$$

with

$$\lambda_i = \mu_i^2 - \beta = \mu_i^2 + P_e^2/4 \quad \text{(MA2.53)}$$

$$c_i = \frac{\int_0^1 \left\{ \frac{P_e}{2\mu_i} \sin(\mu_i z) + \cos(\mu_i z) \right\} e^{-(P_e/2)z} dz}{\int_0^1 \left\{ \frac{P_e}{2\mu_i} \sin(\mu_i z) + \cos(\mu_i z) \right\}^2 dz} \quad \text{(MA2.54)}$$

To conclude this discussion of the analytical solution to eqs. (MA2.47) to (MA2.50), we mention that an infinite series solution of a considerably more difficult problem involving two simultaneous PDEs was derived by Kim, Ma, and Chang (1984) using the concept of a Hilbert space. This approach appears to be effective for handling linear PDE problems that are not amenable to the more classical approach of the preceding analysis.

We now proceed to the programming of the series solution, eqs. (MA2.51) to (MA2.54), for comparison with a numerical solution to eqs. (MA2.47) to (MA2.50). The infinite series of eq. (MA2.51) is evaluated in subroutine SERIES listed below

```
      SUBROUTINE SERIES(NI,NO)
      IMPLICIT DOUBLE PRECISION (A-H,O-Z)
      PARAMETER(NZ=81,NA=200)
      COMMON/T/      T,    NSTOP,     NORUN
     +      /Y/   C(NZ)
     +      /F/   CT(NZ)
     +      /S/   CZ(NZ), CZZ(NZ)
     +      /C/       ZL,       ZU,       ZLU,       PE,
     +            ZMU(NA),  CF(NA)
     +      /I/       NS,       NQ,       NP,        IP,
     +            NCASE,    IN,       JN
C...
C...  DEFINE A FILE FOR MATLAB PLOTTING OF THE ANALYTICAL
C...  SOLUTION
      OPEN(2,FILE='ma2a.out')
C...
C...  NUMBER OF EIGENVALUES
      NS=10
C...
C...  NUMBER OF PANELS IN THE NUMERICAL QUADRATURE
C...  (INTEGRATION)
```

```fortran
          NQ=1000
C...
C...  LIMITS OF INTEGRATION
          ZL=0.0D0
          ZU=1.0D0
C...
C...  SYSTEM LENGTH
          ZLU=1.0D0
C...
C...  SYSTEM PARAMETERS
          PE=10.0D0
C...
C...  COMPUTE AND PRINT NS EIGENVALUES
          CALL EIGVAL
          WRITE(NO,1)
          WRITE( *,1)
1         FORMAT(//,' Eigenvalues',/)
          WRITE(NO,2)(I,ZMU(I),I=1,NS)
2         FORMAT(I5,F10.5)
C...
C...  TEST ORTHOGONALITY OF EIGENFUNCTIONS BY COMPUTING THE
C...  INTEGRAL OF THE PRODUCT OF TWO EIGENFUNCTIONS FOR I
C...  AND J = 1, 2,..., NS
          WRITE(NO,13)
          WRITE( *,13)
13        FORMAT(//,' Product integrals',/)
          DO 10 I=1,NS
          DO 11 J=1,NS
             CALL  ORTHO(I,J,PIJ)
             WRITE(NO,12)I,J,PIJ
12           FORMAT(' i = ',I3,'   j = ',I3,'   Pij = ',F10.5)
             IF(J.EQ.NS)WRITE(NO,14)
14        FORMAT(/)
11        CONTINUE
10        CONTINUE
C...
C...  COMPUTE AND PRINT NS FOURIER COEFFICIENTS
          NS=200
          CALL EIGVAL
          WRITE(NO,22)
          WRITE( *,22)
22        FORMAT(//,' Fourier coefficients',/)
          DO 20 I=1,NS
             CALL COEFF(I)
             WRITE(NO,21)I,CF(I)
21        FORMAT(I5,F10.7)
```

```
20        CONTINUE
C...
C...   FOURIER SERIES FOR UNITY WITH NS TERMS AT A SERIES OF
C...   VALUES OF Z
       NS=200
       WRITE(NO,30)
       WRITE( *,30)
30     FORMAT(//,' Fourier series for unity',/)
C...
C...   STEP THROUGH NZV VALUES OF Z
       NZV=11
       DO 32 J=1,NZV
C...
C...      SET Z
          Z=DFLOAT(J-1)/DFLOAT(NZV-1)
C...
C...      COMPUTE AND PRINT SERIES FOR UNITY
          ONE=UNITY(Z)
          WRITE(NO,31)Z,ONE
31        FORMAT(F10.2,F10.5)
32     CONTINUE
C...
C...   EVALUATE AND PRINT THE SERIES SOLUTION AT Z = 1 FOR A
C...   SERIES OF VALUES OF T USING NS TERMS IN THE SERIES
       WRITE(NO,42)
       WRITE( *,42)
42     FORMAT(//,' Series solution at z = 1, phi1(1,t)',/)
       Z=1.0D0
       DT=0.0025D0
       NTM=81
       DO 40 I=1,NTM
C...
C...      T AND SERIES SOLUTION
          T=DT*DFLOAT(I-1)
          UZT=PHI1(Z,T)
C...
C...      PRINT SOLUTION
          WRITE(NO,41)Z,T,UZT
41        FORMAT(F10.3,F10.4,F10.5)
C...
C...      WRITE SOLUTION FOR MATLAB PLOTTING
          WRITE(2,43)T,UZT
43        FORMAT(2F10.4)
40     CONTINUE
       RETURN
       END
```

We can note the following points about subroutine SERIES:

(1) SERIES has two arguments, NI and NO, to define unit numbers for the input/output. Also, double precision coding is used, and a COMMON block is used to establish communication with other routines

```
      SUBROUTINE SERIES(NI,NO)
      IMPLICIT DOUBLE PRECISION (A-H,O-Z)
      PARAMETER(NZ=81,NA=200)
      COMMON/T/        T,    NSTOP,      NORUN
     +      /Y/   C(NZ)
     +      /F/   CT(NZ)
     +      /S/   CZ(NZ), CZZ(NZ)
     +      /C/        ZL,       ZU,        ZLU,        PE,
     +            ZMU(NA),   CF(NA)
     +      /I/        NS,       NQ,         NP,        IP,
     +            NCASE,      IN,         JN
```

The contents of this COMMON block are used primarily for the numerical integration of the model, eqs. (MA2.47) to (MA2.50), so that discussion of the COMMON block will be essentially deferred until the numerical solution is considered.

(2) File ma2a.out is opened for Matlab plotting of the analytical solution

```
C...
C...  DEFINE A FILE FOR MATLAB PLOTTING OF THE ANALYTICAL
C...  SOLUTION
      OPEN(2,FILE='ma2a.out')
```

(3) The first calculation in SERIES is to test numerically the orthogonality of the basis functions, $\psi(z)$, of eq. (MA2.35) to substantiate numerically eq. (MA2.43). This is done for the first 10 eigenvalues, using Simpson's rule to perform the integration of eq. (MA2.43)

```
C...
C...  NUMBER OF EIGENVALUES
      NS=10
C...
C...  NUMBER OF PANELS IN THE NUMERICAL QUADRATURE
C...  (INTEGRATION)
      NQ=1000
C...
C...  LIMITS OF INTEGRATION
      ZL=0.0D0
      ZU=1.0D0
```

Note that 1,000 panels are used in the Simpson's rule quadrature. This large number of panels was used because of the oscillatory character of the sine and cosine functions of eq. (MA2.35).

(4) Some problem parameters are then defined numerically

```
C...
C...   SYSTEM LENGTH
       ZLU=1.0D0
C...
C...   PROBLEM PARAMETERS
       PE=10.0D0
```

Note in particular that the Peclet number in eq. (MA2.47), vL/D, is 10. This value of the Peclet number corresponds to a substantial axial dispersion so that the solution to eqs. (MA2.47) to (MA2.50) will be smooth (and therefore relatively easy to compute analytically and numerically)

(5) The first 10 eigenvalues from eq. (MA2.52) are then computed by a call to subroutine EIGEN

```
C...
C...   COMPUTE AND PRINT NS EIGENVALUES
       CALL EIGVAL
       WRITE(NO,1)
       WRITE( *,1)
1      FORMAT(//,' Eigenvalues',/)
       WRITE(NO,2)(I,ZMU(I),I=1,NS)
2      FORMAT(I5,F10.5)
```

We will consider the programming in subroutine EIGEN subsequently. The 10 computed eigenvalues are

```
Eigenvalues

    1     2.28445
    2     4.76129
    3     7.46368
    4    10.32661
    5    13.28624
    6    16.30313
    7    19.35516
    8    22.42981
    9    25.51969
   10    28.62025
```

(6) The orthogonality of the first 10 eigenfunctions is then tested numerically by a call to subroutine ORTHO, which performs a Simpson's rule integration according to eq. (MA2.43)

```
C...
C...   TEST ORTHOGONALITY OF EIGENFUNCTIONS BY COMPUTING THE
```

```
C...   INTEGRAL OF THE PRODUCT OF TWO EIGENFUNCTIONS FOR I
C...   AND  J = 1, 2,..., NS
       WRITE(NO,13)
       WRITE( *,13)
13     FORMAT(//,' Product integrals',/)
       DO 10 I=1,NS
       DO 11 J=1,NS
          CALL  ORTHO(I,J,PIJ)
          WRITE(NO,12)I,J,PIJ
12        FORMAT(' i = ',I3,'   j = ',I3,'   Pij = ',F10.5)
          IF(J.EQ.NS)WRITE(NO,14)
14        FORMAT(/)
11     CONTINUE
10     CONTINUE
```

Note that DO loops 10 and 11 cycle through 10 values of i and j in eq. (MA2.43).

Abbreviated output from the WRITE statements of the preceding code is listed below:

```
Product integrals
i =   1    j =    1   Pij =      3.85331
i =   1    j =    2   Pij =       .00000
i =   1    j =    3   Pij =       .00000
i =   1    j =    4   Pij =       .00000
i =   1    j =    5   Pij =       .00000
i =   1    j =    6   Pij =       .00000
i =   1    j =    7   Pij =       .00000
i =   1    j =    8   Pij =       .00000
i =   1    j =    9   Pij =       .00000
i =   1    j =   10   Pij =       .00000

i =   2    j =    1   Pij =       .00000
i =   2    j =    2   Pij =      1.27195
i =   2    j =    3   Pij =       .00000
i =   2    j =    4   Pij =       .00000
i =   2    j =    5   Pij =       .00000
i =   2    j =    6   Pij =       .00000
i =   2    j =    7   Pij =       .00000
i =   2    j =    8   Pij =       .00000
i =   2    j =    9   Pij =       .00000
i =   2    j =   10   Pij =       .00000
                .          .
                .          .
                .          .
i =   9    j =    1   Pij =       .00000
i =   9    j =    2   Pij =       .00000
i =   9    j =    3   Pij =       .00000
```

```
i =   9    j =    4    Pij =    .00000
i =   9    j =    5    Pij =    .00000
i =   9    j =    6    Pij =    .00000
i =   9    j =    7    Pij =    .00000
i =   9    j =    8    Pij =    .00000
i =   9    j =    9    Pij =    .52687
i =   9    j =   10    Pij =    .00000
i =  10    j =    1    Pij =    .00000
i =  10    j =    2    Pij =    .00000
i =  10    j =    3    Pij =    .00000
i =  10    j =    4    Pij =    .00000
i =  10    j =    5    Pij =    .00000
i =  10    j =    6    Pij =    .00000
i =  10    j =    7    Pij =    .00000
i =  10    j =    8    Pij =    .00000
i =  10    j =    9    Pij =    .00000
i =  10    j =   10    Pij =    .52136
```

Note that the integral of eq. (MA2.43) is zero to five figures for $i \neq j$, and, of course, is nonzero for $i = j$ (the latter follows immediately from eq. (MA2.43) since the integral of a squared (positive) function is computed when $i = j$). The programming in subroutine ORTHO will be considered subsequently.

(7) The first 200 Fourier coefficients given by eq. (MA2.54) are then computed by calls to subroutines EIGVAL and COEFF in preparation for the evaluation of the series of eq. (MA2.51)

```
C...
C...     COMPUTE AND PRINT NS FOURIER COEFFICIENTS
         NS=200
         CALL EIGVA
         WRITE(NO,22)
         WRITE( *,22)
22       FORMAT(//,' Fourier coefficients',/)
         DO 20 I=1,NS
            CALL COEFF(I)
            WRITE(NO,21)I,CF(I)
21          FORMAT(I5,F10.7)
20       CONTINUE
```

The first and last several Fourier coefficients computed by the preceding code are listed below:

```
Fourier coefficients

     1    .0858796
     2    .1649248
```

3	.1521910
4	.1143876
5	.0828222
.	
.	
.	
195	.0000538
196	.0000532
197	.0000527
198	.0000522
199	.0000516
200	.0000511

Note in particular how the coefficients decrease in magnitude, which produces the convergence of the series of eq. (MA2.51). However, this decrease in the coefficients is not rapid, so that a relatively large number of terms in the series of eq. (MA2.51) for example, 200, is required to achieve acceptable accuracy in the numerical value of $\phi_1(z,t)$. The programming in subroutine COEFF will be considered subsequently.

(8) As a check on the programming of eq. (MA2.51), the series is computed for the initial condition of eq. (MA2.29), after it is written in the alternate form

$$u(z,0)e^{(P_e/2)z} = e^{(P_e/2)z}e^{-(P_e/2)z} = 1$$

$$= e^{(P_e/2)z}\sum_{i=1}^{\infty} c_i \left\{ \frac{Pe}{2\mu_i}\sin(\mu_i z) + \cos(\mu_i z) \right\} \quad \text{(MA2.55)}$$

```
C...
C...   FOURIER SERIES FOR UNITY WITH NS TERMS AT A SERIES OF
C...   VALUES OF Z
       NS=200
       WRITE(NO,30)
       WRITE( *,30)
30     FORMAT(//,' Fourier series for unity',/)
C...
C...   STEP THROUGH NZV VALUES OF Z
       NZV=11
       DO 32 J=1,NZV
C...
C...      SET Z
          Z=DFLOAT(J-1)/DFLOAT(NZV-1)
C...
C...      COMPUTE AND PRINT SERIES FOR UNITY
          ONE=UNITY(Z)
          WRITE(NO,31)Z,ONE
31        FORMAT(F10.2,F10.5)
32     CONTINUE
```

Eleven values of z are used in eq. (MA2.55), and for each value of z, 200 terms are used in the series of eq. (MA2.55). The programming in function UNITY will be considered subsequently. The output from the preceding code is listed below:

```
Fourier series for unity
       .00     .98984
       .10     .99995
       .20     .99993
       .30     .99988
       .40     .99981
       .50     .99969
       .60     .99949
       .70     .99916
       .80     .99861
       .90     .99771
      1.00     .99624
```

Note that 200 terms in the series of eq. (MA2.55) produce about 2+ figures of accuracy due to the slow convergence of the series.

(9) Finally, the series of eq. (MA2.51) is evaluated at $z = 1$ for 81 values of t in the interval $0 \leq t \leq 0.2$.

```
C...
C...      EVALUATE AND PRINT THE SERIES SOLUTION AT Z = 1 FOR A
C...      SERIES OF VALUES OF T USING NS TERMS IN THE SERIES
          WRITE(NO,42)
          WRITE( *,42)
42        FORMAT(//,' Series solution at z = 1, phi1(1,t)',/)
          Z=1.0D0
          DT=0.0025D0
          NTM=81
          DO 40 I=1,NTM
C...
C...         T AND SERIES SOLUTION
             T=DT*DFLOAT(I-1)
             UZT=PHI1(Z,T)
C...
C...         PRINT SOLUTION
             WRITE(NO,41)Z,T,UZT
41           FORMAT(F10.3,F10.4,F10.5)
C...
C...         WRITE SOLUTION FOR MATLAB PLOTTING
             WRITE(2,43)T,UZT
43           FORMAT(2F10.4)
40        CONTINUE
```

The output from the preceding code is listed below:

```
Series solution at z = 1, phi1(1, t)

    1.000         .0000        .00000
    1.000         .0025       -.00377
    1.000         .0050       -.00376
    1.000         .0075       -.00376
    1.000         .0100       -.00376
    1.000         .0125       -.00376
    1.000         .0150       -.00376
    1.000         .0175       -.00376
    1.000         .0200       -.00373
    1.000         .0225       -.00364
    1.000         .0250       -.00336
    1.000         .0275       -.00273
    1.000         .0300       -.00148
    1.000         .0325        .00068
    1.000         .0350        .00409
    1.000         .0375        .00904
    1.000         .0400        .01582
    1.000         .0425        .02464
    1.000         .0450        .03564
    1.000         .0475        .04888
    1.000         .0500        .06436
                    .
                    .
                    .
    1.000         .0800        .36520
    1.000         .0825        .39354
    1.000         .0850        .42154
    1.000         .0875        .44906
    1.000         .0900        .47602
    1.000         .0925        .50231
    1.000         .0950        .52787
    1.000         .0975        .55264
    1.000         .0000        .57658
    1.000         .1025        .59964
    1.000         .1050        .62181
    1.000         .1075        .64308
    1.000         .1100        .66343
    1.000         .1125        .68288
    1.000         .1150        .70142
    1.000         .1175        .71907
    1.000         .1200        .73585
                    .
                    .
                    .
```

```
     1.000       .1900      .95818
     1.000       .1925      .96084
     1.000       .1950      .96331
     1.000       .1975      .96562
     1.000       .2000      .96777
```

This numerical evaluation of the analytical solution for $\phi_1(z, t)$ from eq. (MA2.51) is subsequently compared with a numerical solution. Also, the programming in function PHI1, which implements the solution of eq. (MA2.51), will be considered subsequently.

We now return to the series of subroutines and functions called by the preceding subroutine SERIES. Subroutine EIGVAL, which computes the eigenvalues defined by eq. (MA2.52), is listed below:

```
      SUBROUTINE EIGVAL
      IMPLICIT DOUBLE PRECISION (A-H,O-Z)
      PARAMETER(NZ=81,NA=200)
      COMMON/T/         T,    NSTOP,    NORUN
     +      /Y/    C(NZ)
     +      /F/    CT(NZ)
     +      /S/    CZ(NZ), CZZ(NZ)
     +      /C/        ZL,      ZU,      ZLU,      PE,
     +            ZMU(NA),  CF(NA)
     +      /I/        NS,      NQ,       NP,      IP,
     +             NCASE,      IN,       JN
C...
C... FUNCTION FOR THE TRANSCENDENTAL EQUATION THAT DEFINES
C... THE EIGENVALUES
      EXTERNAL F1
C...
C... PI
      PI=4.0*DATAN(1.0D0)
C...
C... ERROR TOLERANCE
      EPS=1.0D-06
C...
C... COMPUTE NS EIGENVALUES
      DO 1 I=1,NS
C...
C...    BOUND ROOT (EIGENVALUE)
        ZP=DFLOAT(I-1)*PI
        ZN=DFLOAT(I  )*PI
C...
C...    COMPUTE ROOT AND STORE AS AN EIGENVALUE
        ZMU(I)=ROOT(F1,ZN,ZP,EPS)
C...
```

```
C...    NEXT EIGENVALUE
1       CONTINUE
        RETURN
        END
```

We can note the following points about EIGVAL:

(1) The COMMON block is the same as in subroutine SERIES.

(2) The transcendental function to be zeroed to compute the eigenvalues is declared in an EXTERNAL statement

```
C...
C...    FUNCTION FOR THE TRANSCENDENTAL EQUATION THAT DEFINES
C...    THE EIGENVALUES
        EXTERNAL F1
```

In the present case, function F1 contains the left hand side (LHS) of eq. (MA2.52), as indicated subsequently.

(3) The number π is computed, and the error tolerance for the computed eigenvalues is specified

```
C...
C...    PI
        PI=4.0*DATAN(1.0D0)
C...
C...    ERROR TOLERANCE
        EPS=1.0D-06
```

(4) NS eigenvalues are then computed by cycling through DO loop 1 (NS is set in SERIES before EIGVAL is called, and passed through COMMON /I/)

```
C...
C...    COMPUTE NS EIGENVALUES
        DO 1 I=1,NS
C...
C...       BOUND ROOT (EIGENVALUE)
           ZP=DFLOAT(I-1)*PI
           ZN=DFLOAT(I  )*PI
C...
C...       COMPUTE ROOT AND STORE AS AN EIGENVALUE
           ZMU(I)=ROOT(F1,ZN,ZP,EPS)
C...
C...    NEXT EIGENVALUE
1       CONTINUE
```

The first requirement in computing an eigenvalue is to bound its value. This is done in the present case by making use of the known property of the eigenvalues that they are spaced at regular intervals on a scale defined as a multiple of π. Each eigenvalue is then computed by a call to function ROOT, which will be discussed subsequently. DO loop 1 then concludes subroutine EIGVAL.

Function F1 is listed below:

```
      DOUBLE PRECISION FUNCTION F1(Z)
C...
C...  F1 COMPUTES THE TRANSCENDENTAL FUNCTION THAT DEFINES
C...  THE EIGENVALUES.  F1 IS CALL BY FUNCTION ROOT VIA
C...  SUBROUTINE EIGVAL TO FIND THE ZEROS OF THE
C...  TRANSCENDENTAL FUNCTION.
C...
      IMPLICIT DOUBLE PRECISION (A-H,O-Z)
      PARAMETER(NZ=81,NA=200)
      COMMON/T/       T,    NSTOP,      NORUN
     +      /Y/    C(NZ)
     +      /F/    CT(NZ)
     +      /S/    CZ(NZ), CZZ(NZ)
     +      /C/       ZL,       ZU,        ZLU,       PE,
     +            ZMU(NA),  CF(NA)
     +      /I/       NS,       NQ,         NP,       IP,
     +             NCASE,       IN,         JN
C...
C...  COMPUTE FUNCTION TO BE ZEROED
      F1=1.0D0/DTAN(Z)-Z/PE+PE/(4.0D0*Z)
      RETURN
      END
```

Note that F1 is a straightforward calculation of the LHS of eq. (MA2.52).

```
      F1=1.0D0/DTAN(Z)-Z/PE+PE/(4.0D0*Z)
```

The value of z is set through the argument of F1, and the Peclet number, P_e, comes into F1 through COMMON/C/.

Function ROOT is listed below:

```
      DOUBLE PRECISION FUNCTION ROOT(F,ZN,ZP,EPS)
C...
C...  FUNCTION ROOT COMPUTES THE ROOT ON AN EQUATION (ZERO
C...  OF A FUNCTION) BY THE SECANT METHOD.  THE CODE WAS
C...  ORIGINALLY PROVIDED BY YOUNG-SOON CHOE, AND MODIFIED
C...  SLIGHTLY BY W. E. SCHIESSER.
C...
C...  ARGUMENTS
C...
C...     F      EXTERNAL TO COMPUTE THE FUNCTION TO BE
C...         ZEROED (INPUT)
```

```
C...
C...      ZN      LOWER BOUND ON THE ROOT (INPUT)
C...
C...      ZP      UPPER BOUND ON THE ROOT (INPUT)
C...
C...      EPS     ABSOLUTE ERROR TOLERANCE FOR THE ROOT
C...              (INPUT)
C...
          IMPLICIT DOUBLE PRECISION (A-H,O-Z)
C...
C...      EXTERNAL FOR THE FUNCTION TO BE ZEROED
          EXTERNAL F

C...
C...      BEGIN SEARCH FOR THE ROOT BY THE SECANT METHOD
          IP=0
          IN=0
C...
C...      INITIAL ESTIMATE OF THE ROOT
2         Z=(ZP+ZN)/2.0D0
C...
C...      COMPUTE FUNCTION TO BE ZEROED
          FZ=F(Z)
C...
C...      CONTINUE SEARCH
          IF(FZ.LE.0.0D0)THEN
             ZN=Z
             FM=FZ
             IN=1
          ELSE
             ZP=Z
             FP=FZ
             IP=1
          END IF
          IF((IP*IN).EQ.0)     GO TO 2
          IF((FP-FM).GT.0.1D0)GO TO 2
C...
C...      UPDATE Z AND COMPUTE FUNCTION
3         SL=(FP-FM)/(ZP-ZN)
          Z=ZP-FP/SL
          FZ=F(Z)
C...
C...      TEST FOR CONVERGENCE
          IF((DABS(FZ).LT.EPS).OR.(DABS(ZP-ZN).LT.EPS))THEN
C...
C...           CONVERGENCE ACHIEVED SO STORE ROOT AND RETURN TO
```

```
C...         CALLING PROGRAM
             ROOT=Z
C...
C...         CONTINUE SEARCH
             ELSE
             IF(FZ.GE.0.0D0)THEN
                 FM=FZ
                 ZN=Z
                 GO TO 3
             ELSE
                 FP=FZ
                 ZP=Z
                 GO TO 3
                 END IF
             END IF
         RETURN
         END
```

We can note the following points about ROOT:

(1) Double precision coding is used. The function for which a root is to be computed is declared as an EXTERNAL, in this case the function defined by the previous F1 (the LHS of eq. (MA2.52))

```
             IMPLICIT DOUBLE PRECISION (A-H,O-Z)
C...
C...         EXTERNAL FOR THE FUNCTION TO BE ZEROED
             EXTERNAL F
```

(2) Two integer variables, which control the search for the root, are initialzed. The estimated root is then evaluated at the midpoint of the interval defined in the call to ROOT (the interval consisting of the lower bound to the upper bound)

```
C...
C...         BEGIN SEARCH FOR THE ROOT BY THE SECANT METHOD
             IP=0
             IN=0
C...
C...         INITIAL ESTIMATE OF THE ROOT
2            Z=(ZP+ZN)/2.0D0
C...
C...         COMPUTE FUNCTION TO BE ZEROED
             FZ=F(Z)
```

The function is then evaluated at this initial estimate of the root.

(3) If the function is nonpositive, the lower bound is replaced by the root at the midpoint and the function is stored as FM. If the function is positive, the upper bound is replaced by the midpoint value, and the function is stored as FP

```
C...
C...    CONTINUE SEARCH
        IF(FZ.LE.0.0D0)THEN
            ZN=Z
            FM=FZ
            IN=1
        ELSE
            ZP=Z
            FP=FZ
            IP=1
        END IF
```

(4) Tests are then performed to determine if the preceding interval-halving algorithm should continue to further refine the root (by going to statement 2)

```
        IF((IP*IN).EQ.0)      GO TO 2
        IF((FP-FM).GT.0.1D0)GO TO 2
```

If the root is refined sufficiently, a switch is made to the secant method (a discrete form of Newton's method).

(5) Within the secant method, a finite difference approximation of the slope of the function is first computed

```
C...
C...    UPDATE Z AND COMPUTE FUNCTION
3       SL=(FP-FM)/(ZP-ZN)
        Z=ZP-FP/SL
        FZ=F(Z)
```

The root is then updated by the secant method (note the direct analog with Newton's method). Finally, the function is evaluated at this new estimate of the root.

(6) A convergence test is then made to determine if the function at this new estimate is below the tolerance, or if the change in the root is below the tolerance.

```
C...
C...    TEST FOR CONVERGENCE
        IF((DABS(FZ).LT.EPS).OR.(DABS(ZP-ZN).LT.EPS))THEN
C...
C...        CONVERGENCE ACHIEVED SO STORE ROOT AND RETURN TO
C...        CALLING PROGRAM
            ROOT=Z
```

If either of these tests is passed, function ROOT is set to the value of the new root and the RETURN at the end of ROOT is executed

(7) If the root does not pass the convergence test, the secant method is repeated by a return to statement 3

```
C...
C...      CONTINUE SEARCH
          ELSE
          IF(FZ.GE.0.0D0)THEN
              FM=FZ
              ZN=Z
              GO TO 3
          ELSE
              FP=FZ
              ZP=Z
              GO TO 3
          END IF
```

In summary, the basic organization of ROOT is to compute an approximate root using interval-halving, then refine the root using the secant method (which has near quadratic convergence even though it does not require the derivative of the function as does Newton's method). Note that this programming does not preclude the possibility of an infinite loop (the execution of the RETURN occurs only if one of the convergence tests is passed), but function ROOT proved to be reliable for this particular application.

Subroutine COEFF to test the orthogonality of eq. (MA2.43) is listed below:

```
          SUBROUTINE ORTHO(I,J,PIJ)
          IMPLICIT DOUBLE PRECISION (A-H,O-Z)
          PARAMETER(NZ=81,NA=200)
          COMMON/T/         T,     NSTOP,     NORUN
         +        /Y/    C(NZ)
         +        /F/    CT(NZ)
         +        /S/    CZ(NZ), CZZ(NZ)
         +        /C/       ZL,       ZU,       ZLU,      PE,
         +              ZMU(NA),   CF(NA)
         +        /I/       NS,       NQ,        NP,      IP,
         +              NCASE,       IN,        JN
C...
C...      EXTERNAL THE FUNCTION FOR THE INTEGRAND OF THE PRODUCT
C...      INTEGRAL
          EXTERNAL F2
C...
C...      SET THE INDICES OF THE EIGENFUNCTIONS
          IN=I
          JN=J
```

```
C...
C...    PERFORM THE INTEGRATION OF THE PRODUCT OF TWO
C...    EIGENFUNCTIONS FROM ZL = 0 TO ZU = 1
        PIJ=SIMP(F2,ZL,ZU,NQ)
        RETURN
        END
```

We can note the following points about COEFF:

(1) The COMMON area is the same as in subroutine SERIES.

(2) The function of eq. (MA2.35), which appears in the integrand of eq. (MA2.43), is declared in an EXTERNAL statement

```
C...
C...    EXTERNAL THE FUNCTION FOR THE INTEGRAND OF THE PRODUCT
C...    INTEGRAL
        EXTERNAL F2
```

(3) The indices of the two eigenfunctions in eq. (MA2.43) are then set, and the integral of eq. (MA2.43) is computed by a call to function SIMP, which performs the Simpson's rule quadrature

```
C...
C...    SET THE INDICES OF THE EIGENFUNCTIONS
        IN=I
        JN=J
C...
C...    PERFORM THE INTEGRATION OF THE PRODUCT OF TWO
C...    EIGENFUNCTIONS FROM ZL = 0 TO ZU = 1
        PIJ=SIMP(F2,ZL,ZU,NQ)
```

Note that the indices are passed from ORTHO to function F2 through COMMON/I/.

Function SIMP is listed below:

```
        DOUBLE PRECISION FUNCTION SIMP(F,A,B,NQ)
C...
C...    FUNCTION SIMP PERFORMS A NUMERICAL QUADRATURE
C...    (INTEGRATION) BY SIMPSON'S RULE
C...
C...    ARGUMENTS
C...
C...       F        EXTERNAL TO COMPUTE THE INTEGRAND (INPUT)
C...
C...       A        LOWER LIMIT OF THE INTEGRAL (INPUT)
C...
C...       B        UPPER LIMIT OF THE INTEGRAL (INPUT)
```

```
C...
C...       NQ        NUMBER OF PANELS USED IN SIMPSON'S RULE
C...                 (MUST BE EVEN) (INPUT)
C...
       IMPLICIT DOUBLE PRECISION (A-H,O-Z)
C...
C...   EXTERNAL FUNCTION FOR INTEGRAND
       EXTERNAL F
C...
C...   CHECK FOR AN EVEN NUMBER OF PANELS
       IF((NQ/2*2).NE.NQ)THEN
          WRITE(*,2)NQ
2         FORMAT(//,' NQ = ',I3,' (NOT EVEN)')
          SIMP=0.0D0
          RETURN
       END IF
C...
C...   INTEGRATION INTERVAL
       DZ=(B-A)/DFLOAT(NQ)
C...
C...   INITIALIZE SUM
       SIMP=(F(A)+4.0D0*F(A+DZ)+F(B))
C...
C...   CONTINUE SUM
       IF(NQ.GT.2)THEN
          DO 1 I=2,NQ-1,2
             SIMP=SIMP
     +            +2.0D0*F(A+DFLOAT(I  )*DZ)
     +            +4.0D0*F(A+DFLOAT(I+1)*DZ)

1         CONTINUE
       END IF
C...
C...   INTEGRAL
       SIMP=(DZ/3.0D0)*SIMP
       RETURN
       END
```

We can note the following points about SIMP:

(1) The function that defines the integrand of the integral to be evaluated, in this case $\psi_i \psi_j$ of eq. (MA2.43), is declared as an external

```
C...
C...   EXTERNAL FUNCTION FOR INTEGRAND
       EXTERNAL F
```

(2) A test is then made to determine if SIMP is called with an even number of panels specified (an odd number of quadrature points)

```
C...
C...    CHECK FOR AN EVEN NUMBER OF PANELS
        IF((NQ/2*2).NE.NQ)THEN
            WRITE(*,2)NQ
2           FORMAT(//,' NQ = ',I3,' (NOT EVEN)')
            SIMP=0.0D0
            RETURN
        END IF
```

In the present case, 1,000 panels are specified when SIMP is called (originally specified in SERIES).

(3) The integration interval is then computed as the difference between the upper and lower limits divided by the number of panels. Note that the integration interval can be negative (if the upper limit is less than the lower limit); the following code still executes correctly

```
C...
C...    INTEGRATION INTERVAL
        DZ=(B-A)/DFLOAT(NQ)
```

(4) The code for Simpson's rule is then executed

```
C...
C...    INITIALIZE SUM
        SIMP=(F(A)+4.0D0*F(A+DZ)+F(B))
C...
C...    CONTINUE SUM
        IF(NQ.GT.2)THEN
            DO 1 I=2,NQ-1,2
                SIMP=SIMP
     +              +2.0D0*F(A+DFLOAT(I  )*DZ)
     +              +4.0D0*F(A+DFLOAT(I+1)*DZ)
1           CONTINUE
        END IF
C...
C...    INTEGRAL
        SIMP=(DZ/3.0D0)*SIMP
```

This code is a straightforward implementation of Simpson's rule

$$\int_a^b f(x)dx \approx (\Delta x/3)\{f(a) + 2f(a + \Delta x) + 4f(a + 2\Delta x) + 2f(a + 3\Delta x)$$
$$+ \cdots + 2f(b - 2\Delta x) + 4f(b - \Delta x) + f(b)\} \quad \text{(MA2.56)}$$

Function F2 to compute the integrand of eq. (MA2.43) is listed below:

```
      DOUBLE PRECISION FUNCTION F2(Z)
C...
C...  FUNCTION F2 COMPUTES THE PRODUCT OF TWO
C...  EIGENFUNCTIONS. F2 IS CALLED BY FUNCTION
C...  SIMP VIA SUBROUTINE ORTHO.
C...
      IMPLICIT DOUBLE PRECISION (A-H,O-Z)
      PARAMETER(NZ=81,NA=200)
      COMMON/T/     T,      NSTOP,    NORUN
     +      /Y/   C(NZ)
     +      /F/   CT(NZ)
     +      /S/   CZ(NZ), CZZ(NZ)
     +      /C/       ZL,      ZU,      ZLU,       PE,
     +            ZMU(NA),  CF(NA)
     +      /I/       NS,      NQ,       NP,       IP,
     +            NCASE,       IN,       JN
C...
C...  EIGENFUNCTION PSI(I)
      PSII=PE/(2.0D0*ZMU(IN))*DSIN(ZMU(IN)*Z)
     +     +DCOS(ZMU(IN)*Z)
C...
C...  EIGENFUNCTION PSI(J)
      PSIJ=PE/(2.0D0*ZMU(JN))*DSIN(ZMU(JN)*Z)
     +     +DCOS(ZMU(JN)*Z)
C...
C...  PRODUCT OF THE EIGENFUNCTIONS
      F2=PSII*PSIJ
      RETURN
      END
```

We can note the following points about F2:

(1) The COMMON area is the same as in subroutine SERIES.

(2) The eigenfunction $\psi_i(z)$ in eq. (MA2.43) is computed first

```
C...
C...  EIGENFUNCTION PSI(I)
      PSII=PE/(2.0D0*ZMU(IN))*DSIN(ZMU(IN)*Z)
     +     +DCOS(ZMU(IN)*Z)
```

Note that this coding follows directly from eq. (MA2.35).

(3) Then the eigenfunction ψ_j in eq. (MA2.43) is computed

```
C...
C...  EIGENFUNCTION PSI(J)
```

```
      PSIJ=PE/(2.0D0*ZMU(JN))*DSIN(ZMU(JN)*Z)
     +      +DCOS(ZMU(JN)*Z)
```

(4) Finally, the product of the two eigenfunctions is computed to form the integrand of eq. (MA2.43)

```
C...
C...  PRODUCT OF THE EIGENFUNCTIONS
      F2=PSII*PSIJ
```

Function COEFF, which computes the Fourier coefficients according to eq. (MA2.54), is listed below:

```
      SUBROUTINE COEFF(I)
      IMPLICIT DOUBLE PRECISION (A-H,O-Z)
      PARAMETER(NZ=81,NA=200)
      COMMON/T/       T,     NSTOP,    NORUN
     +      /Y/    C(NZ)
     +      /F/    CT(NZ)
     +      /S/    CZ(NZ), CZZ(NZ)
     +      /C/       ZL,      ZU,       ZLU,      PE,
     +           ZMU(NA),   CF(NA)
     +      /I/       NS,      NQ,        NP,      IP,
     +             NCASE,     IN,        JN
C...
C...  EXTERNAL THE FUNCTION FOR THE INTEGRAND OF THE
C...  NUMERATOR AND DENOMINATOR INTEGRALS OF THE FOURIER
C...  COEFFICIENT
      EXTERNAL F3, F4
C...
C...  SET THE INDEX OF THE FOURIER COEFFICIENT
      IN=I
C...
C...  PERFORM THE INTEGRATION FOR THE NUMERATOR AND
C...  DENOMINATOR INTEGRALS FROM ZL = 0 TO ZU = 1
      ZNUM=SIMP(F3,ZL,ZU,NQ)
      ZDEN=SIMP(F4,ZL,ZU,NQ)
C...
C...  FOURIER COEFFICIENT
      CF(I)=ZNUM/ZDEN
      RETURN
      END
```

We can note the following points about COEFF:

(1) The COMMON area is the same as in subroutine SERIES.

(2) The functions that compute the integrands in the numerator and denominator of eq. (MA2.54) are defined as externals

```
C...
C...   EXTERNAL THE FUNCTION FOR THE INTEGRAND OF THE
C...   NUMERATOR AND DENOMINATOR INTEGRALS OF THE FOURIER
C...   COEFFICIENT
       EXTERNAL F3, F4
```

(3) The index of the Fourier coefficient is set to IN, which is then passed through COMMON/I/ to functions F3 and F4

```
C...
C...   SET THE INDEX OF THE FOURIER COEFFICIENT
       IN=I
C...
C...   PERFORM THE INTEGRATION FOR THE NUMERATOR AND
C...   DENOMINATOR INTEGRALS FROM ZL = 0 TO ZU = 1
       ZNUM=SIMP(F3,ZL,ZU,NQ)
       ZDEN=SIMP(F4,ZL,ZU,NQ)
```

The integrals in the numerator and denominator of eq. (MA2.54) are then computed by calls to function SIMP.

(4) Finally, the Fourier coefficient is computed according to eq. (MA2.54)

```
C...
C...   FOURIER COEFFICIENT
       CF(I)=ZNUM/ZDEN
```

Function F3 to compute the integrand in the numerator of eq. (MA2.54) is listed below:

```
       DOUBLE PRECISION FUNCTION F3(Z)
C...
C...   FUNCTION F3 COMPUTES THE NUMERATOR INTEGRAND OF THE
C...   FOURIER COEFFICIENT.  F3 IS CALLED BY FUNCTION SIMP
C...   VIA SUBROUTINE COEFF.
C...
       IMPLICIT DOUBLE PRECISION (A-H,O-Z)
       PARAMETER(NZ=81,NA=200)
       COMMON/T/        T,    NSTOP,    NORUN
      +      /Y/    C(NZ)
      +      /F/    CT(NZ)
      +      /S/    CZ(NZ), CZZ(NZ)
      +      /C/        ZL,       ZU,      ZLU,       PE,
      +             ZMU(NA),  CF(NA)
      +      /I/        NS,       NQ,       NP,       IP,
      +             NCASE,        IN,       JN
C...
C...   EIGENFUNCTION PSI(I)
```

```
      PSII=PE/(2.0D0*ZMU(IN))*DSIN(ZMU(IN)*Z)
     +     +DCOS(ZMU(IN)*Z)
C...
C... PRODUCT OF EXPONENTIAL AND EIGENFUNCTION
      F3=DEXP((-PE/2.0D0)*Z)*PSII
      RETURN
      END
```

We can note the following points about F3:

(1) The COMMON area is the same as in subroutine SERIES.

(2) The computation of the numerator integrand of eq. (MA2.54) is straightforward

```
C...
C... EIGENFUNCTION PSI(I)
      PSII=PE/(2.0D0*ZMU(IN))*DSIN(ZMU(IN)*Z)
     +     +DCOS(ZMU(IN)*Z)
C...
C... PRODUCT OF EXPONENTIAL AND EIGENFUNCTION
      F3=DEXP((-PE/2.0D0)*Z)*PSII
```

Function F4, listed below, is a direct analog of F3 (refer to the denominator of eq. (MA2.54)):

```
      DOUBLE PRECISION FUNCTION F4(Z)
C...
C... FUNCTION F4 COMPUTES THE DENOMINATOR INTEGRAND OF THE
C... FOURIER COEFFICIENT.  F4 IS CALLED BY FUNCTION SIMP
C... VIA SUBROUTINE COEFF.
C...
      IMPLICIT DOUBLE PRECISION (A-H,O-Z)
      PARAMETER(NZ=81,NA=200)
      COMMON/T/        T,    NSTOP,     NORUN
     +      /Y/    C(NZ)
     +      /F/    CT(NZ)
     +      /S/    CZ(NZ), CZZ(NZ)
     +      /C/        ZL,       ZU,       ZLU,        PE,
     +           ZMU(NA),   CF(NA)
     +      /I/        NS,       NQ,        NP,        IP,
     +             NCASE,       IN,        JN
C...
C... EIGENFUNCTION PSI(I)
      PSII=PE/(2.0D0*ZMU(IN))*DSIN(ZMU(IN)*Z)
     +     +DCOS(ZMU(IN)*Z)
C...
C... EIGENFUNCTION SQUARED
```

```
      F4=PSII**2
      RETURN
      END
```

Note in particular that the eigenfunction is squared in computing the denominator integrand of eq. (MA2.54).

To complete this discussion of the analytical solution, we have to consider only two functions that evaluate the infinite series solutions of eq. (MA2.55) and (MA2.51). Function UNITY computes the Fourier series expansion of unity according to eq. (MA2.55)

```
      DOUBLE PRECISION FUNCTION UNITY(Z)
C...
C...  FUNCTION UNITY COMPUTES THE SERIES EXPANSION OF UNITY
C...  (THE NUMBER ONE) AT Z
C...
      IMPLICIT DOUBLE PRECISION (A-H,O-Z)
      PARAMETER(NZ=81,NA=200)
      COMMON/T/      T,     NSTOP,     NORUN
     +      /Y/   C(NZ)
     +      /F/   CT(NZ)
     +      /S/   CZ(NZ),  CZZ(NZ)
     +      /C/       ZL,      ZU,      ZLU,      PE,
     +            ZMU(NA),  CF(NA)
     +      /I/       NS,      NQ,       NP,      IP,
     +            NCASE,      IN,       JN
C...
C...  INITIALIZE SERIES
      SUM=0.0D0
C...
C...  PRECOMPUTE THE COMMON EXPONENTIAL FACTOR
      EXPZ=DEXP((PE/2.0D0)*Z)
C...
C...  COMPUTE SERIES WITH NS TERMS
      DO 1 I=1,NS
C...
C...     EIGENFUNCTION PSI(I)
         PSII=PE/(2.0D0*ZMU(I))*DSIN(ZMU(I)*Z)
     +        +DCOS(ZMU(I)*Z)
C...
C...     RUNNING SUM
         SUM=SUM+CF(I)*PSII
C...
C...  CONTINUE SUMMATION
1     CONTINUE
```

```
C...
C...    INCLUDE EXPONENTIAL FACTOR
        UNITY=EXPZ*SUM
C...
C...    CALCULATION IS COMPLETE
        RETURN
        END
```

We can note the following points about function UNITY:

(1) The COMMON area is the same as in subroutine SERIES.

(2) The series of eq. (MA2.55) is initialized first, and the common exponential factor is computed

```
C...
C...    INITIALIZE SERIES
        SUM=0.0D0
C...
C...    PRECOMPUTE THE COMMON EXPONENTIAL FACTOR
        EXPZ=DEXP((PE/2.0D0)*Z)
```

(3) NS (= 200) terms of the series in eq. (M2.55) are then computed in DO loop 1

```
C...
C...    COMPUTE SERIES WITH NS TERMS
        DO 1 I=1,NS
C...
C...        EIGENFUNCTION PSI(I)
            PSII=PE/(2.0D0*ZMU(I))*DSIN(ZMU(I)*Z)
     +          +DCOS(ZMU(I)*Z)
C...
C...        RUNNING SUM
            SUM=SUM+CF(I)*PSII
C...
C...    CONTINUE SUMMATION
1       CONTINUE
```

Note that in computing each term of the series, the corresponding eigenvalue, μ_i, given by eq. (MA2.52), and computed previously by the call to EIGVAL, is used. These eigenvalues, in array ZMU(I), are passed through COMMON/C/ to UNITY.

(3) The series of eq. (MA2.55) is then completed by including the common exponential factor

```
C...
C...    INCLUDE EXPONENTIAL FACTOR
        UNITY=EXPZ*SUM
```

Finally, we consider function PHI1, which implements the series of eq. (MA2.51)

```fortran
      DOUBLE PRECISION FUNCTION PHI1(Z,T)
C...
C...  FUNCTION PHI1 COMPUTES THE SERIES SOLUTION AT Z AND T
C...
      IMPLICIT DOUBLE PRECISION (A-H,O-Z)
      PARAMETER(NZ=81,NA=200)
      COMMON/T/      TM,     NSTOP,     NORUN
     +      /Y/    C(NZ)
     +      /F/    CT(NZ)
     +      /S/    CZ(NZ),  CZZ(NZ)
     +      /C/       ZL,       ZU,        ZLU,       PE,
     +            ZMU(NA), CF(NA)
     +      /I/       NS,       NQ,        NP,        IP,
     +            NCASE,       IN,        JN
C...
C...  INITIALIZE SERIES
      SUM=0.0D0
C...
C...  PRECOMPUTE THE COMMON EXPONENTIAL FACTOR
      EXPZ=DEXP((PE/2.0D0)*Z)
C...
C...  COMPUTE SERIES WITH NS TERMS
      DO 1 I=1,NS
C...
C...     EXPONENTIAL IN T
         TLAM=ZMU(I)**2+PE**2/4.0D0
C...
C...     TEST MAGNITUDE OF TLAM*T TO AVOID UNDERFLOW OF EXP
C...     FUNCTION
         TLAM=-TLAM*T
         IF(TLAM.LT.-1.0D+20)THEN
C...
C...     TERMS ARE NEGLIGIBLY SMALL SO TERMINATE SUMMING
C...     SERIES
            GO TO 2
         ELSE
C...
C...     CONTINUE TO SUM SERIES
            EXPT=DEXP(TLAM)
         END IF
C...
C...     EIGENFUNCTION PSI(I)
         PSII=PE/(2.0D0*ZMU(I))*DSIN(ZMU(I)*Z)
     +        +DCOS(ZMU(I)*Z)
C...
```

```
C...        RUNNING SUM
            SUM=SUM+CF(I)*(1.0D0-EXPT)*PSII
C...
C...        CONTINUE SUMMATION
1           CONTINUE
C...
C...        INCLUDE EXPONENTIAL FACTOR
2           PHI1=EXPZ*SUM
C...
C...        CALCULATION IS COMPLETE
            RETURN
            END
```

We can note the following points about function PHI1:

(1) The COMMON area is the same as in subroutine SERIES.

(2) The series of eq. (MA2.51) is initialized first, and the common exponential factor is computed

```
C...
C...        INITIALIZE SERIES
            SUM=0.0D0
C...
C...        PRECOMPUTE THE COMMON EXPONENTIAL FACTOR
            EXPZ=DEXP((PE/2.0D0)*Z)
```

(3) NS (= 200) terms of the series in eq. (MA2.51) are computed in DO loop 1. First the exponential in t of eq. (MA2.51)

$$e^{-\lambda_i t}$$

is computed. Note that λ_i is given by eq. (MA2.53)

```
C...
C...        COMPUTE SERIES WITH NS TERMS
            DO 1 I=1,NS
C...
C...           EXPONENTIAL IN T
               TLAM=ZMU(I)**2+PE**2/4.0D0
C...
C...           TEST MAGNITUDE OF TLAM*T TO AVOID UNDERFLOW OF EXP
C...           FUNCTION
               TLAM=-TLAM*T
               IF(TLAM.LT.-1.0D+02)THEN
C...
C...              TERMS ARE NEGLIGIBLY SMALL SO TERMINATE SUMMING
C...              SERIES
                  GO TO 2
               ELSE
```

```
C...
C...      CONTINUE TO SUM SERIES
              EXPT=DEXP(TLAM)
          END IF
```

Note that t is available as the second argument of PHI for the calculation of the exponential function. A test is included in this code to determine if the exponent, $-\lambda_i t$, is small enough to cause a possible underflow (i.e., $-\lambda_i t < -10^2$). If this condition occurs, the corresponding term, and all subsequent terms in the series of eq. (MA2.51) are considered negligibly small, and the summation is terminated (by a transfer to statement 2).

(4) If $-\lambda_i t \geq -10^2$, the next term in the series is computed and added to the running sum

```
          ELSE
C...
C...      CONTINUE TO SUM SERIES
              EXPT=DEXP(TLAM)
          END IF
C...
C...      EIGENFUNCTION PSI(I)
          PSII=PE/(2.0D0*ZMU(I))*DSIN(ZMU(I)*Z)
     +         +DCOS(ZMU(I)*Z)
C...
C...      RUNNING SUM
          SUM=SUM+CF(I)*(1.0D0-EXPT)*PSII
```

The coding in the segment follows directly from the summation of eq. (MA2.51).

(5) Finally, after DO loop 1 is completed, or the branch to statement 2 as discussed previously occurs, the sum is multiplied by the common exponential factor to complete the numerical solution (assigned to PHI)

```
C...
C...   INCLUDE EXPONENTIAL FACTOR
2      PHI1=EXPZ*SUM
```

This completes the programming of the analytical solution, eqs. (MA2.51) to (MA2.54). We now consider the programming of the numerical solution, starting with subroutine INITAL, which defines initial condition (MA2.48)

```
          SUBROUTINE INITAL
          IMPLICIT DOUBLE PRECISION (A-H,O-Z)
          PARAMETER(NZ=81,NA=200)
          COMMON/T/     T,    NSTOP,    NORUN
     +         /Y/    C(NZ)
```

```
     +          /F/   CT(NZ)
     +          /S/   CZ(NZ), CZZ(NZ)
     +          /C/       ZL,         ZU,       ZLU,           PE,
     +                ZMU(NA),     CF(NA)
     +          /I/       NS,         NQ,        NP,           IP,
     +                  NCASE,         IN,        JN
C...
C...   SPATIAL DIFFERENTIATOR
C...
C...       NCASE = 1 - DSS020
C...
C...       NCASE = 2 - DSS024
C...
       NCASE=1
C...
C...   SYSTEM LENGTH
       ZL=1.0D0
C...
C...   SYSTEM PARAMETERS
       PE=10.0D0
C...
C...   INITIAL CONDITIONS
       DO 1 I=1,NZ
          C(I)=0.0D0
1      CONTINUE
C...
C...   INITIAL DERIVATIVES
       CALL DERV
C...
C...   INITIALIZE COUNTERS FOR PRINTING AND PLOTTING
       IP=0
       NP=81
       RETURN
       END
```

We can note the following points about subroutine INITAL:

(1) The COMMON area is the same as used in the routines for the analytical solutions, for example, subroutine SERIES. Note that an 81-point grid in z is used for the numerical solution. $\phi_1(z, t)$ of eq. (MA2.47) is stored in array C(NZ) in COMMON/Y/, and the temporal derivative, $\partial \phi_1 / \partial t$, is in array CT(NZ) in COMMON/F/. The first- and second-order derivatives, $\partial \phi_1 / \partial z$ and $\partial^2 \phi_1 / \partial z^2$, in eq. (MA2.47) are in arrays CZ(NZ) and CZZ(NZ), in COMMON/S/, respectively, and time, $t \, (= T)$, is in COMMON/T/.

(2) Two cases can be executed corresponding to the use of subroutines DSS020 and DSS024 for the computation of the first-order spatial derivative, $\partial \phi_1 / \partial z$, in

eq. (MA2.47)

```
C...
C...    SPATIAL DIFFERENTIATOR
C...
C...        NCASE = 1 - DSS020
C...
C...        NCASE = 2 - DSS024
C...
        NCASE=1
```

Subroutine DSS020 implements five-point, biased upwind approximations (Schiesser 1991, 135), while subroutine DSS024 implements a combination of two-point and five-point, biased upwind approximations.

(3) The problem parameters are then defined as before in the analytical solution

```
C...
C...    SYSTEM LENGTH
        ZL=1.0D0
C...
C...    SYSTEM PARAMETERS
        PE=10.0D0
```

(4) Initial condition (MA2.48) is then set in DO loop 1

```
C...
C...    INITIAL CONDITIONS
        DO 1 I=1,NZ
            C(I)=0.0D0
1       CONTINUE
```

(5) Finally, a call to DERV is used to test the calculations of the temporal derivative, $\partial \phi_1 / \partial t$, in eq. (MA2.47)

```
C...
C...    INITIAL DERIVATIVES
        CALL DERV
C...
C...    INITIALIZE COUNTERS FOR PRINTING AND PLOTTING
        IP=0
        NP=81
```

Counters for the printed and plotted solution are also initialized at the end of INITAL.

Equation (MA2.47) and boundary conditions (MA2.49) and (MA2.50) are programmed in subroutine DERV listed below:

```fortran
      SUBROUTINE DERV
      IMPLICIT DOUBLE PRECISION (A-H,O-Z)
      PARAMETER(NZ=81,NA=200)
      COMMON/T/      T,    NSTOP,    NORUN
     +     /Y/   C(NZ)
     +     /F/   CT(NZ)
     +     /S/   CZ(NZ), CZZ(NZ)
     +     /C/       ZL,      ZU,     ZLU,      PE,
     +           ZMU(NA),  CF(NA)
     +     /I/       NS,      NQ,      NP,      IP,
     +            NCASE,      IN,      JN
C...
C...  CZ
      IF(NCASE.EQ.1)CALL DSS020(0.0D0,ZL,NZ,C,CZ,1.0D0)
      IF(NCASE.EQ.2)CALL DSS024(0.0D0,ZL,NZ,C,CZ,1.0D0)
C...
C...  BOUNDARY CONDITION, Z = 0
          NL=2
          CZ(1)=PE*(C(1)-FB(T))
C...
C...  BOUNDARY CONDITION, Z = 1
          NU=2
          CZ(NZ)=0.0D0
C...
C...  CZZ
      CALL DSS044(0.0D0,ZL,NZ,C,CZ,CZZ,NL,NU)
C...
C...  PDE
      DO 1 I=1,NZ
          CT(I)=CZZ(I)-PE*CZ(I)
 1    CONTINUE
      RETURN
      END

      DOUBLE PRECISION FUNCTION FB(T)
      IMPLICIT DOUBLE PRECISION (A-H,O-Z)
C...
C...  UNIT STEP FUNCTION
      IF(T.LT.0.0D0)THEN
          FB=0.0D0
      ELSE
     +IF(T.GE.0.0D0)THEN
          FB=1.0D0
      END IF
      RETURN
      END
```

We can note the following points about DERV and subordinate function FB:

(1) The COMMON block is the same as in INITAL.

(2) The first-order spatial derivative in eq. (MA2.47), $\partial \phi_1/\partial z$, is computed by either a call to DSS020 (NCASE = 1) or DSS024 (NCASE = 2)

```
C...
C...   CZ
       IF(NCASE.EQ.1)CALL DSS020(0.0D0, ZL, NZ, C, CZ, 1.0D0)
       IF(NCASE.EQ.2)CALL DSS024(0.0D0, ZL, NZ, C, CZ, 1.0D0)
```

Note that ϕ_1 (= C) is an input to DSS020 or DSS024, and $\partial \phi_1/\partial z$ (= CZ) is an output.

(3) Boundary condition (MA2.49) is then implemented

```
C...
C...   BOUNDARY CONDITION, Z = 0
       NL=2
       CZ(1)=PE*(C(1)-FB(T))
```

This is a Neumann boundary condition that is specified with NL = 2 (for subsequent calculation of the second-order derivative, $\partial^2 \phi_1/\partial z^2$, by subroutine DSS044). Function FB implements a unit step function at $t = 0$.

(4) Boundary condition (MA2.50) is implemented next

```
C...
C...   BOUNDARY CONDITION, Z = 1
       NU=2
       CZ(NZ)=0.0D0
```

Note again that this is a Neumann boundary condition.

(5) The second-order derivative, $\partial^2 \phi_1/\partial z^2$, in eq. (MA2.47) is then computed by a call to DSS044

```
C...
C...   CZZ
       CALL DSS044(0.0D0,ZL,NZ,C,CZ,CZZ,NL,NU)
```

Note that $\partial \phi_1/\partial z$ (= CZ) at $z = 0$ and $z = 1$ (as set by the boundary conditions as CZ(1) and CZ(NZ), respectively) are inputs to DSS044, and $\partial^2 \phi_1/\partial z^2$ (= CZZ) is the output.

(6) Finally, eq. (MA2.47) is implemented in DO loop 1.

```
C...
C...   PDE
       DO 1 I=1,NZ
```

```
          CT(I)=CZZ(I)-PE*CZ(I)
1      CONTINUE
```

Note the close correspondence between eq. (MA2.47) and its programming in DO loop 1, which is one of the major advantages of this approach to the numerical integration of PDEs, termed the *Numerical Method of Lines* or MOL (Schiesser 1991). The essential function of subroutine DERV is to compute the temporal derivative of eq. (MA2.47), $\partial \phi_1/\partial t$ ($=$ CT), which is sent to an ODE integrator through COMMON/F/. The integrator then returns the dependent variable, ϕ_1, through COMMON/Y/ so that it can be used to compute $\partial \phi_1/\partial t$ in DERV.

The numerical solution produced by subroutines INITAL and DERV is printed and plotted by subroutine PRINT listed below:

```
      SUBROUTINE PRINT(NI,NO)
      IMPLICIT DOUBLE PRECISION (A-H,O-Z)
      PARAMETER(NZ=81,NA=200)
      COMMON/T/      T,    NSTOP,       NORUN
     +     /Y/   C(NZ)
     +     /F/   CT(NZ)
     +     /S/   CZ(NZ), CZZ(NZ)
     +     /C/       ZL,      ZU,       ZLU,         PE,
     +         ZMU(NA), CF(NA)
     +     /I/       NS,      NQ,        NP,         IP,
     +           NCASE,      IN,        JN
C...
C...  MONITOR THE OUTPUT
      WRITE(*,*)IP
C...
C...  WRITE NUMERICAL AND SERIES SOLUTIONS EVERY 40TH CALL
C...  TO PRINT
      IP=IP+1
      IF (((IP-1)/40*40).EQ.(IP-1))THEN
         WRITE(NO,1)T,(C(I),I=1,NZ)
1        FORMAT(/,' t = ',F7.2,/,F10.5,/,(5F10.5))
      END IF
C...
C...  MATLAB PLOTTING OF THE SOLUTION
      CALL PLOTM
C...
C...  SERIES SOLUTION AT END OF NUMERICAL SOLUTION
      IF(IP.EQ.NP)CALL SERIES(NI,NO)
      RETURN
      END
```

We can note the following points about PRINT:

(1) The COMMON area is the same as in INITAL and DERV.

(2) After the counter IP is written to the screen to monitor the progress of the solution, $\phi_1(z, t)$ (= C) is printed as a function of z and t every 40th call to PRINT, (PRINT is called 81 times to produce enough points for a smooth plot of the solution; to limit the output, the WRITE statement for the solution is executed only three times corresponding to $t = 0, 0.1$ and 0.2).

```
C...
C...   MONITOR THE OUTPUT
       WRITE(*,*)IP
C...
C...   WRITE NUMERICAL AND SERIES SOLUTIONS EVERY 40TH CALL
C...   TO PRINT
       IP=IP+1
       IF(((IP-1)/40*40).EQ.(IP-1))THEN
          WRITE(NO,1)T,(C(I),I=1,NZ)
1         FORMAT(/,' t = ',F5.2,/,F10.5,/,(5F10.5))
       END IF
```

(3) The solution $\phi_1(1, t)$ is plotted versus t by Matlab through a call to subroutine PLOTM

```
C...
C...   MATLAB PLOTTING OF THE SOLUTION
       CALL PLOTM
```

(4) Finally, at the end of the numerical solution (when IP = NP = 201), subroutine SERIES is called to compute and write the analytical solution, as discussed previously.

```
C...
C...   SERIES SOLUTION AT END OF NUMERICAL SOLUTION
       IF(IP.EQ.NP)CALL SERIES(NI,NO)
```

To complete the discussion of the programming of the numerical solution, subroutine PLOTM is listed below:

```
       SUBROUTINE PLOTM
       IMPLICIT DOUBLE PRECISION (A-H,O-Z)
       PARAMETER(NZ=81,NA=200)
       COMMON/T/       T,     NSTOP,    NORUN
     +      /Y/    C(NZ)
     +      /F/    CT(NZ)
     +      /S/    CZ(NZ),  CZZ(NZ)
     +      /C/        ZL,       ZU,      ZLU,       PE,
     +           ZMU(NA),   CF(NA)
     +      /I/        NS,       NQ,       NP,       IP,
     +             NCASE,       IN,       JN
```

```
C...
C...   DEFINE A FILE FOR MATLAB PLOTTING OF THE NUMERICAL
C...   SOLUTION
       IF(IP.EQ.1)OPEN(1,FILE='ma2b.out')
C...
C...   WRITE SOLUTION AT Z = 1 FOR MATLAB PLOTTING (EVERY
C...   FIFTH POINT IS USED SO THAT THE INDIVIDUAL PLOTTED
C...   POINTS CAN BE DISTINGUISHED)
       IF(((IP-1)/5*5).EQ.(IP-1))WRITE(1,1)T,C(NZ)
1      FORMAT(2F10.4)
       RETURN
       END
```

We can note the following points about PLOTM:

(1) The COMMON area is the same as in subroutines INITAL, DERV, and PRINT.

(2) At the beginning of the solution (when IP = 1), file ma2b.out is opened to store the solution, which is then subsequently sent to Matlab for plotting of the solution

```
C...
C...   DEFINE A FILE FOR MATLAB PLOTTING
       IF(IP.EQ.1)OPEN(1,FILE='ma2b.out')
```

(3) Finally, the solution $\phi_1(1, t)$ is written as a function of t to file ma2b.out

```
C...
C...   WRITE SOLUTION AT Z = 1 FOR MATLAB PLOTTING (EVERY
C...   FIFTH POINT IS USED SO THAT THE INDIVIDUAL PLOTTED
C...   POINTS CAN BE DISTINGUISHED)
       IF(((IP-1)/5*5).EQ.(IP-1))WRITE(1,1)T,C(NZ)
1      FORMAT(2F10.4)
```

The time integration of the 81 ODEs (with the derivatives CT(NZ) in COMMON/F/) was performed by subroutine RKF45 called by a main program similar to the main program in Appendix A. The main program reads a data file listed below:

```
One-dimensional, convective-diffusion
0.         0.2        0.0025
    81                 0.0001
END OF RUNS
```

t covers the interval $0 \leq t \leq 0.2$ with an increment of 0.0025, that is, 81 calls to subroutine PRINT. Eighty one ODEs are integrated with an error tolerance of 0.0001.

The output from subroutine PRINT is abbreviated below:

```
RUN NO. -    1  One-dimensional, convective-diffusion

INITIAL T -    .000D+00

  FINAL T -    .200D+00

  PRINT T -    .250D-02

NUMBER OF DIFFERENTIAL EQUATIONS -   81

MAXIMUM INTEGRATION ERROR -     .100D-03

t =    .00
     .00000
     .00000     .00000     .00000     .00000     .00000
     .00000     .00000     .00000     .00000     .00000
     .00000     .00000     .00000     .00000     .00000
     .00000     .00000     .00000     .00000     .00000
                  .              .
                  .              .
                  .              .
     .00000     .00000     .00000     .00000     .00000
     .00000     .00000     .00000     .00000     .00000
     .00000     .00000     .00000     .00000     .00000
     .00000     .00000     .00000     .00000     .00000
t =    .10
     .99437
     .99363     .99284     .99198     .99106     .99006
     .98899     .98784     .98660     .98528     .98387
     .98236     .98075     .97904     .97723     .97529
     .97325     .97108     .96878     .96635     .96379
                  .              .
                  .              .
                  .              .
     .71026     .70099     .69173     .68251     .67334
     .66428     .65536     .64662     .63811     .62988
     .62200     .61452     .60753     .60110     .59533
     .59032     .58618     .58304     .58104     .58033
t =    .20
     .99978
     .99976     .99973     .99969     .99966     .99962
     .99957     .99953     .99948     .99943     .99937
     .99931     .99924     .99917     .99909     .99901
```

Mass transfer with simultaneous convection and diffusion

.99892	.99883	.99873	.99862	.99850
.		.		
.		.		
.		.		
.98242	.98169	.98095	.98021	.97946
.97872	.97798	.97725	.97653	.97583
.97516	.97451	.97391	.97335	.97285
.97241	.97204	.97177	.97159	.97153

We can now make a comparison of the analytical and numerical solutions. At $z = 1$, we have

```
analytical (taken from file ma2a.out):
  1 (z=1,t=0.0975) = 0.5526
  1 (z=1,t=0.1000) = 0.5766
  1 (z=1,t=0.1025) = 0.5996

numerical (taken from file ma2b.out):
  1 (z=1,t=0.0975) = 0.5564
  1 (z=1,t=0.1000) = 0.5803
  1 (z=1,t=0.1025) = 0.6034
```

The discrepancy between the analytical and numerical solutions is not necessarily due to the numerical solution. Because only a finite number of terms ($= 200$) was used in the evaluation of the series of eq. (MA2.51), the analytical solution as computed is only approximately correct. Note, for example, the negative values for $\phi_1(1, t)$ in the preceding output (in file ma2a.out) for small t

```
Series solution at z = 1, phi1(1,t)

     1.000          .000        .00000
     1.000          .0025      -.00377
     1.000          .0050      -.00376
     1.000          .0075      -.00376
     1.000          .0100      -.00376
     1.000          .0125      -.00376
     1.000          .0150      -.00376
     1.000          .0175      -.00376
     1.000          .0200      -.00373
     1.000          .0225      -.00364
     1.000          .0250      -.00336
     1.000          .0275      -.00273
     1.000          .0300      -.00148
     1.000          .0325       .00068
     1.000          .0350       .00409
     1.000          .0375       .00904
```

1.000	.0400	.01582
1.000	.0425	.02464
1.000	.0450	.03564
1.000	.0475	.04888
1.000	.0500	.06436

On the other hand, we did not establish convergence of the numerical solution with respect to z by varying, for example, the number of spatial grid points, that is, the preceding numerical solution is only for 81 grid points in z. The sources of error in the analytical and numerical solutions could therefore be investigated further, but we end the discussion at this point by referring to Figure MA2.2, which generally shows good agreement between the two solutions. Figure MA2.2 was produced from file ma2b.out (the numerical solution written in subroutine PLOTM) and file ma2a.out (the analytical solution written in subroutine SERIES).

In summary, we have worked through the details of the analytical and numerical solutions for a one-dimensional, hyperbolic-parabolic PDE, eq. (MA2.47). The level of effort clearly indicates that the numerical solution was easier to obtain. Also, the numerical solution could easily be extended to include other terms in the PDE, which could be linear or nonlinear; specifically, only the one line in subroutine

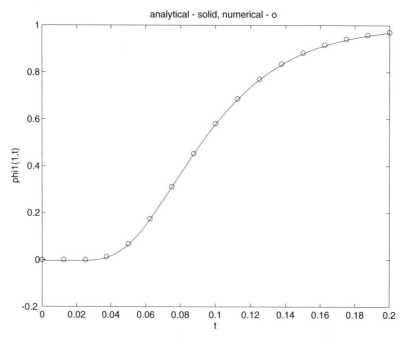

Figure MA2.2. A Comparison of the Analytical and Numerical Solutions to Eqs. (MA 2.47) to (MA 2.50)

DERV

```
CT(I)=CZZ(I)-PE*CZ(I)
```

would require modification to include other terms in the PDE. On the other hand, the derivation of the analytical solution for a variant of eq. (MA2.51) would be difficult (for a linear modification) to essentially impossible (for a nonlinear modification). Thus, we conclude that the numerical approach is the better approach for computing a solution to a spectrum of PDE problems with reasonable effort.

References

Kim, D. H., G. Y. Ma, and K. S. Chang. 1984. The Unsteady-State Solution by an Operator Theory for a Dispersion-Type Tubular Reactor with an Immobile Zone, *Chem. Eng. Commun.* 28:283–95.

Morse, P. M., and H. Feshbach. 1953. *Methods of Theoretical Physic.* New York: McGraw-Hill.

Schiesser, W. E. 1991. *The Numerical Method of Lines Integration of Partial Differential Equations.* San Diego: Academic Press.

MA3

Transient multicomponent diffusion

Consider the system consisting of two bulbs connected by a tube depicted in Figure MA3.1. The gas in the system consists of three components: H_2 ($i = 1$), N_2 ($i = 2$), and CO_2 ($i = 3$).

Develop a mathematical model that can be used to compute the concentrations of the three components in the bulbs as a function of time. Program the model and compute the time-dependent component concentrations and component fluxes.

Solution

The system depicted in Figure MA3.1 is modeled with the following equations, starting with the component balances for the left bulb:

$$\frac{dx_{io}}{dt} = -\frac{n_i A}{C_t V_o}, \quad i = 1, 2 \tag{MA3.1}$$

where

- x_{io} mole fraction of component i in the left bulb
- t time (s)
- n_i flux of component i in the tube (gm moles i/s-cm^2)
- A cross sectional area of the tube (cm^2)
- C_t total molar concentration of gas (gm moles gas/cm^3)
- V_o volume of the left bulb (cm^3)

The index i indicates that eq. (MA3.1) is written for two of the three components; that is, $i = 1$ is for H_2, and $i = 2$ is for N_2. The mole fraction of the third component, $i = 3$ for CO_2, can then be obtained "by difference," that is, the sum of the mole fractions must be unity.

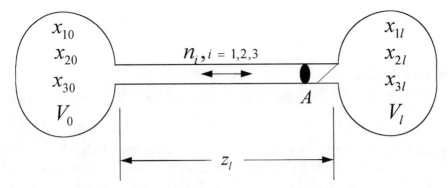

Figure MA3.1. Transient Multicomponent Diffusion

A similar set of component balances applies to the right bulb (with subscript o replaced by l) gives

$$\frac{dx_{il}}{dt} = -\frac{n_i A}{C_t V_l}, \quad i = 1, 2 \tag{MA3.2}$$

Then, the concentration and flux of the third component are available from the equations

$$x_{1o} + x_{2o} + x_{3o} = 1 \tag{MA3.3}$$

$$x_{1l} + x_{2l} + x_{3l} = 1 \tag{MA3.4}$$

$$n_1 + n_2 + n_3 = 0 \tag{MA3.5}$$

The fluxes n_1 and n_2 can be calculated from the Stefan–Maxwell equation (Bird, Stewart and Lightfoot, 1960; Taylor and Krishna 1993)

$$\nabla x_i = \sum_{j=1}^{3} \frac{x_i n_j - x_j n_i}{C_t D_{ij}}, \quad i = 1, 2 \tag{MA3.6}$$

We also use the following assumptions:

(1) The temperature and pressure in the system are uniform. Also, the ideal gas law can be used to calculate the total molar concentration, C_t.

(2) The diffusion in the tube is essentially at steady state, that is, the concentration profiles in the tube adjust instantaneously to changes in the concentrations in the bulbs according to the so-called *quasi-steady state* assumption.

(3) The concentration profiles in the tube are linear, which is based on the *quasi-steady state* assumption.

(4) The fluxes of the three components in the tubes are evaluated at the midpoint.

(5) The concentrations in the bulbs are uniform (i.e., dependent on time only, not position).

The parameters are $V_o = V_l = 80$ cm^3, $z_l = 10$ cm, $A = 0.04$ cm^2, $D_{12} = 0.833$ cm^2/s, $D_{13} = 0.68$ cm^2/s, $D_{23} = 0.168$ cm^2/s (these are binary diffusivities), $T = 35.2°$C, $P = 101.3$ kPa.

The initial conditions for eqs. (MA3.1) and (MA3.2) are

$$x_{1o}(0) = 0, \qquad x_{2o}(0) = 0.5 \qquad \text{(MA3.7)}$$

$$x_{1l}(0) = 0.5, \qquad x_{2l}(0) = 0.5 \qquad \text{(MA3.8)}$$

Because of the relatively slow diffusion, the analysis is in hours rather than seconds.

We now consider a Fortran program for the numerical integration of eqs. (MA3.1) and (MA3.2). Initial conditions (MA3.7) and (MA3.8) are first set in subroutine INITAL

```
      SUBROUTINE INITAL
      IMPLICIT DOUBLE PRECISION (A-H,O-Z)
      COMMON/T/    T,    NSTOP,   NORUN
     +      /Y/    X10,  X20,     X1L,     X2L
     +      /F/    X10T, X20T,    X1LT,    X2LT
     +      /C/    A,    ZL,      VO,      VL,
     +             TC,   P,       R,       CT,
     +      D12,   D13,  D23,     X1,      X2,
     +             X30,  X3L,     DELX1,   DELX2,
     +             N1,   N2,      N3
     +      /I/    IP
      DOUBLE PRECISION    N1,      N2,      N3
C...
C...  MODEL PARAMETERS
C...
C...     TUBE LENGTH (CM)
         ZL=10.0D0
C...
C...     TUBE CROSS SECTIONAL AREA (CM**2)
         A=0.04D0
C...
C...     TERMINAL VOLUMES (CM**3)
         VO=80.0D0
         VL=80.0D0
C...
C...     TEMPERATURE (C)
         TC=35.2D0
C...
C...     PRESSURE (PA)
         P=101300.0D0
C...
C...     GAS CONSTANT (PA-CM**3/K-GM MOL)
         R=101300.0D0*22400.0D0/273.16D0
```

```
C...
C...      MOLAR DENSITY (GM MOL/CM**3)
          CT=P/(R*(TC+273.16D0))
C...
C...      BINARY DIFFUSIVITIES (CM**2/HR)
          D12=8.33D-01*3600.0D0
          D13= 6.8D-01*3600.0D0
          D23=1.68D-01*3600.0D0
C...
C...      INITIAL CONDITIONS
          X1O=0.0D0
          X2O=0.5D0
          X1L=0.5D0
          X2L=0.5D0
C...
C...      INITIAL DERIVATIVES
          CALL DERV
          IP=0
          RETURN
          END
```

We can note the following points about INITAL:

(1) The coding is in double precision format

```
          IMPLICIT DOUBLE PRECISION (A-H,O-Z)
          COMMON/T/     T,   NSTOP,   NORUN
         +      /Y/    X1O,    X2O,     X1L,    X2L
         +      /F/   X1OT,   X2OT,    X1LT,   X2LT
         +      /C/     A,     ZL,      VO,     VL,
         +              TC,     P,       R,     CT,
         +      D12,   D13,    D23,     X1,     X2,
         +             X3O,    X3L,   DELX1,  DELX2,
         +             N1,     N2,      N3
         +      /I/    IP
```

The dependent variables of eq. (MA3.1), x_{1o} and x_{2o}, and of eq. (MA3.2), x_{1l} and x_{2l}, are in COMMON/Y/ (X1O, X2O, X1L, and X2L, respectively). The temporal derivatives of these four dependent variables, defined by the right hand side (RHS) of eqs. (MA3.1) and (MA3.2), are in COMMON/F/ (X1OT, X2OT, X1LT, and X2LT, respectively). The model parameters and the algebraic variables are in COMMON/C/.

(2) The given model parameters are then set numerically

```
C...
C...      MODEL PARAMETERS
C...
```

```
C...          TUBE LENGTH (CM)
              ZL=10.0D0
C...
C...          TUBE CROSS SECTIONAL AREA (CM**2)
              A=0.04D0
C...
C...          TERMINAL VOLUMES (CM**3)
              V0=80.0D0
              VL=80.0D0
C...
C...          TEMPERATURE (C)
              TC=35.2D0
C...
C...          PRESSURE (PA)
              P=101300.0D0
```

The pressure of 101300 Pa (Pascals or Newtons/m^2) corresponds to one atmosphere.

(3) The ideal gas constant is computed next from the conditions at standard temperature and pressure, and this gas constant is then used to calculate the molar density of the gas at the prescribed temperature and pressure

```
C...
C...          GAS CONSTANT (PA-CM**3/K-GM MOL)
              R=101300.0D0*22400.0D0/273.16D0
C...
C...          MOLAR DENSITY (GM MOL/CM**3)
              CT=P/(R*(TC+273.16D0))
```

(4) The binary gas diffusivities are defined and converted from seconds to hours

```
C...
C...          BINARY DIFFUSIVITIES (CM**2/HR)
              D12=8.33D-01*3600.0D0
              D13= 6.8D-01*3600.0D0
              D23=1.68D-01*3600.0D0
```

(5) Initial conditions (MA3.7) and (MA3.8) are then set

```
C...
C...   INITIAL CONDITIONS
       X10=0.0D0
       X20=0.5D0
       X1L=0.5D0
       X2L=0.5D0
```

(6) Finally, subroutine DERV is called to calculate the initial derivatives (in COMMON/F/), and an integer counter is initialized for the plotted solution

```
C...
C...   INITIAL DERIVATIVES
       CALL DERV
       IP=0
```

Subroutine DERV is listed below:

```
       SUBROUTINE DERV
       IMPLICIT DOUBLE PRECISION (A-H,O-Z)
       COMMON/T/      T,     NSTOP,   NORUN
      +      /Y/      X10,   X20,     X1L,     X2L
      +      /F/      X10T,  X20T,    X1LT,    X2LT
      +      /C/      A,     ZL,      V0,      VL,
      +               TC,    P,       R,       CT,
      +      D12,     D13,   D23,     X1,      X2,
      +               X30,   X3L,     DELX1,   DELX2,
      +               N1,    N2,      N3
      +      /I/      IP
       DOUBLE PRECISION     N1,    N2,     N3
C...
C...   GRADIENTS
       DELX1=(X1L-X10)/ZL
       DELX2=(X2L-X20)/ZL
C...
C...   MIDPOINT CONCENTRATIONS
       X1=(X1L+X10)/2.0D0
       X2=(X2L+X20)/2.0D0
C...
C...   DEFINE LINEAR ALGEBRAIC EQUATIONS FOR N1 AND N2
       X3=1.0D0-X1-X2
       A11=1.0D0/(CT*D12)*(-X2)+(1.0D0/(CT*D13))*(-X1-X3)
       A12=1.0D0/(CT*D12)*( X1)+(1.0D0/(CT*D13))*(-X1)
        B1=DELX1
       A21=1.0D0/(CT*D12)*( X2)+(1.0D0/(CT*D23))*(-X2)
       A22=1.0D0/(CT*D12)*(-X1)+(1.0D0/(CT*D23))*(-X2-X3)
        B2=DELX2
C...
C...   SOLUTION OF LINEAR ALGEBRAIC EQUATIONS BY KRAMER'S
C...   RULE
C...
C...      DISCRIMINANT
          DIS=A11*A22-A12*A21
C...
C...      CHECK FOR NEAR-SINGULAR SYSTEM
          IF(DABS(DIS).LT.1.0D-20)THEN
             WRITE(*,1)DIS,T
1            FORMAT(' Discriminant = ',D12.3,' at t = ',F6.2)
```

```
              STOP
          END IF
C...
C...      FLUXES
          N1=(A22*B1-A12*B2)/DIS
          N2=(A11*B2-A21*B1)/DIS
C...
C...      ODES
          X10T=-N1*A/(CT*V0)
          X20T=-N2*A/(CT*V0)
          X1LT= N1*A/(CT*VL)
          X2LT= N2*A/(CT*VL)
          RETURN
          END
```

We can note the following points about DERV:

(1) The COMMON area is the same as in INITAL.

(2) The gradients of components 1 and 2, ∇x_1 and ∇x_2, are first calculated by simple difference approximations, which are exact only if the gradients are linear (in z)

```
C...
C...      GRADIENTS
          DELX1=(X1L-X10)/ZL
          DELX2=(X2L-X20)/ZL
```

(3) Midpoint concentrations (at $z = z_l/2$) are computed next as the average of the terminal concentrations

```
C...
C...      MIDPOINT CONCENTRATIONS
          X1=(X1L+X10)/2.0D0
          X2=(X2L+X20)/2.0D0
```

(4) The coefficients in a 2×2 system of linear algebraic equations obtained from the Stefan–Maxwell equation, eq. (MA3.6), are then defined

```
C...
C...   DEFINE LINEAR ALGEBRAIC EQUATIONS FOR N1 AND N2
       X3=1.0D0-X1-X2
       A11=1.0D0/(CT*D12)*(-X2)+(1.0D0/(CT*D13))*(-X1-X3)
       A12=1.0D0/(CT*D12)*( X1)+(1.0D0/(CT*D13))*(-X1)
        B1=DELX1
       A21=1.0D0/(CT*D12)*( X2)+(1.0D0/(CT*D23))*(-X2)
       A22=1.0D0/(CT*D12)*(-X1)+(1.0D0/(CT*D23))*(-X2-X3)
        B2=DELX2
```

These coefficients follow from writing eq. (MA3.6) for $i = 1, 2$

$$\nabla x_1 = \frac{x_1 n_1 - x_1 n_1}{C_t D_{11}} + \frac{x_1 n_2 - x_2 n_1}{C_t D_{12}} + \frac{x_1 n_3 - x_3 n_1}{C_t D_{13}} \quad (MA3.9)$$

$$\nabla x_2 = \frac{x_2 n_1 - x_1 n_2}{C_t D_{21}} + \frac{x_2 n_2 - x_2 n_2}{C_t D_{22}} + \frac{x_2 n_3 - x_3 n_2}{C_t D_{23}} \quad (MA3.10)$$

Also,

$$n_3 = -n_1 - n_2 \quad (MA3.11)$$

Equations (MA3.9) to (MA3.11) are a 2×2 linear system in the two fluxes, n_1 and n_2

$$\frac{x_{1l} - x_{10}}{\Delta z} = \frac{x_1 n_2 - x_2 n_1}{C_t D_{12}} + \frac{x_1 n_3 - x_3 n_1}{C_t D_{13}}$$

$$\frac{x_{2l} - x_{20}}{\Delta z} = \frac{x_2 n_1 - x_1 n_2}{C_t D_{21}} + \frac{x_2 n_3 - x_3 n_2}{C_t D_{23}}$$

or

$$\left\{\frac{-x_2}{C_t D_{12}} + \frac{-x_3}{C_t D_{13}} + \frac{-x_1}{C_t D_{13}}\right\} n_1 + \left\{\frac{x_1}{C_t D_{12}} + \frac{-x_1}{C_t D_{13}}\right\} n_2 = \frac{x_{1l} - x_{10}}{\Delta z}$$

$$\left\{\frac{x_2}{C_t D_{21}} + \frac{-x_2}{C_t D_{23}}\right\} n_1 + \left\{\frac{-x_1}{C_t D_{21}} + \frac{-x_3}{C_t D_{23}} + \frac{-x_2}{C_t D_{23}}\right\} n_2 = \frac{x_{2l} - x_{20}}{\Delta z}$$

or

$$a_{11} n_1 + a_{12} n_2 = b_1 \quad (MA3.12)$$

$$a_{21} n_1 + a_{22} n_2 = b_2 \quad (MA3.13)$$

where

$$a_{11} \quad \frac{-x_2}{C_t D_{12}} + \frac{-x_3}{C_t D_{13}} + \frac{-x_1}{C_t D_{13}}$$

$$a_{12} \quad \frac{x_1}{C_t D_{12}} + \frac{-x_1}{C_t D_{13}}$$

$$b_1 \quad \frac{x_{1l} - x_{10}}{\Delta z}$$

$$a_{21} \quad \frac{x_2}{C_t D_{21}} + \frac{-x_2}{C_t D_{23}}$$

$$a_{22} \quad \frac{-x_1}{C_t D_{21}} + \frac{-x_3}{C_t D_{23}} + \frac{-x_2}{C_t D_{23}}$$

$$b_2 \quad \frac{x_{2l} - x_{20}}{\Delta z}$$

Also, $D_{ji} = D_{ij}$.

(5) The linear system of eqs. (MA3.12) and (MA3.13) is then solved for n_1 and n_2 by Kramer's rule. First the discriminant is calculated and checked for a near-singular system

```
C...
C...      SOLUTION OF LINEAR ALGEBRAIC EQUATIONS BY KRAMER'S
C...      RULE
C...
C...         DISCRIMINANT
             DIS=A11*A22-A12*A21
C...
C...         CHECK FOR NEAR-SINGULAR SYSTEM
             IF(DABS(DIS).LT.1.0D-20)THEN
                WRITE(*,1)DIS,T
1               FORMAT(' Discriminant = ',D12.3,' at t = ',F6.2)
                STOP
             END IF
```

The algebraic system is then solved (using determinants)

```
C...
C...         FLUXES
             N1=(A22*B1-A12*B2)/DIS
             N2=(A11*B2-A21*B1)/DIS
```

(6) Finally, the temporal derivatives of eqs. (MA3.1) and (MA3.2) are computed

```
C...
C...         ODES
             X10T=-N1*A/(CT*VO)
             X20T=-N2*A/(CT*VO)
             X1LT= N1*A/(CT*VL)
             X2LT= N2*A/(CT*VL)
```

This set of four derivatives is passed to the ODE integrator, RKF45, through COMMON/F/, and the dependent variables, x_{1o}, x_{2o}, x_{1l}, and x_{2l}, are returned by the ODE integrator through COMMON/Y/ for use in the calculation of the derivatives.

Subroutine PRINT to display the solution is listed below:

```
      SUBROUTINE PRINT(NI,NO)
      IMPLICIT DOUBLE PRECISION (A-H,O-Z)
      PARAMETER(NT=101)
      COMMON/T/    T,    NSTOP,   NORUN
     +      /Y/   X10,    X20,     X1L,     X2L
     +      /F/   X10T,   X20T,    X1LT,    X2LT
     +      /C/    A,     ZL,      VO,      VL,
     +             TC,    P,       R,       CT,
```

```
     +          D12,       D13,       D23,        X1,         X2,
     +                     X30,       X3L,      DELX1,      DELX2,
     +                     N1,        N2,         N3
     +          /I/        IP
            DOUBLE PRECISION          N1,         N2,         N3
C...
C...    MONITOR THE OUTPUT
        IP=IP+1
        WRITE(*,*)IP,T
C...
C...    HEADING FOR THE SOLUTION
        IF(IP.EQ.1)WRITE(NO,1)
1       FORMAT(9X,'t',7X,'X10',7X,'X20',7X,'X30',/,
     +                  17X,'X1L',7X,'X2L',7X,'X3L',/,
     +                  17X,' N1',7X,' N2',7X,' N3',/,
     + 3X,'dX10/dt',3X,'dX20/dt',3X,'dX1L/dt',3X,'dX2L/dt')
C...
C...    WRITE THE SOLUTION EVERY 10TH CALL TO PRINT
        X30=1.0D0-X10-X20
        X3L=1.0D0-X1L-X2L
        IF(((IP-1)/10*10).EQ.(IP-1))THEN
            WRITE(NO,2)T,X10,X20,X30,X1L,X2L,X3L,
     +                  N1,N2,N3,X10T,X20T,X1LT,X2LT
2           FORMAT(F10.2,3F10.5,/,
     +                  10X,3F10.5,/,
     +                  10X,3D10.2,/,
     +                      4D10.2,//)
        END IF
C...
C...    PLOT THE SOLUTION WITH MATLAB
        CALL MPLOT
        RETURN
        END
```

We can note the following points about PRINT:

(1) The COMMON area is the same as in INITAL and DERV.

(2) The integer counter IP and time are displayed to monitor the progress of the solution

```
C...
C...    MONITOR THE OUTPUT
        IP=IP+1
        WRITE(*,*)IP,T
```

(3) A heading is printed next at the beginnng of the solution (IP = 1)

```
C...
C...   HEADING FOR THE SOLUTION
       IF(IP.EQ.1)WRITE(NO,1)
1      FORMAT(9X,'t',7X,'X10',7X,'X20',7X,'X30',/,
      +             17X,'X1L',7X,'X2L',7X,'X3L',/,
      +             17X,' N1',7X,' N2',7X,' N3',/,
      + 3X,'dX10/dt',3X,'dX20/dt',3X,'dX1L/dt',3X,'dX2L/dt')
```

(4) The solution, consisting of time, the six terminal concentrations, the three fluxes, and the four temporal derivatives computed in DERV, is then printed

```
C...
C...   WRITE THE SOLUTION EVERY 10TH CALL TO PRINT
       X30=1.0D0-X10-X20
       X3L=1.0D0-X1L-X2L
       IF(((IP-1)/10*10).EQ.(IP-1))THEN
           WRITE(NO,2)T,X10,X20,X30,X1L,X2L,X3L,
      +               N1,N2,N3,X10T,X20T,X1LT,X2LT
2          FORMAT(F10.2,3F10.5,/,
      +             10X,3F10.5,/,
      +             10X,3D10.2,/,
      +                 4D10.2,//)
       END IF
```

(5) Finally, subroutine MPLOT is called to write the solution to a file that can be called by Matlab to plot the solution

```
C...
C...   PLOT THE SOLUTION WITH MATLAB
       CALL MPLOT
```

Subroutine MPLOT is listed below:

```
       SUBROUTINE MPLOT
       IMPLICIT DOUBLE PRECISION (A-H,O-Z)
       PARAMETER(NT=101)
       COMMON/T/      T,    NSTOP,   NORUN
      +       /Y/    X10,    X20,     X1L,      X2L
      +       /F/   X10T,   X20T,    X1LT,     X2LT
      +       /C/      A,     ZL,      VO,       VL,
      +              TC,      P,       R,       CT,
      +      D12,   D13,    D23,      X1,       X2,
      +             X30,    X3L,   DELX1,    DELX2,
      +              N1,     N2,      N3
      +       /I/     IP
       DOUBLE PRECISION    N1,     N2,       N3
C...
```

```
C...    OPEN FILES FOR MATLAB PLOTTING
        IF(IP.EQ.1)THEN
            OPEN(1,FILE='ma3c.out')
            OPEN(2,FILE='ma3f.out')
        END IF
C...
C...    WRITE THE SOLUTION FOR PLOTTING
C...
C...        CONCENTRATIONS
            X30=1.0D0-X10-X20
            X3L=1.0D0-X1L-X2L
            WRITE(1,1)T, X10, X20, X30, X1L, X2L, X3L
1           FORMAT(7F9.5)
C...
C...        FLUXES
            N3=-N1-N2
            WRITE(2,2)T, N1, N2, N3
2           FORMAT(F8.4,3F13.7)
        RETURN
        END
```

We can note the following points about MPLOT:

(1) The COMMON area is the same as in INITAL, DERV, and PRINT.

(2) Files ma3c.out and ma3f.out are opened at the beginning of the solution (IP = 1) to store the numerical solution

```
C...
C...    OPEN FILES FOR MATLAB PLOTTING
        IF(IP.EQ.1)THEN
            OPEN(1,FILE='ma3c.out')
            OPEN(2,FILE='ma3f.out')
        END IF
```

(3) The concentrations are then written to file ma3c.out

```
C...
C...    WRITE THE SOLUTION FOR PLOTTING
C...
C...        CONCENTRATIONS
            X30=1.0D0-X10-X20
            X3L=1.0D0-X1L-X2L
            WRITE(1,1)T, X10, X20, X30, X1L, X2L, X3L
1           FORMAT(7F9.5)
```

(4) The fluxes are written next to file ma3f.out

```
C...
C...       FLUXES
           N3=-N1-N2
           WRITE(2,2)T, N1, N2, N3
2          FORMAT(F8.4,3F13.7)
```

The main program, RKF45M, that calls RKF45, INITAL, and PRINT is listed in Appendix A; RKF45, in turn, calls DERV. RKF45M also reads the following data file to control the numerical integration

```
Three component diffusion
0.           50.0         0.5
     4                    0.0001
END OF RUNS
```

The data lines specify:

(1) The initial, final, and output intervals of time. Thus, time ranges over $0 \leq t \leq 50$, and PRINT is called at intervals of 0.5 so there are 101 calls to PRINT (counting $t = 0$).

(2) Four ordinary differential equations (ODEs) are integrated with an accuracy of 0.0001.

(3) One run is programmed (three lines of data), so execution is terminated after the first run with the characters END OF RUNS (read by main program RKF45M to execute a STOP statement).

The tabular output from PRINT (written to file "output" opened in the main program RKF45M) is summarized below:

```
RUN NO. -   1  Three component diffusion

INITIAL T -  0.000D+00

  FINAL T -  0.500D+02

  PRINT T -  0.500D+00

NUMBER OF DIFFERENTIAL EQUATIONS -    4

MAXIMUM INTEGRATION ERROR -  0.100D-03

              t         X10          X20          X30
                        X1L          X2L          X3L
                        N1           N2           N3
         dX10/dt     dX20/dt      dX1L/dt      dX2L/dt
            0.00     0.00000      0.50000      0.50000
                     0.50000      0.50000      0.00000
```

	-0.55D-02	0.27D-02	0.00D+00
0.69D-01	-0.34D-01	-0.69D-01	0.34D-01
5.00	0.18666	0.42573	0.38761
	0.31334	0.57427	0.11239
	-0.14D-02	0.24D-03	0.12D-02
0.17D-01	-0.30D-02	-0.17D-01	0.30D-02
10.00	0.23333	0.43053	0.33615
	0.26667	0.56947	0.16385
	-0.34D-03	-0.24D-03	0.62D-03
0.43D-02	0.31D-02	-0.43D-02	-0.31D-02
15.00	0.24519	0.44766	0.30715
	0.25481	0.55234	0.19285
	-0.89D-04	-0.27D-03	0.37D-03
0.11D-02	0.34D-02	-0.11D-02	-0.34D-02
20.00	0.24834	0.46287	0.28879
	0.25166	0.53713	0.21121
	-0.25D-04	-0.21D-03	0.24D-03
0.31D-03	0.26D-02	-0.31D-03	-0.26D-02
	.	.	
	.	.	
	.	.	
50.00	0.24992	0.49600	0.25408
	0.25008	0.50400	0.24592
	-0.46D-06	-0.24D-04	0.25D-04
0.59D-05	0.30D-03	-0.59D-05	-0.30D-03

We can note the following points about this output:

(1) The preceding data (in the data file) are confirmed, for example, $0 \leq t \leq 50$, 4 ODEs integrated with an accuracy of 0.0001.

(2) The initial conditions at $t = 0$ are in agreement with eqs. (MA3.7) and (MA3.8). Of course, the initial value for component 3 is obtained by difference in PRINT.

t	X10	X20	X30
	X1L	X2L	X3L
	N1	N2	N3
dX10/dt	dX20/dt	dX1L/dt	dX2L/dt
0.00	0.00000	0.50000	0.50000
	0.50000	0.50000	0.00000
	-0.55D-02	0.27D-02	0.00D+00
0.69D-01	-0.34D-01	-0.69D-01	0.34D-01

Note that we have an initial driving force for diffusion (i.e., differences in the terminal concentrations) for components 1 and 3, but not for component 2. The interaction of the diffusion of the three components, including component 2, is an interesting aspect of this study.

(3) The transient behavior of the three components can be observed in the output over the interval $0 \leq t \leq 50$. In particular, note that the concentration of component 2 goes through a minimum at the left end ($z = 0$) and through a maximum at the right end ($z = z_l$) at approximately $t = 5$

t	X10 X1L	X20 X2L	X30 X3L
0.00	0.00000 0.50000	0.50000 0.50000	0.50000 0.00000
5.00	0.18666 0.31334	0.42573 0.57427	0.38761 0.11239
10.00	0.23333 0.26667	0.43053 0.56947	0.33615 0.16385

(4) Also, the flux of component 2 undergoes a sign change (change in direction) at approximately $t = 5$

t	N1	N2	N3
0.00	-0.55D-02	0.27D-02	0.00D+00
5.00	-0.14D-02	0.24D-03	0.12D-02
10.00	-0.34D-03	-0.24D-03	0.62D-03

Initially, component 2 is diffusing left to right, that is, $n_2 > 0$; at $t = 10$, the diffusive flux has switched to right to left.

(5) The system approaches a steady state for $t \to \infty$. At $t = 50$, the changes in the terminal concentrations are small. This is reflected in small fluxes, and from eqs. (MA3.1) and (MA3.2), the temporal derivatives (in COMMON/F/) are also small

t	X10 X1L N1	X20 X2L N2	X30 X3L N3
	dX10/dt	dX20/dt	dX1L/dt	dX2L/dt
50.00	0.24992 0.25008	0.49600 0.50400	0.25408 0.24592
	-0.46D-06	-0.24D-04	0.25D-04
0.59D-05	0.30D-03	-0.59D-05	-0.30D-03

This point can be appreciated by comparing the magnitudes of the fluxes and temporal derivatives at $t = 50$ with those at the beginning of the solution (and shortly after $t = 0$). This is a typical characteristic of a stable transient or dynamic system, that is, the rates of change (and thus the temporal derivatives) tend to be largest at the beginning of the solution. Of course, there are possible exceptions to this generalization. For example:

(5.1) If the initial conditions were selected as

$$x_{1o}(0) = 0.25, \quad x_{2o}(0) = 0.5 \quad \text{(MA3.14)}$$

$$x_{1l}(0) = 0.25, \quad x_{2l}(0) = 0.5 \quad \text{(MA3.15)}$$

the four temporal derivatives given by eqs. (MA3.1) and (MA3.2) would be zero initially and, in fact, would remain at zero; thus, the fluxes would also remain at zero and the concentrations would stay at their initial values for $t > 0$, that is, the system would stay at a steady state condition corresponding to the initial conditions of eqs. (MA3.14) and (MA3.15).

We might ask about the origin of initial conditions (MA3.14) and (MA3.15); in fact, they follow immediately from the preceding solution at $t = 50$, for which the system has closely approached a steady state. This suggests two ideas: (a) a transient mathematical model can be used to conveniently compute a steady state solution (formulate a dynamic model, then let it run to essentially steady state), and (b) initial conditions (MA3.14) and (MA3.15) could be used in place of initial conditions (MA3.7) and (MA3.8) in INITAL; we would expect that all of the temporal derivatives in COMMON/F/ would remain at zero, and therefore the component concentrations would not change with time.

(5.2) If the ODE system were unstable (this would not be possible with just the diffusion of this model, but if an unstable chemical reaction were also included in the analysis, the system might be unstable), then the initial temporal derivatives might not have the largest values. For example, an unstable system could have unbounded growth, with increasing temporal derivatives, as in the case of an explosive mixture of gases. However, for the present system, the instance of the greatest rate of change is near $t = 0$.

These properties of the transient diffusion can be seen also in a plot of the solution (produced by the output from subroutine MPLOT sent to Matlab) in Figures MA3.2 to MA3.4. Note in particular the maximum and minimum concentrations of component 2 (Figures MA3.2 and MA3.3) and the change in the sign of the flux of component 2 (Figure MA3.4).

In summary, we have observed some interesting, and possibly unexpected, behavior of this multicomponent diffusion system. The analysis was based on approximate computation of the concentration gradients, ∇x_1 and ∇x_2, (according to the calculations of DELX1 and DELX2 at the beginning of DERV) and use of midpoint

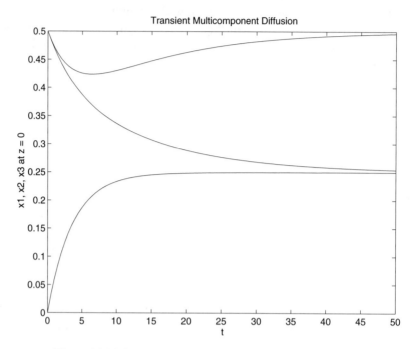

Figure MA3.2. Terminal Concentrations at $z = 0$ versus t

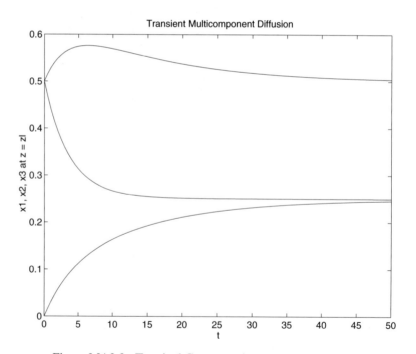

Figure MA3.3. Terminal Concentrations at $z = z_l$ versus t

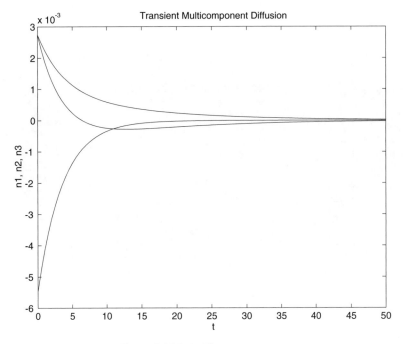

Figure MA3.4. Fluxes versus t

concentrations (according to the calculation of X1 and X2 at the beginning of DERV).

A more rigorous approach would be to represent the connecting tube in Figure MA3.1 as a system distributed over space (z), then calculate the concentration gradients, ∇x_1 and ∇x_2, as a function of z using, for example, finite difference approximations. Thus, these concentration gradients would be a function of z. The corresponding fluxes (from the Stefan–Boltzman equation (MA3.6)) would be used in an unsteady state material balance for the tube. The fluxes $n_1(0, t)$ and $n_2(0, t)$ would also be used in eq. (MA3.1) while the fluxes $n_1(z_l, t)$ and $n_2(z_l, t)$, would be used in eq. (MA3.2). An interesting excerise, then, would be to do this more rigorous calculation and compare the solution with the previously computed solution.

References

Bird, R. B., W. E. Stewart, and E. N. Lightfoot. 1960. *Transport Phenomena*. New York: John Wiley and Sons.

Taylor, R., and R. Krishna. 1993. *Multicomponent Mass Transfer*. New York: John Wiley and Sons.

Appendix A

Main program RKF45M to call ODE integrator RKF45

```
      PROGRAM RKF45M
C...
C...  PROGRAM RKF45M CALLS: (1) SUBROUTINE INITAL TO DEFINE
C...  THE ODE INITAL CONDITIONS, (2) SUBROUTINE RKF45 TO
C...  INTEGRATE THE ODES, AND (3) SUBROUTINE PRINT TO PRINT
C...  THE SOLUTION.
C...
C...  THE FOLLOWING CODING IS FOR 1000 ODES. IF MORE ODES
C...  ARE TO BE INTEGRATED, ALL OF THE 1000'S SHOULD BE
C...  CHANGED TO THE REQUIRED NUMBER (THIS CAN EASILY BE
C...  DONE BY USING AN EDITOR TO SEARCH FOR AND CHANGE 1000
C...  TO THE MAXIMUM NUMBER OF ODES).
C...
C...  THE SIZING OF THE WORK ARRAYS, RWORK AND IWORK, IS IN
C...  ACCORDANCE WITH THE FOLLOWING COMMENTS TAKEN FROM
C...  RKF45
C...
C     WORK(*) -- ARRAY TO HOLD INFORMATION INTERNAL TO
C             RKF45 WHICH IS NECESSARY FOR SUBSEQUENT CALLS.
C             MUST BE DIMENSIONED AT LEAST   3+6*NEQN
C     IWORK(*) -- INTEGER ARRAY USED TO HOLD INFORMATION
C             INTERNAL TO RKF45 WHICH IS NECESSARY FOR
C             SUBSEQUENT CALLS. MUST BE DIMENSIONED AT LEAST 5
C...
      PARAMETER (NODE=1000,IRWA=3+6*NODE,IIWA=5)
C...
C...  DOUBLE PRECISION CODING IS USED
      IMPLICIT DOUBLE PRECISION (A-H,O-Z)
      COMMON/T/          T,       NSTOP,      NORUN
     1      /Y/    Y(NODE)
     2      /F/    F(NODE)
C...
```

```
C...   THE NUMBER OF DIFFERENTIAL EQUATIONS IS IN COMMON/N/
C...   FOR USE IN SUBROUTINE FCN
       COMMON/N/     NEQN
C...
C...   COMMON AREA TO PROVIDE THE INPUT/OUTPUT UNIT NUMBERS
C...   TO OTHER SUBROUTINES
       COMMON/IO/        NI,         NO
C...
C...   ABSOLUTE DIMENSIONING OF THE ARRAYS REQUIRED BY RKF45
       DIMENSION YV(NODE), WORK(IRWA), IWORK(IIWA)
C...
C...   EXTERNAL THE DERIVATIVE ROUTINE CALLED BY RKF45
       EXTERNAL FCN
C...
C...   ARRAY FOR THE TITLE (FIRST LINE OF DATA), CHARACTERS
C...   END OF RUNS
       CHARACTER TITLE(20)*4, ENDRUN(3)*4
C...
C...   DEFINE THE CHARACTERS  END OF RUNS
       DATA ENDRUN/'END ','OF R','UNS '/
C...
C...   DEFINE THE INPUT/OUTPUT UNIT NUMBERS
       NI=5
       NO=6
C...
C...   OPEN INPUT AND OUTPUT FILES
       OPEN(NI,FILE='DATA',   STATUS='OLD')
       OPEN(NO,FILE='OUTPUT',STATUS='NEW')
C...
C...   INITIALIZE THE RUN COUNTER
       NORUN=0
C...
C...   BEGIN A RUN
1      NORUN=NORUN+1
C...
C...   INITIALIZE THE RUN TERMINATION VARIABLE
       NSTOP=0
C...
C...   READ THE FIRST LINE OF DATA
       READ(NI,1000,END=999)(TITLE(I),I=1,20)
C...
C...   TEST FOR  END OF RUNS  IN THE DATA
       DO 2 I=1,3
       IF(TITLE(I).NE.ENDRUN(I))GO TO 3
2      CONTINUE
C...
```

```
C...   AN END OF RUNS HAS BEEN READ, SO TERMINATE EXECUTION
999    STOP
C...
C...   READ THE SECOND LINE OF DATA
3      READ(NI,1001,END=999)T0,TF,TP
C...
C...   READ THE THIRD LINE OF DATA
       READ(NI,1002,END=999)NEQN,ERROR
C...
C...   PRINT A DATA SUMMARY
       WRITE(NO,1003)NORUN,(TITLE(I),I=1,20),
     1                    T0,TF,TP,
     2                    NEQN,ERROR
C...
C...   INITIALIZE TIME
       T=T0
C...
C...   SET THE INITIAL CONDITIONS
       CALL INITAL
C...
C...   SET THE INITIAL DERIVATIVES (FOR POSSIBLE PRINTING)
       CALL DERV
C...
C...   PRINT THE INITIAL CONDITIONS
       CALL PRINT(NI,NO)
C...
C...   SET THE INITIAL CONDITIONS FOR SUBROUTINE RKF45
       TV=T0
       DO 5 I=1,NEQN
       YV(I)=Y(I)
5      CONTINUE
C...
C...   SET THE PARAMETERS FOR SUBROUTINE RKF45
       RELERR=ERROR
       ABSERR=ERROR
       IFLAG=1
       TOUT=T0+TP
C...
C...   CALL SUBROUTINE RKF45 TO START THE SOLUTION FROM THE
C...   INITIAL CONDITION (IFLAG = 1) OR COMPUTE THE SOLUTION
C...   TO THE NEXT PRINT POINT (IFLAG = 2)
4      CALL RKF45(FCN,NEQN,YV,TV,TOUT,RELERR,ABSERR,IFLAG,
     +            WORK,IWORK)
C...
C...   PRINT THE SOLUTION AT THE NEXT OUTPUT POINT
       T=TV
```

```
            DO 6 I=1,NEQN
            Y(I)=YV(I)
6           CONTINUE
            CALL DERV
            CALL PRINT(NI,NO)
C...
C...    TEST FOR AN ERROR CONDITION
        IF(IFLAG.NE.2)THEN
C...
C...       PRINT A MESSAGE INDICATING AN ERROR CONDITION
           WRITE(NO,1004)IFLAG
C...
C...       GO ON TO THE NEXT RUN
           GO TO 1
        END IF
C...
C...    CHECK FOR A RUN TERMINATION
        IF(NSTOP.NE.0)GO TO 1
C...
C...    CHECK FOR THE END OF THE RUN
        TOUT=TV+TP
        IF(TV.LT.(TF-0.5D0*TP))GO TO 4
C...
C...    THE CURRENT RUN IS COMPLETE, SO GO ON TO THE NEXT RUN
        GO TO 1
C...
C...    ***********************************************************
C...
C...    FORMATS
C...
1000    FORMAT(20A4)
1001    FORMAT(3D10.0)
1002    FORMAT(I5,20X,D10.0)
1003    FORMAT(1H1,
       1 ' RUN NO. - ',I3,2X,20A4,//,
       2 ' INITIAL T - ',D10.3,//,
       3 '   FINAL T - ',D10.3,//,
       4 '   PRINT T - ',D10.3,//,
       5 ' NUMBER OF DIFFERENTIAL EQUATIONS - ',I3,//,
       6 ' MAXIMUM INTEGRATION ERROR - ',D10.3,//,
       7 1H1)
1004    FORMAT(1H ,//,' IFLAG = ',I3,//,
       1 ' INDICATING AN INTEGRATION ERROR, SO THE CURRENT
       2   RUN',/,
       3 ' IS TERMINATED.  PLEASE REFER TO THE DOCUMENTATION'
       4   FOR',/,
```

```
     5 ' SUBROUTINE',//,25X,'RKF45',//,
     6 ' FOR AN EXPLANATION OF THESE ERROR INDICATORS' )
       END

       SUBROUTINE FCN(TV,YV,YDOT)
C...
C...   SUBROUTINE FCN IS AN INTERFACE ROUTINE BETWEEN
C...   SUBROUTINES RKF45 AND DERV
C...
C...   NOTE THAT THE SIZE OF ARRAYS Y AND F IN THE FOLLOWING
C...   COMMON AREA IS ACTUALLY SET BY THE CORRESPONDING
C...   COMMON STATEMENT IN MAIN PROGRAM RKF45M
       PARAMETER (NODE=1000)
       IMPLICIT DOUBLE PRECISION (A-H,O-Z)
       COMMON/T/       T,       NSTOP,      NORUN
     1       /Y/      Y(1)
     2       /F/      F(1)
C...
C...   THE NUMBER OF DIFFERENTIAL EQUATIONS IS AVAILABLE
C...   THROUGH COMMON/N/
       COMMON/N/      NEQN
C...
C...   ABSOLUTE DIMENSION THE DEPENDENT VARIABLE, DERIVATIVE
C...   VECTORS
       DIMENSION YV(NODE), YDOT(NODE)
C...
C...   TRANSFER THE INDEPENDENT VARIABLE, DEPENDENT VARIABLE
C...   VECTOR FOR USE IN SUBROUTINE DERV
       T=TV
       DO 1 I=1,NEQN
       Y(I)=YV(I)
1      CONTINUE
C...
C...   EVALUATE THE DERIVATIVE VECTOR
       CALL DERV
C...
C...   TRANSFER THE DERIVATIVE VECTOR FOR USE BY SUBROUTINE
C...   RKF45
       DO 2 I=1,NEQN
       YDOT(I)=F(I)
2      CONTINUE
       RETURN
       END

       DOUBLE PRECISION FUNCTION DFLOAT(I)
C...
```

```
C...   FUNCTION DFLOAT CONVERTS A SINGLE PRECISION INTEGER
C...   INTO A DOUBLE PRECISION FLOATING POINT.  THIS FUNCTION
C...   IS PROVIDED IN CASE THE USER'S FORTRAN COMPILER DOES
C...   NOT INCLUDE DFLOAT.
C...
       DFLOAT=DBLE(FLOAT(I))
       RETURN
       END
```

Appendix B

Main program MEM and ODE integrator EULERM

```
      PROGRAM MEM
C...
C...  THIS MAIN PROGRAM CALLS A FIXED STEP MODIFIED EULER
C...  INTEGRATION ROUTINE, EULERM, FOR USE WHEN THE STEP
C...  SIZE OF  AN ODE INTEGRATOR WITH AUTOMATIC STEP SIZE
C...  ADJUSTMENT IS CONSTRAINED BECAUSE OF A SMALL OUTPUT
C...  INTERVAL
C...
C...  DOUBLE PRECISION CODING IS RECOMMENDED FOR SHORT WORD
C...  LENGTH
C...  COMPUTERS
      IMPLICIT DOUBLE PRECISION (A-H,O-Z)
C...
C...  SET THE MAXIMUM NUMBER OF ODES
      PARAMETER (NEQ=101)
C...
C...  ABSOLUTE DIMENSION THE ARRAYS
      COMMON/T/         T,    NSTOP,     NORUN
     +      /Y/     Y(NEQ)
     +      /F/     F(NEQ)
      DIMENSION     YV(NEQ)
C...
C...  COMMON AREA TO PROVIDE THE INPUT/OUTPUT UNIT NUMBERS
C...  TO OTHER SUBROUTINES
      COMMON/IO/       NI,        NO
C...
C...  EXTERNAL THE DERIVATIVE SUBROUTINE CALLED BY EULERM
      EXTERNAL FCN
C...
C...  ARRAY FOR THE TITLE (FIRST LINE OF DATA), CHARACTERS
C...  END OF RUNS
      CHARACTER TITLE(20)*4, ENDRUN(3)*4
```

```
C...
C...   DEFINE THE CHARACTERS   END OF RUNS
       DATA ENDRUN/'END ','OF R','UNS '/
C...
C...   DEFINE THE INPUT/OUTPUT UNIT NUMBERS
       NI=5
       NO=6
C...
C...   OPEN INPUT AND OUTPUT FILES
       OPEN(NI,FILE='DATA',   STATUS='OLD')
       OPEN(NO,FILE='OUTPUT',STATUS='NEW')
C...
C...   INITIALIZE THE RUN TERMINATION VARIABLE
       NSTOP=0
C...
C...   INITIALIZE THE RUN COUNTER
       NORUN=0
C...
C...   START THE FIRST (OR NEXT) RUN BY CALLING SUBROUTINE
C...   INITAL TO SET THE INITIAL CONDITIONS
4      NORUN=NORUN+1
C...
C...   READ THE FIRST LINE OF DATA
       READ(NI,1000,END=999)(TITLE(I),I=1,20)
C...
C...   TEST FOR  END OF RUNS  IN THE DATA
       DO 2 I=1,3
       IF(TITLE(I).NE.ENDRUN(I))GO TO 3
2      CONTINUE
C...
C...   AN END OF RUNS HAS BEEN READ, SO TERMINATE EXECUTION
999    STOP
C...
C...   READ THE SECOND LINE OF DATA
3      READ(NI,1001,END=999)T0,TF,TP
C...
C...   READ THE THIRD LINE OF DATA
       READ(NI,1002,END=999)NEQN,NSTEPS
C...
C...   PRINT A DATA SUMMARY
       WRITE(NO,1003)NORUN,(TITLE(I),I=1,20),
     1                     T0,TF,TP,
     2                     NEQN,NSTEPS
C...
C...   INTEGRATION STEP
       H=TP/DFLOAT(NSTEPS)
```

```
C...
C...      CALL INITAL TO SET THE INITIAL CONDITIONS
          T =T0
          CALL INITAL
C...
C...      SET THE INITIAL CONDITIONS FOR SUBROUTINE EULERM
          TV=T0
          DO 5 I=1,NEQN
             YV(I)=Y(I)
5         CONTINUE
          TOUT=T0
C...
C...      PRINT THE INITIAL CONDITION OR THE SOLUTION AT THE
C...      NEXT OUTPUT
C...      POINT
1         CALL PRINT(NI,NO)
C...
C...      AT THE FINAL VALUE OF T, GO ON TO THE NEXT RUN
          TOUT=TOUT+TP
          IF(TOUT.GT.(TF+0.5D0*TP))GO TO 4
C...
C...      CALL SUBROUTINE EULERM TO COMPUTE THE SOLUTION AT THE
C...      NEXT OUTPUT POINT
          CALL EULERM(FCN,H,NEQN,TV,TOUT,YV)
C...
C...      UPDATE THE DEPENDENT AND DERIVATIVE VECTORS
          T=TOUT
          DO 6 I=1,NEQN
             Y(I)=YV(I)
6         CONTINUE
          CALL DERV
C...
C...      CHECK FOR A RUN TERMINATION
          IF(NSTOP.NE.0)GO TO 4
C...
C...      PRINT THE SOLUTION AND CONTINUE THE INTEGRATION
          GO TO 1
C...
C...      **********************************************************
C...
C...      FORMATS
C...
1000      FORMAT(20A4)
1001      FORMAT(3D10.0)
1002      FORMAT(I5,20X,I5)
1003      FORMAT(1H1,
```

```
      1 ' RUN NO. - ',I3,2X,20A4,//,
      2 ' INITIAL T - ',D10.3,//,
      3 '   FINAL T - ',D10.3,//,
      4 '   PRINT T - ',D10.3,//,
      5 ' NUMBER OF DIFFERENTIAL EQUATIONS - ',I3,//,
      6 ' INTEGRATION STEPS/PRINT INTERVAL - ',I5,//,
      7 1H1)
      END


      SUBROUTINE EULERM(FCN,H,NEQN,TV,TOUT,YV)
C...
C...  SUBROUTINE EULERM PERFORMS A FIXED-STEP EXPLICIT
C...  MODIFIED EULER INTEGRATION
C...
C...  DIMENSION THE ARRAYS CONTAINING THE DEPENDENT VARIABLE
C...  VECTOR (T AND TB) AND THE VECTOR OF DERIVATIVES (TT
C...  AND TTP).  AS AN INCIDENTAL, BUT IMPORTANT, POINT, THE
C...  FOLLOWING ABSOLUTE DIMENSIONING ILLUSTRATES WHY
C...  LIBRARY ROUTINES OFTEN HAVE WORK ARRAYS IN THEIR
C...  ARGUMENTS, E.G., YB, YDOT AND YDOTP (DIMENSIONED AS
C...  250) ARE REQUIRED FOR INTERMEDIATE STORAGE, AND THEY
C...  OBVIOUSLY LIMIT THE SIZE OF THE ODE PROBLEM THAT CAN
C...  BE ACCOMMODATED UNLESS THEY ARE REDIMENSIONED TO A
C...  LARGER SIZE. SUCH REDIMENSIONING IS UNDESIRABLE IN A
C...  LIBRARY ROUTINE  (THE USER SHOULD NOT HAVE TO GO INTO
C...  A LIBRARY ROUTINE TO MAKE CHANGES)
      IMPLICIT DOUBLE PRECISION (A-H,O-Z)
      DIMENSION YV(NEQN), YB(250), YDOT(250), YDOTP(250)
C...
C...  EXTERNAL THE DERIVATIVE SUBROUTINE CALLED BY EULERM
      EXTERNAL FCN
C...
C...  NUMBER OF MODIFIED EULER STEPS
      NSTEPS=INT((TOUT-TV)/H-0.01D0)+1
C...
C...  STEP THROUGH NSTEPS MODIFIED EULER STEPS
      DO 3 NS=1,NSTEPS
C...
C...  EVALUATE THE TEMPORAL DERIVATIVES AT THE BASE POINT
      CALL FCN(NEQN,TV,YV,YDOT)
C...
C...  ADVANCE THE DEPENDENT VARIABLE VECTOR USING ONE EULER
C...  STEP AFTER STORING THIS VECTOR AT THE BASE POINT
      DO 1 I=1,NEQN
         YB(I)=YV(I)
         YV(I)=YV(I)+YDOT(I)*H
```

```
1     CONTINUE
C...
C...  UPDATE TIME FOR THE PRECEDING EULER STEP
      TV=TV+H
C...
C...  EVALUATE THE TEMPORAL DERIVATIVES AT THE PREDICTED
C...  POINT
      CALL FCN(NEQN,TV,YV,YDOTP)
C...
C...  ADVANCE THE DEPENDENT VARIABLE VECTOR FROM THE BASE
C...  POINT USING THE MODIFIED EULER METHOD
      DO 2 I=1,NEQN
         YV(I)=YB(I)+(YDOT(I)+YDOTP(I))*H/2.0D0
2     CONTINUE
C...
C...  TAKE THE NEXT MODIFIED EULER STEP
3     CONTINUE
      RETURN
      END


      SUBROUTINE FCN(NEQN,TV,YV,YDOT)
C...
C...  SUBROUTINE FCN IS AN INTERFACE ROUTINE BETWEEN
C...  SUBROUTINES EULERM AND DERV
C...
C...  NOTE THAT THE SIZE OF ARRAYS Y AND F IN THE FOLLOWING
C...  COMMON AREA IS ACTUALLY SET BY THE CORRESPONDING
C...  COMMON   STATEMENT IN MAIN PROGRAM MEM
      IMPLICIT DOUBLE PRECISION (A-H,O-Z)
      COMMON/T/       T,      NSTOP,       NORUN
     1      /Y/       Y(1)
     2      /F/       F(1)
C...
C...  VARIABLE DIMENSION THE DEPENDENT VARIABLE, DERIVATIVE
C...  VECTORS
      DIMENSION YV(NEQN), YDOT(NEQN)
C...
C...  TRANSFER THE INDEPENDENT VARIABLE, DEPENDENT VARIABLE
C...  VECTOR FOR USE IN SUBROUTINE DERV
      T=TV
      DO 1 I=1,NEQN
         Y(I)=YV(I)
1     CONTINUE
C...
C...  EVALUATE THE DERIVATIVE VECTOR
      CALL DERV
```

```
C...
C...   TRANSFER THE DERIVATIVE VECTOR FOR USE BY SUBROUTINE
C...   EULERM
       DO 2 I=1,NEQN
          YDOT(I)=F(I)
2      CONTINUE
       RETURN
       END
```

Appendix C

Subroutine DSS024

```
          SUBROUTINE DSS024(XL,XU,N,U,UX,V)
C...
C...   SUBROUTINE DSS024 IS AN APPLICATION OF FOURTH-ORDER
C...   DIRECTIONAL DIFFERENCING IN THE NUMERICAL METHOD OF
C...   LINES.  IT IS INTENDED SPECIFICALLY FOR THE ANALYSIS
C...   OF CONVECTIVE SYSTEMS MODELLED BY FIRST-ORDER
C...   HYPERBOLIC PARTIAL DIFFERENTIAL EQUATIONS AS DISCUSSED
C...   IN SUBROUTINE  DSS012. THE COEFFICIENTS OF THE FINITE
C...   DIFFERENCE APPROXIMATIONS USED HEREIN ARE TAKEN FROM
C...   BICKLEY, W. G., FORMULAE FOR NUMERICAL DIFFERENTIATION,
C...   THE MATHEMATICAL GAZETTE, PP. 19-27,
C...   1941, N = 4, M = 1, P = 0, 1, 2, 3, 4.  THE
C...   IMPLEMENTATION IS THE **FIVE-POINT  BIASED UPWIND
C...   FORMULA** OF M. B. CARVER AND H. W. HINDS, THE METHOD
C...   OF LINES AND THE ADVECTION EQUATION, SIMULATION,
C...   VOL. 31, NO. 2, PP. 59-69, AUGUST, 1978.
C...
C...   DSS024 IS AN EXTENSION OF DSS020.  THESE ROUTINES HAVE
C...   THE FOLLOWING APPROXIMATIONS:
C...
C...      DSS020 - FIVE POINT APPROXIMATIONS OVER THE ENTIRE
C...               SPATIAL GRID.  AT THE INTERIOR POINTS, THE
C...               FIVE POINT, BIASED UPWIND APPROXMTAION IS
C...               USED.
C...
C...      DSS024 - A COMBINATION OF TWO POINT, UPWIND AND
C...               FIVE POINT, BIASED UPWIND APPROXIMATIONS
C...               IS USED.  THE FIVE POINT APPROXIMATION IS
C...               USED AT THE INTERIOR POINTS TO PROVIDE
C...               RESOLUTION OF MOVING FRONTS (I.E., TO
C...               MINIMIZE NUMERICAL DIFFUSION).  THE TWO
C...               POINT APPROXIMATION IS USED NEAR THE
```

```
C...                 BOUNDARIES TO REDUCE NUMERICAL
C...                 OSCILLATIONS FROM THE FIVE POINT
C...                 APPROXIMATION.
C...
C...  TYPE REAL VARIABLES AS DOUBLE PRECISION
      IMPLICIT DOUBLE PRECISION (A-H,O-Z)
      DIMENSION U(N),UX(N)
C...
C...  NUMBER OF GRID POINTS AT THE LEFT AND RIGHT ENDS WITH
C...  TWO POINT UPWIND APPROXIMATIONS
      NL=5
      NU=5
C...
C...  COMPUTE THE COMMON FACTOR FOR EACH FINITE DIFFERENCE
C...  APPROXIMATION CONTAINING THE SPATIAL INCREMENT, THEN
C...  SELECT THE FINITE DIFFERENCE APPROXIMATION DEPENDING
C...  ON THE SIGN OF V (SIXTH ARGUMENT).
      DX=(XU-XL)/DFLOAT(N-1)
      R4FDX=1.D+00/(12.D+00*DX)
      IF(V.GE.0.D+00)THEN
C...
C...     (1)  FINITE DIFFERENCE APPROXIMATION FOR POSITIVE V
C...
C...  TWO POINT UPWIND APPROXIMATIONS AT LEFT END (FOR NL
C...  GRID POINTS)
      UX(  1)=(1.0D0/DX)*
     1(  -1.D+00      *U(1)
     2   +1.D+00      *U(2))
      DO 2 I=2,NL
      UX(  I)=(1.0D0/DX)*
     1(  -1.D+00      *U(I-1)
     2   +1.D+00      *U(I  ))
2     CONTINUE
C...
C...  FIVE POINT BIASED UPWIND APPROXIMATIONS FOR INTERIOR
C...  GRID POINTS NL+1 TO N-NU-1
      DO 1 I=NL+1,N-NU-1
      UX(  I)=R4FDX*
     1(  -1.D+00      *U(I-3)
     2   +6.D+00      *U(I-2)
     3  -18.D+00      *U(I-1)
     4  +10.D+00      *U(I  )
     5   +3.D+00      *U(I+1))
1     CONTINUE
C...
C...  TWO POINT UPWIND APPROXIMATIONS AT RIGHT END (FOR NU
```

```
C...    GRID POINTS)
        DO 3 I=N-NU,N
        UX(I  )=(1.0D0/DX)*
     1(   -1.D+00        *U(I-1)
     2    +1.D+00        *U(I  ))
3       CONTINUE
C...
C...       (2)  FINITE DIFFERENCE APPROXIMATION FOR NEGATIVE V
C...
        ELSE IF(V.LT.0.D+00)THEN
C...
C...    TWO POINT UPWIND APPROXIMATIONS AT LEFT END (FOR NL
C...    GRID POINTS)
        DO 12 I=1,NL
        UX(  I)=(1.0D0/DX)*
     1(   -1.D+00        *U(I  )
     2    +1.D+00        *U(I+1))
12      CONTINUE
C...
C...    FIVE POINT BIASED UPWIND APPROXIMATIONS FOR INTERIOR
C...    GRID POINTS NL+1 TO N-NU-1
        DO 11 I=NL+1,N-NU-1
        UX(  I)=R4FDX*
     1(   -3.D+00        *U(I-1)
     2   -10.D+00        *U(I  )
     3   +18.D+00        *U(I+1)
     4    -6.D+00        *U(I+2)
     5    +1.D+00        *U(I+3))
11      CONTINUE
C...
C...    TWO POINT UPWIND APPROXIMATIONS AT RIGHT END (FOR NU
C...    GRID POINTS)
        DO 13 I=N-NU,N-1
        UX(  I)=(1.0D0/DX)*
     1(   -1.D+00        *U(I  )
     2    +1.D+00        *U(I+1))
13      CONTINUE
        UX(  N)=(1.0D0/DX)*
     1(   -1.D+00        *U(N-1)
     2    +1.D+00        *U(N  ))
        END IF
        RETURN
        END
```

Index

A

Analytical solution(s), 25
  convective systems, 56–57, 289–294
  convective-diffusion systems, 167, 183, 207, 280–282, 370–372
  heat conduction
    Bessel function, 149–163
    infinite series, 127–131
    Laplace transform, 92–93
    similarity transformation, 90–92
  heat convection, 290–294
  laminar flow in a tube, 38–47
  singularity, 73
  see also Series solution

B

Backward differentiation formulas *see* BDF
Barrett, L. C., 149
Basis functions, 381, 398–399
BDF, 62
Bessel function
  first kind, 38, 42–44
    BESSJ1, 44–45
  formulas from Watson *see* Watson
  general ODE solution, 149
  large argument limit, 163
  modified, first kind, 149
    BESSI0, BESSI1, 157–161
  modified, second kind, 149
    BESSK0, BESSK1, 157–161
  order, 38, 149
  roots of J0, 38, 40, 42–44
    ZERO, 43
  second kind, 149
  series solution, 38
    convergence, 38
Biased approximation, *see* Finite difference approximations, DSS020, DSS024, DSS034
Biot number, 108, 111, 120, 131–134
Bird, R. B., 25

Blasius equation, 4
  series solution, 4, 17–22
Boundary conditions, 5, 25, 89–90, 107–108, 136, 167, 207, 289, 342, 371
  alternate formulations, 62, 140, 147–148, 211
  axial, 245–246, 268, *see also* Boundary conditions, outflow
  constant flux, 167, 169
  Dirichlet, 112–113, 122–124, 147–149
  discontinuous, 292
    effect on numerical solutions, 305–325
  fictitious points, 80, 141, 177
  from minimum reduction of PDEs, 244
  insulated, 108, 136
  Neumann, 112–113, 122–124, 147–149, 410
  outflow, 211, 222–225, 264
    distortion of solution, 233–234, 256–257
    effect on solution, 227–271
  smooth, 308–309
  symmetry, 108, 169, 211, 223, 241, 255
  time varying, 289–294
  see also Finite difference approximations, ODEs
Boundary layer, 3, 23
Boundary-value problems, 5, 23
  see also ODEs
Burgers' equation 56
  inviscid form, 56
  front-flattening, 55, 69–86
  front-sharpening, 55, 69–86
  nonlinearity, 56, 86
  shock formation, 86
  solution singularity, 73
  upwind approximations, 63

C

Cartesian coordinates, 89, 107, 115–121, 131–133
Cauchy's equation, 149–150
Characteristic equation, 92, 374, 378
Component balance, 330, 370, *see also* Mass transfer
Continuity equation, 3, 86

Index

Convective systems
    centered approximations, 78, 80, 84–86, 172, 177
    effect of axial conduction, 212, 214–215, 244–248
    effect of radial conduction, 214–215
    outflow boundary conditions, *see* Boundary conditions
    modeled by first-order hyperbolic PDEs, 289
    upwind approximations, 63, 172, 178
    with diffusion, 167, 172, 207–212, 370–372
    see also Burger's equation, DSS034
Convergence to boundary condition, 12–14, 23
Cylindrical coordinates, 107, 121–126, 133, 136, 167–169, 207–210
    centerline condition, *see* Boundary conditions, symmetry

D

DAEs, 62
DASSL, 62
DFLOAT, 22, 27, 45, 104–105, 161, 441–442
Differential algebraic equations
    see DAEs
Dirichlet boundary conditions
    see Boundary conditions
Discontinuous solutions, 292–294
    approximated numerically, 294–325
    see also DSS020, DSS024, DSS034
Double precision, 6, 22
    see also DFLOAT
Drake, R. M., 167
DSS004, 121–122
DSS020, 298–299, 306–308, 317–322, 343, 407–410
DSS024, 298–299, 312–317, 322–325, 449–451, 407–410
DSS034, 172, 195–199, 215, 249–271
DSS044, 112, 409

E

Eckert, E. R. G., 167
Eigenfunctions, 38, 127, 374, 398–399
Eigenvalues, 38, 40, 127, 374, 378, 382, 388–389
    calculation, 382, 388–390
    defined by characteristic equation, 374, 378, 390
    dominant, 286
Energy equation, 86, 89, 107, 135, 167–170, 207–210, 290
Equilibrium relationships, *see* Mass transfer
Error function, 91–93
    Hasting's approximation, 103–104, 106
    numerical evaluation, 93–106
Euler's equation, 149–150
Euler's method, 50–51
    explicit, 51
    implicit, 54
    stability limit, 53–54
    time step, 50
    see also Modified Euler method
Explicit integration, 53
    stability limit, 53–54

F

Feshbach, H., 375
Fictitious points, *see* Boundary conditions
Finite difference approximations, 29, 63, 141, 144
    accuracy, 53, 78
    backward, 63, 172, 221, 236
    biased, upwind approximation, 198, 215, 254–256, 298–299, *see also* DSS020, DSS024, DSS034
        oscillation, 256, 261, 266, 306–308
    centered, 172, 177, 222, 224–226, 252–254
    discontinuous solutions, 294–325, *see also* DSS020, DSS024, DSS034
    explicit programming, 172, 175–176, 214–215
    numerical diffusion, 261, 311–314, 325
    numerical oscillation, 256, 261, 266, 306–308, 325
    order, 63, 144, 203
    stability limit, 53
    upwind, 63, 172, 178, 221, *see also* DSS020, DSS024, DSS034
    see also Boundary conditions
Finlayson, B. A., 207
Five-point, biased upwind approximation, 198, 215, 254–256
    oscillation, 256, 261
    see also DSS020, DSS024, DSS034
Forsythe, G. E., 4
Fourier coefficients, 128, 376, 379, 384–385, 399–401
Fourier's first law, 143–144
Fourier's second law, 89, 107
    solution by
        Bessel functions, 149–166
        method of lines, 109–127, 136–151
        numerical quadrature, 102–106
    time constant, 136, 138
Fourier series, 127–131, 380, 385–386, 402–406
    rate of convergence, 386

G

Graetz problem
    with constant wall heat flux, 167–206
    with constant wall temperature, 207–288
Gurney Lurie chart, 121

H

Hand calculation, 19
Heat conduction equation
    see Fourier's second law
Heat exchanger dynamics, 289–325
Height of a transfer unit, *see* Mass transfer
Henry's law, *see* Mass transfer
Hyperbolic PDEs
    first-order, 289
Hyperbolic-parabolic PDEs, 371
    conversion to parabolic PDEs, 372–373

I

Ill-conditioned systems, 360, 423, 426
Implicit integration, 54
    stability limit, 54

Infinite series, *see* Series solutions
Initial conditions, 5, 25, 89–90, 107–108, 136, 169, 210, 289, 335–338, 370, 420
Initial-value problems, 5
   see also ODEs
Interval-halving, *see* Root finding
Isotropic systems, 209
$I(x)$ modified Bessel function of first kind, *see* Bessel functions

J

Jakob, M., 207
$J(x)$ Bessel functions of first kind, *see* Bessel functions

K

Kramer's rule, 360, 423, 426
$K(x)$ modified Bessel functions of second kind, *see* Bessel functions

L

Laminar flow, 3, 25
Laplace transforms, 92–93, 106, 290–294
   convolution, 292
   double, 291
   unit step function, 292
Least squares analysis, 350–361
l'Hospital's rule, 29, 125, 178–179, 222, 265
Library routines
   advantages, 206
   see also DASSL, DSS034, DSS004, DSS044, ODEPACK, LSODE, LSODES, RKF45
Lightfoot, E. N., 25
Log-mean driving force, *see* Mass transfer
LSODE, 62
LSODES, 62

M

Malcolm, M. A., 4
Mass transfer
   ammonia absorber, 334
      detailed design, 334–369
   coefficient, 330, 341
   comparison of design methods, 369
   component balance, 330
   dynamic model, 329, 342–343
      advantages, 351–352
      boundary conditions, 342
      initial conditions, 342
      spatial derivatives, 343
      see also PDEs
   equilibrium relationship, 331, 337, 341, 350, 352–361
   height of a transfer unit, 333, 363–364
   Henry's law, 331
   log-mean driving force, 333–335, 338, 341, 350, 361–363
   number of transfer units, 333, 338, 351, 363–368
   operating line, 333–334, 337, 340–341, 350, 352
   rate, 330–331
   resistance, 335
   steady state model, 331–334, 351
Matlab
   plotting, 31–33, 64–65, 114, 183–184, 227, 279, 302, 305, 380, 412–413, 428–429
   Method of lines, 25–30, 63, 109–127, 136–148, 170–206, 335–347, 406–416
      advantages, 113, 255, 300, 411, 416–417
      approximation of discontinuous solutions, 294–325, *see also* DSS020, DSS024
      finite difference approximations, 29–30, 47, 78, 144, 177–179, 212–271
         see also DSS034, DSS004, DSS044
      general introduction, 207–288
      initial-value spatial variable, 272–278
      numerical diffusion, 261
      numerical oscillation, 84, 127, 256, 261, 266, 306–308
      spatial grid, 26, 110, 136, 173
         two-dimensional, 173, 214, 251
      steady state formulation, 272–278
      time integration, 47
         Euler's method, 50–51
         see also RKF45
      "time-like" spatial variable, 272–278
Minimum reduction of PDEs, 244
Modified Bessel functions, *see* Bessel functions
Modified Euler method, 443–448
Moler, C. B., 4
Momentum equation, 3, 25, 55, 86
   see also Navier–Stokes equations
Morse, P. M., 375
Multicomponent diffusion
   approach to steady state, 432
   binary diffusivities, 420, 422
   fluxes from Stefan-Maxwell equation, 419, 424–426
   transient model, 420
Multidimensional solutions, *see* PDEs

N

Navier-Stokes equations, 3
   Burgers' equation, 56
   precursor, 55
   stress tensor, 55
Near-singular system, 360, 423, 426
Neumann boundary conditions
   see Boundary conditions
Newton's law for fluids, 55, 168, 209
Newton's method, 14
   see also Root finding
Normalization temperature, 169, 173
Number of transfer units, *see* Mass transfer
Numerical quadrature
   see Simpson's rule
Numerical solution(s)
   see Method of lines, ODEs, PDEs

O

ODEPACK, 62
ODEs
  boundary conditions, 5
  general Bessel function solution, 149–151
  initial conditions, *see* Initial conditions
  initial-value, 4, 26
  integration, *see* BDF, Euler's method, ODEPACK, LSODE, LSODES, RKF45
  multicomponent diffusion, 418–420
  nth-order, 4
  shooting method, 5
  with variable coefficients, 149
Operating line, *see* Mass transfer
Order of approximations, 63
Order-of-magnitude analysis, 3
Ordinary differential equations
  see ODEs
Orthogonality, 128, 375–376, 379, 381–384, 394–395, 398–399
Outflow boundary conditions, *see* Boundary conditions

P

Parabolic velocity profile, 167, 173, 207
Parabolic-hyperbolic PDEs, 371
Partial differential equation(s), *see* PDE(s)
PDE(s)
  boundary conditions from minimum reduction, 244
  convection and diffusion, 167, 370–371
    analytical solution, 167, 183, 377–406
    numerical solution, 170–206, 406–416
  dimensionless variables, 90, 108, 169, 210, 371–372
  first-order hyperbolic, 289, 331
  heat conduction, 89, 107–108, 135–136
    analytical solution, 90–93, 149–166
    numerical solution, 93, 109–126, 136–149
    series solution, 127–131
  heat convection, 289
    analytical solution, 291–293
    numerical solution, 294–325
  hyperbolic-parabolic, 371
  laminar flow in a tube, 25
    numerical solution, 25
    series solution, 38–47
  mass transfer, 329–331
    numerical solution, 335–347
  method of lines solutions, *see* Method of lines
  multidimensional solutions, 131–134
  parabolic-hyperbolic, 371
  product solutions, 131–134
  subscript notation, 89
  time integration, *see* Method of lines
Peclet number, 169, 207, 210, 372
  effect on solution, 272, 286
  measure of convection and diffusion, 372
Perturbation variables, 289
Product solutions, *see* PDEs

Q

Quadrature, *see* Simpson's rule
QUANC8, 102
Quasi-steady state assumption, 419

R

Root finding, 7
  Bessel functions, 42
  for eigenvalues, 390–394
  initial estimate of root, 14
  interval-halving, 7, 11
    advantages, 14
  Newton's method, 14
  secant method, 390–394
RKF45, 4–6, 15–17, 26, 32–34, 47–48, 63, 115, 143, 184, 203, 342, 430
  calling program, 437–442

S

Schiesser, W. E., 25
Search
  for missing initial-conditions, 9–12, 23
Secant method, *see* Root finding
Separation of variables, 127–131, 373–374
Series solutions, 4, 17–22, 38–47, 127–131, 207, 280–282, 373–374, 378
  convergence, 23, 38, 46
  divergence, 23
  infinite series
    Bessel function, 38
    cosine, 127–131
    sine and cosine, 378
Shock capture, 86
Shock formation, 86
Shooting method, 5, 23–24
Silebi, C. A., 369
Similarity transformation, 4, 90–92, 106
Simpson's rule, 93–106
  applied to
    orthogonal functions, 381–384, 395–397
    polynomials, 96–102
    mass transfer, 325, 338, 351, 363–368
  number of panels, 94, 381
Single precision
  conversion to double precision, *see* DFLOAT
Singularity, 73
Spatial derivatives
  see Finite difference approximations
Spatial grid, 26, 60, 173
  two-dimensional, 173
  see also Finite difference approximations, Method of lines
Spherical coordinates, 107, 121–126
  centerline condition, 125
Stefan-Maxwell equation, 419
Stewart, W. E., 25
Stream function, 3
Stress tensor, 55
Strongly convective PDEs, 372
Strongly hyperbolic PDEs, 372

Sturm-Liouville systems, 375
    orthogonality, 375–376
Subscript notation for PDEs, 89
Superaccurate approximations, 63
    exact integrals, 96–102
Superposition of solutions, 127, 374
Symmetry boundary conditions, *see* Boundary
        conditions, symmetry

T

Thermal conductivity, 107, 168
Thermal diffusivity, 89, 107, 169, 210

U

Unsteady flow, 25
Upwind approximations, 63, 85, 198

V

Velocity profile, 167, 173, 207
Velocity vector, 55
Viscosity, 55

W

Watson, G. N., 44, 54, 151, 163
Welty, J. R., 334
Wicks, C. E., 334
Wilson, R. E., 334
Wylie, C. R., 149

Y

$Y(x)$ Bessel function of second kind, *see* Bessel
        functions